Springer Series on Atomic, Optical, and Plasma Physics

Volume 83

The Springer Series on Atomic, Optical, and Plasma Physics covers in a comprehensive manner theory and experiment in the entire field of atoms and molecules and their interaction with electromagnetic radiation. Books in the series provide a rich source of new ideas and techniques with wide applications in fields such as chemistry, materials science, astrophysics, surface science, plasma technology, advanced optics, aeronomy, and engineering. Laser physics is a particular connecting theme that has provided much of the continuing impetus for new developments in the field, such as quantum computation and Bose-Einstein condensation. The purpose of the series is to cover the gap between standard undergraduate textbooks and the research literature with emphasis on the fundamental ideas, methods, techniques, and results in the field.

More information about this series at http://www.springer.com/series/411

Valentin Igochine
Editor

Active Control of Magneto-hydrodynamic Instabilities in Hot Plasmas

Springer

Editor
Valentin Igochine
Max-Planck-Institut für Plasmaphysik
Garching
Germany

ISSN 1615-5653 ISSN 2197-6791 (electronic)
ISBN 978-3-662-44221-0 ISBN 978-3-662-44222-7 (eBook)
DOI 10.1007/978-3-662-44222-7

Library of Congress Control Number: 2014949367

Springer Heidelberg New York Dordrecht London

Printed on acid-free paper

Springer is part of Springer Science+Business Media (www.springer.com)

Preface

Magneto-hydrodynamic instabilities, that can develop in a hot plasma, strongly limit the range of possible operation parameters of a fusion reactor. This issue was identified from the beginning of fusion research and a lot of attention has been paid in the field toward avoiding, stabilizing, or controlling these instabilities. During the last decades, remarkable progress has been made in practically all areas of instability control. This book aims to provide a concise introduction into this field, with the main focus on basic mechanisms of instability, their identification, and control. Experimental results and theoretical interpretation presented in the book are the current status of our understanding of this subject. One has to note that both the physics and the control of these instabilities are far from completely understood, in many cases, and remain as areas of active research. The research areas are very broad and it is extremely difficult to find an expert who can cover all these instabilities. This was the motivation for our joint work on this book. Different authors, all of them active physicists, working on different fusion devices, describe different subjects, and provide concise overviews of the research area in which they are working as senior researchers or heads of research groups.

Garching, Germany Valentin Igochine

Preface

Acknowledgments

The authors are thankful to "Wilhelm and Else Heraeus Stiftung" for organization of the seminar "Active Control of Instabilities in Hot Plasmas," which was held in June 2011 in Bad Honnef, Germany. Discussions during this meeting identified the need for this book and were the starting point of our joint work.

Valentin Igochine is grateful to T. Hender, H. Zohm, and V. Pustovitov for many important comments and changes of RWM and Operation limit chapters, I. Stepanov and M. Dunn for careful reading of these chapters and useful remarks. He thanks S. Günter, K. Lackner, E. Strumberger, and P.R. Brunsel for many useful discussions on RWMs which helped to better understand the physics.

Tim Hender thanks J. Vega for assisting with preparing the text on the APODIS code in the Disruptions chapter, and G. Pautasso, A. Thornton, R. Granetz and A. Murari for many helpful comments on Chap. 7.

Anja Gude thanks H. Zohm, M. Maraschek, and V. Igochine for many important comments and discussions. She is grateful to V. Igochine, M. Maraschek, and M. Weiland for preparing some of the figures and to I. Chapman for reading and correcting the manuscript of Chap. 3.

Piero Martin acknowledges all the authors of this book, for the great work they have done and that he has extensively used in this chapter. He acknowledges Dave Humphreys for a precious source of inspiration with his lecture given at the 2010 Iter International Summer School held in Austin, TX. A particular thank you to the colleagues of the RFX-mod team and of the ITPA Working Group on Stability and control, from whom he learned a lot on MHD control.

Contents

Main Definitions, Notations and Abbreviations

Coordinate System

Torus coordinates: (r, θ, ϕ) or (R, z, ϕ): R is the major radius, r is the minor radius, θ is the poloidal angle, ϕ is the toroidal angle and z is the height above the midplane. Straight field line coordinates: (ρ, θ^*, ϕ): ρ is the radial coordinate, θ^* is the poloidal straight field line angle and ϕ is the toroidal angle. The radial coordinate ρ can be expressed in terms of the poloidal fluxes, Ψ,

$$\rho_{pol} = \sqrt{\frac{\Psi - \Psi_0}{\Psi_a - \Psi_0}}$$

or toroidal fluxes, Φ,

$$\rho_{tor} = \sqrt{\frac{\Phi - \Phi_0}{\Phi_a - \Phi_0}},$$

where the index a refers to the separatrix (or plasma boundary) and index 0 to the magnetic axis. These coordinates are scaled such that $\rho = 0$ on the magnetic axis and $\rho = 1$ at the plasma boundary.

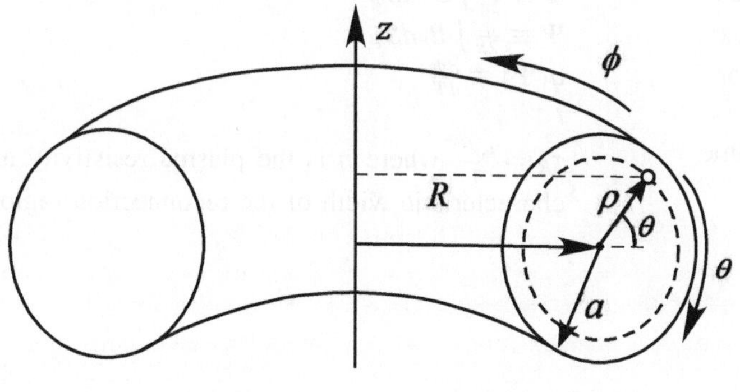

MHD Instability

MHD instability with poloidal mode number m and toroidal mode number n is located at the resonant surface $\rho_{m,n}$ or $\rho_{q=m/n}$. The resonant surface can be also marked as r_{res} or ρ_{res}. Radial displacement due to an instability has displacement amplitude $\hat{\xi}_r(\rho)$ and can be written in the following form:

$$\xi_r = \hat{\xi}_r(\rho) \cdot \cos(m\theta^* - n\phi + \omega t),$$

where ω is a real number representing the angular frequency of the mode. In the linear regime, the mode growth is assumed to be exponential, $\hat{\xi}_r(t) \sim e^{\gamma t}$, with real growth rate γ.

General Notations and Relations

Temperature	T_e and T_i are electron and ion temperatures respectively. In plasma physics, the temperature is often given in units of the thermal energy: $T[eV] = k_B T[K]$, where Boltzmann constant $k_B = 8.6173 \cdot 10^{-5}[eV/K]$.
Density	n_e and n_i are electron and ion densities respectively. Quasi-neutrality condition requires $n_e = n_i = n$ for pure Hydrogen plasmas.
Mass	m_e and m_i are electron and ion masses respectively.
Mass density in MHD	$\rho \equiv m_i n$
Total plasma current	I_p
Plasma current density	$\vec{j}$
Electric field	$\vec{E}$
Adiabatic factor	Γ
Magnetic field	$\vec{B}$ is the total magnetic field. B_ϕ is the toroidal component of the magnetic field and B_θ is the poloidal component of the magnetic field.
Toroidal flux	$\Phi \equiv \frac{1}{2\pi} \int B_\phi dS_\Phi$
Poloidal flux	$\Psi \equiv \frac{1}{2\pi} \int B_\theta dS_\Psi$
Safety factor	$q(\Psi) \equiv \frac{d\Phi}{d\Psi}$
Time	t
Resistive time	$\tau_R = \frac{\mu_0 l^2}{\eta}$, where η is the plasma resistivity and l is the characteristic width of the reconnection region.

Pressure Related Quantities

Pressure	p is the plasma pressure which is the sum of the electron and ion pressures $p = p_e + p_i$.
Beta	is the ratio of the average plasma pressure and magnetic field pressure $\beta = \frac{\langle p \rangle}{\langle B^2 \rangle / 2\mu_0}$
Poloidal beta	$\beta_p = \frac{\langle p \rangle}{\langle B_\theta^2 \rangle / 2\mu_0}$
Local poloidal beta at the resonant surface s	$\beta_p(r_s) = \frac{p(r_s)}{\langle B_\theta^2(r_s) \rangle / 2\mu_0}$
Normalized beta	$\beta_N \equiv \frac{aB_\phi}{I_p} \beta$

Velocities and Frequencies

Velocities	$\vec{v}_e$ and $\vec{v}_i$ are electron and ion velocities respectively and plasma velocity $\vec{v} \equiv \vec{v}_i$ for single fluid MHD. The thermal velocity of the plasma ions is $v_{th,i} = \sqrt{2T_i/m_i}$. $v_\parallel$ and $v_\perp$ are parallel and perpendicular velocities with respect to the magnetic field $\vec{B}$. v_{pol} and v_{tor} are poloidal and toroidal plasma rotation velocities respectively. Alfven velocity is $v_{Alfven} = B_\phi / \sqrt{\mu_0 \rho}$.
Angular frequency	ω_c is the cyclotron frequency, ω_{tor} is angular rotation frequency of the plasma rotation, etc.

General Abbreviations

AE	Alfvén Eigenmodes
BAE	Beta induced Alfvén Eigenmodes
EAE	Ellipticity induced Alfvén Eigenmodes
ECCD	Electron Cyclotron Current Drive
ECE	Electron Cyclotron Emission
ECEI	Electron Cyclotron Emission Imaging
ECRH or ECH	Electron Cyclotron Resonance Heating
EFCC	Error Field Correction Coils
ELM	Edge Localized Mode
FFT	Fast Fourier Transformation
FIR	Frequently Interrupted Regime of neoclassical tearing mode

FLR	Finite Larmor Radius (effect/correction)
GEM	Gas Electron Multiplication detector
H-mode	High confinement mode
HFS	High Field Side
ICCD	Ion Cyclotron Current Drive
ICRH	Ion Cyclotron Resonance Heating
ILW	ITER Like Wall
L-mode	Low confinement mode
LFS	Low Field Side
LHCD	Lower Hybrid Current Drive
LOS	Lines Of Sight
MARFE	Multifaceted Asymmetric Radiation From the Edge
MCCD	Mode Conversion Current Drive
MGI	Massive Gas Injection
MHD	Magneto-hydrodynamics
MP	Magnetic Perturbations
MSE	Motional Stark Effect
N-NBI	Negative Neutral Beam Injection
NAE	Non-circularity induced Alfvén Eigenmodes
NBCD	Neutral Beam Current Drive
NBI	Neutral Beam Injection
NTM	Neoclassical Tearing Mode
NTV	Neoclassical Toroidal Viscosity
QH-mode	Quiescent High confinement mode
RFA	Resonant Field Amplification
RMP	Resonant Magnetic Perturbations
RSAE	Reversed Shear Alfvén Eigenmodes
RWM	Resistive Wall Mode
SVD	Singular Value Decomposition
SXR	Soft X-ray Cameras
TAE	Toroidal Alfvén Eigenmodes
TM	Tearing Mode
VDE	Vertical Displacement Event

Contributors

Ian T. Chapman CCFE, Culham Science Centre, Abingdon, Oxfordshire, UK

Anja Gude Max Planck Institute for Plasma Physics, Garching, Germany

Tim C. Hender Directorate, CCFE Culham Science Centre, Abingdon, Oxfordshire, UK

Valentin Igochine Max Planck Institute for Plasma Physics, Garching, Germany

Yunfeng Liang Forschungszentrum Jülich GmbH, Association EURATOM-FZ Jülich, Institut Für Energieforschung—Plasmaphysik, Trilateral Euregio Cluster, Jülich, Germany

Marc Maraschek Max Planck Institute for Plasma Physics, Garching, Germany

Piero Martin Department of Physics and Astronomy, University of Padova, Padua, Italy; Consorzio RFX, Padua, Italy

Simon D. Pinches ITER Organization, Route de Vinon-sur-Verdon, CS 90 046, 13067 St. Paul-lez-Durance, Cedex, France

Sergei E. Sharapov CCFE, Culham Science Centre, Abingdon, Oxfordshire, UK

Hartmut Zohm Max Planck Institute for Plasma Physics, Garching, Germany

Chapter 1
Introduction to Tokamak Operational Scenarios

Hartmut Zohm

Abstract The goal of nuclear fusion using magnetic confinement is to confine a plasma consisting of hydrogen isotopes and heat it to temperatures that allow the energy released from the fusion reactions to largely compensate the energy loss from the plasma due to convection, conduction and radiation so that stationary nuclear burning of the fuel is achieved with net energy output. Presently, the reaction envisaged is the fusion of Deuterium and Tritium, which releases the fusion power P_{fus} in He nuclei (α–particles) of 3.5 MeV and neutrons of 14.1 MeV. In this chapter, we discuss optimization of power balance for this reaction in conventional and advanced tokamak regimes.

1.1 Plasma Conditions Needed for Nuclear Fusion

The goal of nuclear fusion using magnetic confinement is to confine a plasma consisting of hydrogen isotopes and heat it to temperatures that allow the energy released from the fusion reactions to largely compensate the energy loss from the plasma due to convection, conduction and radiation so that stationary nuclear burning of the fuel is achieved with net energy output. Presently, the reaction envisaged is the fusion of Deuterium and Tritium, which releases the fusion power P_{fus} in He nuclei (α-particles) of 3.5 MeV and neutrons of 14.1 MeV. The α-particles must be confined in the magnetic field so that they can heat the plasma by collisional slowing down, providing the α-heating P_α, while the neutrons will deposit their energy in the first wall. Since the energy is distributed between the two reaction partners according to their mass ratio, the relation $P_\alpha = 1/5\, P_{fus}$ holds.

The plasma parameters needed for this state are determined by the energy balance, i.e. the α-heating should compensate the losses due to convection, conduction and radiation. In the absence of a first principles theory describing the convective

H. Zohm (✉)
Max Planck Institute for Plasma Physics, Boltzmannstr. 2, 85748 Garching, Germany
e-mail: hartmut.zohm@ipp.mpg.de

V. Igochine (ed.), *Active Control of Magneto-hydrodynamic Instabilities in Hot Plasmas*, Springer Series on Atomic, Optical, and Plasma Physics 83,
DOI 10.1007/978-3-662-44222-7_1

$n_{\max} < n_{GW} = I_p/(\pi a^2)$ where a is the plasma minor radius. Hence, this limitation calls for operation at high I_p.

A similar strategy results from the optimization of the energy confinement time τ_E. Although there is no first principles theory to predict the value of τ_E due to the fact that it is mainly determined by gradient driven turbulent transport, empirical scalings derived from a large number of tokamaks clearly indicate an approximately linear scaling $\tau_E \sim I_p$, providing another strong motivation to run a tokamak at high plasma current.

However, for given toroidal field, the plasma current I_p is limited by the Kruskal-Shafranov limit that states that the edge safety factor $q_a \sim a^2/R\ B_t/I_p$ must not be lower than $q_a = 1$ since then, the plasma is unstable to an ideal external kink mode. In praxi, this limit is more restrictive since without active control, an external kink mode will already occur at $q_a = 2$ (see Chap. 2).[3] Moreover, there is a tendency for disruptions to occur more often with higher current, particularly in the window $q_a < 3$. Hence, while high current is definitely desirable from the point of view of optimizing $nT\tau_E$, there will be a maximum allowable current set by the minimum tolerable q_a.

Finally, when optimizing $nT\tau_E$, there is also a limitation to the maximum achievable pressure $p = nT$ in the form of a limit to the dimensionless number β, which is the ratio of kinetic to magnetic pressure. This so-called β-limit, described in its various aspects in Chaps. 2, 6 and 8 of this book, will, for given toroidal field, limit the achievable β to the order of a few per cent. In ideal MHD, it is usually given by an external kink mode that is driven by a combination of pressure and current gradient. It was mentioned above that, due to its ideal nature, this instability will grow too fast to be accessible by feedback control, but its growth can be slowed down by nearby conducting structures that, for sufficient inductive coupling, will slow down the growth rate to the inverse of the resistive time scale of the wall, converting the ideal kink to the Resistive Wall Mode (RWM) treated in Chap. 6 of this book. For ideal MHD modes, the ideal β-limit can be described as a limit to the so-called normalized β, $\beta_N = \beta\ /\ I_p/(aB)$. It has been found that $\beta_{N,\max}$ is a function of the shape of the profile of toroidal current, with peaked current profiles being more stable than broad ones because it is mainly the edge current that drives the kink mode unstable. We mention here that the introduction of finite resistivity effects leads to an additional stability boundary, usually with $\beta_{N,\max}$ values lower than the ideal limit, set by the Neoclassical Tearing Mode (NTM) treated in Chap. 8.

---

[3] Recent experiments on the RFX reversed field pinch, when run as a tokamak, have clearly demonstrated that with adequate feedback control by additional coils, a tokamak discharge can in principle be run at $q_a < 2$.

1.2.2 Conventional Tokamak Scenarios

From these considerations, a commonly used strategy for a tokamak scenario reaching optimum $nT\tau_E$ is the following: for given size, geometry and toroidal field, the plasma current should be as high as allowed by the q_a limit. In praxi, this will be around $q_a = 3$. The density is set to a value that is below the density limit, $n < n_{GW}$,[4] and β is raised to a value $\beta_N < \beta_{N,\max}$. The current profile will be peaked in the center, consistent with the tendency of ohmically driven current to have the largest current density where the plasma is hottest due to the decrease of electrical resistivity with temperature, $\eta \sim T^{3/2}$. Here, the condition that $q > 1$ everywhere in the plasma will limit the central current density due to the occurrence of the sawtooth instability described in Chap. 4 when $q(0)$ is below 1. In fact, the baseline operation scenario for ITER, aimed at achieving $Q = 10$, i.e. dominant α-heating $P_\alpha = 2\,P_{ext}$, follows precisely this strategy, also known as 'conventional tokamak scenario'. We mention here that such a scenario will also have to be examined with respect to fast particle driven instabilities as described in Chap. 9 such that the α-particles are confined well enough to provide central heating by classical slowing down. For all the limits described above, active control will act either to extend the limitation or to allow operation close to the limit with the possibility to recover the operational point should one of the limits be violated transiently.

1.2.3 Advanced Tokamak Scenarios

There is, however, an important change in optimization strategy when also the pulse length is introduced in the optimization. It is clear that a tokamak scenario that has an inductively driven component of the plasma current will be pulsed since at some point, the transformer flux that is used to compensate the resistive loss in the plasma is exhausted. On the other hand, a fusion reactor preferably has to operate under stationary conditions or at least have a reasonable duty cycle and for the reason just given, the boundary condition of long pulses calls for a large fraction of the plasma current to be driven non-inductively. External current drive systems that are also used to heat the plasma have a relatively low current drive efficiency $I_p\,/P_{ext}$ so that this optimization strategy usually relies on a large component of the so-called bootstrap fraction, which is a toroidal current occurring in a toroidal magnetic confinement system due to a thermo-electrical effect. It can be shown that the bootstrap current density is roughly proportional to the radial pressure gradient, so that f_{bs}, the fraction of total current supported by the bootstrap effect, becomes

[4] For typical parameter values of present and future tokamaks, this will roughly be of the order of $10^{20}\ \mathrm{m}^{-3}$.

created in this case, thus only strongly shaped [illegible] providing [illegible] the [illegible] [illegible] the [illegible].

1.4 Summary

In summary, tokamak operational space is restricted [illegible] parameters by the occurrence of MHD instabilities and active control [illegible] and help to optimize the tokamak performance [illegible]. The [illegible] outlined [illegible] along different routes, pushing different [illegible] and need to actively control the [illegible]. The [illegible] [illegible].

Chapter 2
Magneto-Hydrodynamics and Operational Limits

Valentin Igochine

Abstract The main aim of the fusion research is to confine the plasma long enough for fusion power production. The required temperatures are so high that no material can tolerate direct contact. The magnetic field acts on the plasma and gives opportunity for isolation of the burning region from the surrounding materials. Different magnetic systems have different limits for plasma confinement and the plasma is lost if these limits are crossed. Thus, understanding of these limits is of primary importance for stable plasma operation. The present chapter provides a review of density, current and beta limits with particular focus on magneto-hydrodynamic instabilities in tokamaks.

2.1 Introduction to the Main Concepts of Magnetic Confinement

The main goal of fusion research is to achieve a favorable power balance in a fusion reactor, which implies the fusion power to be much higher than the input power. In order to reach this goal, the plasma has to be confined for a sufficiently long time with sufficient heat isolation at high temperatures. Assuming Deuterium-Tritium reaction, the triple product of plasma density $n[m^{-3}]$, confinement time $\tau[s]$ and plasma temperature $T[keV]$[1] should be sufficiently high to maintain the plasma heating solely by products of the reaction: the ignition condition is $nT\tau > 3 \times 10^{21}\ \mathrm{m^{-3}\ keVs}$ [1]. This number is valid for the most energetically favorable reaction, between deuterium and tritium. Even in this case, the optimal reaction

[1] In plasma physics, the temperature is often given in units of the thermal energy: $T[eV] = k_B T[K]$, where Boltzmann constant $k_B = 8.6173 \times 10^{-5}$ [eV/K].

V. Igochine (✉)
Max Planck Institute for Plasma Physics, Boltzmannstr 2, 85748 Garching, Germany
e-mail: valentin.igochine@ipp.mpg.de

V. Igochine (ed.), *Active Control of Magneto-hydrodynamic Instabilities in Hot Plasmas*, Springer Series on Atomic, Optical, and Plasma Physics 83,
DOI 10.1007/978-3-662-44222-7_2

temperature, $T \approx 20$ keV $\left(T \approx 230 \times 10^6 \text{ K}\right)$, is too high to be tolerated by any materials and a magnetic field is the only way to confine the plasma in a reactor under stationary conditions and isolate the burning region from the surrounding materials.

One has to start from the properties of the magnetic field to get an idea about plasma confinement. The primary action of the magnetic field on the plasma can be easily understood from single particle motion in a uniform magnetic field. A charged particle will gyrate in the plane perpendicular to the magnetic field (see Fig. 2.1a). This gyration constrains particle motion perpendicular to the magnetic field. The gyration frequency $\left(\omega_c = q_p B / m_p\right)$ and the gyro-radius $\left(r_L = \frac{m_p v_\perp}{|q_p| B}\right)$ follows directly from the single particle equation of motion in a uniform magnetic field taking into account the Lorentz force $\left(\vec{F}_L = q_p\left(\vec{v} \times \vec{B}\right)\right)$ acting on the particle (here m_p is the particle mass, q_p is the particle charge, $v_\perp$ is the perpendicular component of the particle velocity with respect to the magnetic field $\vec{B}$). The important consequence of this behavior is unconstrained motion along the magnetic field. The combined perpendicular and parallel motion of the charged particle corresponds to a helical trajectory along a magnetic field line with the initial parallel velocity of the particle, $v_\parallel$. Restriction of the freedom of the particle motion along the magnetic field can be achieved in two ways:

(i) variation of the magnetic field strength and/or electrostatic barriers along the field line,
(ii) magnetic configurations with closed field lines.

The first approach is implemented in mirror systems, where variation of the magnetic field strength creates a magnetic bottle with higher magnetic field strength at both ends (see Fig. 2.1b). In this case, a particle with sufficient perpendicular velocity $(v_\perp)$ will be confined in a region with minimal magnetic field. The particle is reflected from the high field region due to conversion of parallel velocity of the particle into perpendicular velocity. The physical reason for this behavior is the conservation of the flux enclosed by the particle trajectory over a gyro period. The conversion itself is described by the first adiabatic invariant which is approximately $\mu = \frac{m_p v_\perp^2(t)}{2B} = const$. Particles with small perpendicular velocities are not confined in the system. They move freely along the magnetic field lines and become lost; the loss condition is $v_\parallel > v_\perp \sqrt{(B_{\max} - B_{\min})/B_{\min}}$. This produces an unpopulated "loss cone" in the velocity space (Fig. 2.1c). Making the so called "mirror ratio" $(R_M = B_{\max}/B_{\min})$ high, one could reduce the loss cone. Unfortunately, Coulomb collisions change particle velocities and move new portions of the particles permanently into the loss cone. This produces a constant outflow of particles from the device along the magnetic field lines. Further complications come from anomalous losses of electrons and turbulent processes. Present configurations are able to reduce losses along the magnetic field lines but not at the level required for a reactor.

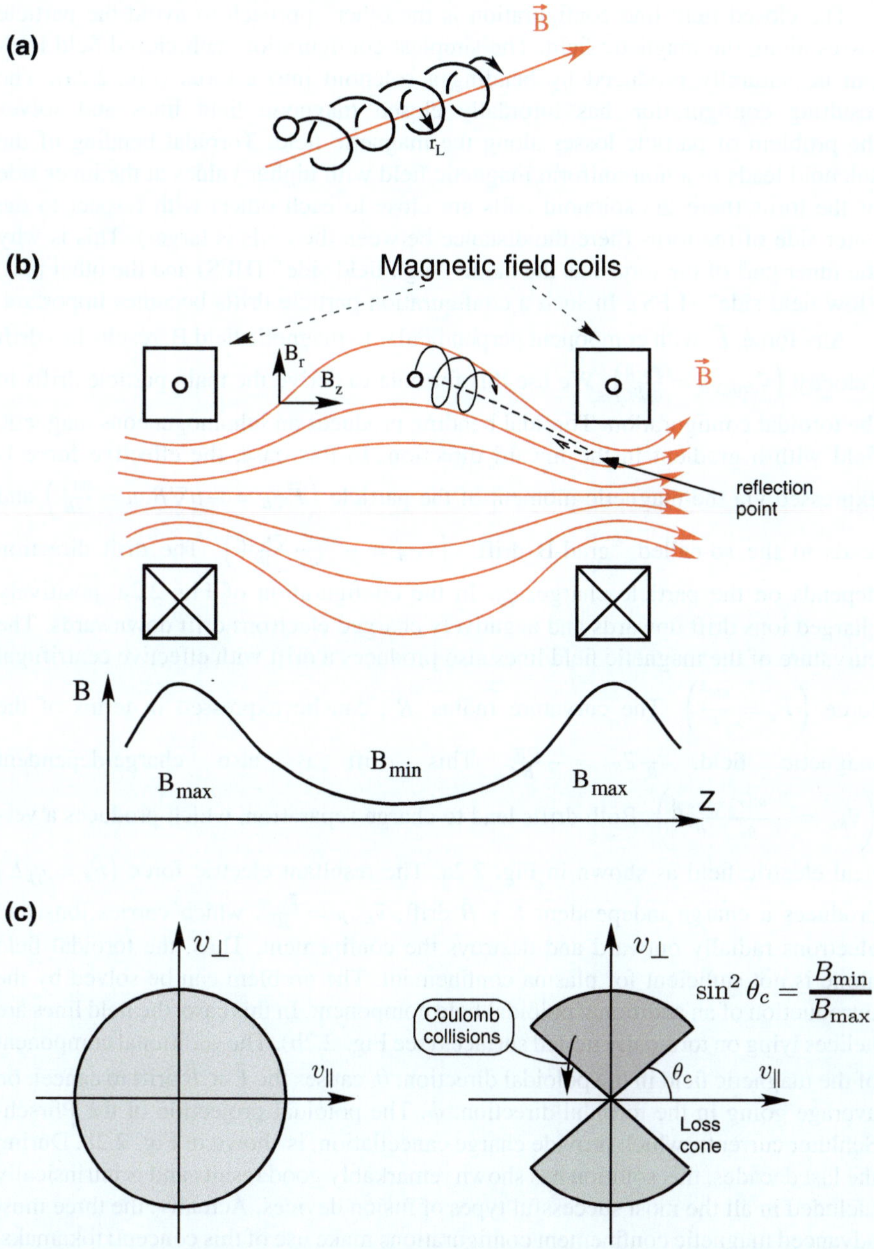

Fig. 2.1 **a** Single charge particle motion in a uniform magnetic field. **b** Magnetic bottle. Particle is reflected in high field region. **c** Velocity space before and after losses in the "loss cone". Collisions move new particles permanently into the loss cone

The closed field line configuration is the other approach to avoid the particle losses along the magnetic field. The simplest configuration with closed field lines can be naturally produced by bending a solenoid into a torus (Fig. 2.2a). The resulting configuration has toroidally closed magnetic field lines and solves the problem of particle losses along the magnetic field. Toroidal bending of the solenoid leads to a non-uniform magnetic field with higher values at the inner side of the torus (here the solenoid coils are close to each other) with respect to the outer side of the torus (here the distance between the coils is larger). This is why the inner part of the torus has the name "high field side" (HFS) and the other part, "low field side" (LFS). In such a configuration particle drifts becomes important.

Any force, $\vec{F}$, with component perpendicular to magnetic field B, results in a drift velocity $\left(\vec{v}_{drift \perp B} = \frac{\vec{F}\times\vec{B}}{q_p B^2}\right)$. We use this formula to derive the main particle drifts in the toroidal configuration. Toroidal bending produces an inhomogeneous magnetic field with a gradient in the inward direction. In this case, the effective force is expressed via the magnetic moment of the particle $\left(\vec{F}_{\nabla B} = -\mu \nabla B, \mu = \frac{mv_\perp^2}{2B}\right)$ and leads to the so-called "grad-B drift" $\left(\vec{v}_{\nabla B} = -\frac{m_p v_\perp^2}{2q_p}\frac{\nabla B\times\vec{B}}{B^3}\right)$. The drift direction depends on the particle charge, q_p. In the configuration of Fig. 2.2a, positively charged ions drift upwards and negatively charged electrons drift downwards. The curvature of the magnetic field lines also produces a drift with effective centrifugal force $\left(F_c = \frac{mv_\parallel^2}{r}\right)$. The curvature radius, R_c, can be expressed in terms of the magnetic field, $\frac{1}{R_c}\vec{e}_{Rc} = -\frac{\nabla B}{B}$. This drift is also charge-dependent $\left(\vec{v}_{Rc} = -\frac{m_p v_\parallel^2}{q_p}\frac{\nabla B\times\vec{B}}{B^3}\right)$. Both drifts lead to charge separation, which produces a vertical electric field as shown in Fig. 2.2a. The resultant electric force $\left(\vec{F}_e = q_p\vec{E}\right)$ produces a charge independent $\vec{E}\times\vec{B}$ drift, $\vec{v}_{E\times B} = \frac{\vec{E}\times\vec{B}}{B^2}$, which carries ions and electrons radially outward and destroys the confinement. Thus, the toroidal field alone is not sufficient for plasma confinement. The problem can be solved by the introduction of an additional poloidal field component. In this case, the field lines are helices lying on toroidally nested surfaces (see Fig. 2.2b). The additional component of the magnetic field in the poloidal direction, θ, causes the $\vec{E}\times\vec{B}$ drift to cancel, on average going in the toroidal direction, ϕ. The poloidal projection of the Pfirsch-Schlüter currents, which provide charge cancellation, is shown in Fig. 2.2b. During the last decades, this solution has shown remarkably good results and is intrinsically included in all the most successful types of fusion devices. Actually, the three most advanced magnetic confinement configurations make use of this concept: tokamaks, stellarators and reversed field pinches. The difference between these devices is in the way the poloidal field is produced, and its magnitude.

- *Tokamaks* The poloidal magnetic field is produced by the plasma current and is much smaller compared to the primary toroidal magnetic field: $B_\theta \ll B_\phi$ (see Fig. 2.2c).

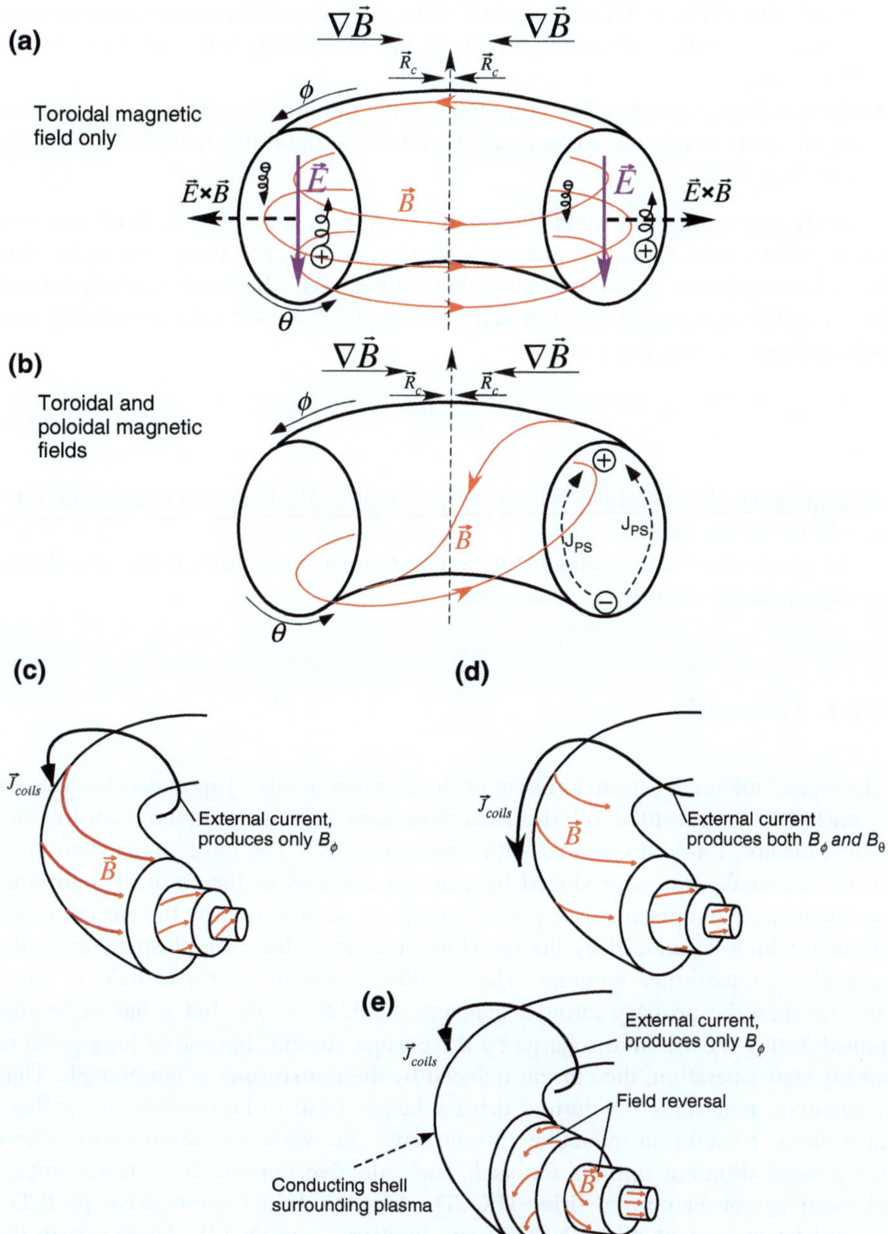

Fig. 2.2 **a** A simple toroidal magnetic field produces particle drifts, charge separation and ultimately confinement loss due to the $\vec{E} \times \vec{B}$ drift. **b** A helical magnetic field removes charge separation. **c** Tokamak. **d** Stellarator. **e** Reversed Field Pinch (RFP)

- *Stellarators* The poloidal magnetic field is produced by currents in external conductors. Thus, plasma current is not necessary for confinement (see Fig. 2.2d).
- *Reversed field pinches* The amplitude of the poloidal magnetic field produced by the plasma current is comparable with the toroidal magnetic field: $B_\theta \approx B_\phi$ (see Fig. 2.2e).

In all cases, magnetic field lines describe helices around nested toroidal surfaces, which form sequence around a single closed curve (magnetic axis). The helical winding of a field line is a topological quantity which can be characterized by the relation between the toroidal winding of the field line $\Delta\phi$ during one poloidal turn around the torus:

$$q = \frac{\Delta\phi}{2\pi} \tag{2.1}$$

This quantity, the so-called "safety factor", is strongly linked to plasma stability, as will be shown later.

All these magnetic confinement configurations have advantages and disadvantages, which we briefly discuss here.

2.1.1 Tokamak

The word "tokamak" is an acronym of the Russian words "**то**роидальная **ка**мера с **ма**гнитными **к**атушками" (toroidal'naya kamera s magnitnymi katushkami), which means "toroidal chamber with magnetic coils". The toroidal magnetic field in the tokamak, B_ϕ, is produced by external currents in the toroidal field coils which encircle the plasma. The poloidal field, B_θ, is produced by the current in the plasma, which is induced by the transformer action where the plasma acts as the secondary transformer winding. The resulting poloidal magnetic field is much smaller than the primary toroidal magnetic field, $B_\theta \ll B_\phi$, but it has to be sustained during the whole discharge by the plasma current. In case of long-pulse or steady state operation, the current induced by the transformer is not enough. Thus alternative, non-inductive current drive schemes have to be considered for these operations, because an inductive current drive can work only temporarily. There are several different options for such non-inductive current drive in tokamaks: electron cyclotron current drive (ECCD), lower hybrid current drive (LHCD), neutral beam current drive (NBCD) and bootstrap current [2]. Among them the bootstrap current [3] is the most attractive option, which can produce a sufficiently large amount of toroidal current in big tokamaks without additional costs for the current drive system. The physical mechanism behind the bootstrap current is directly related to the inhomogeneity of the magnetic field strength which was the problem for the particle confinement in the previous section, $B \propto 1/R$. This inhomogeneity separates particles into "trapped particles", which are reflected by the higher magnetic field on the high field side, and "passing particles", which are able to make a full poloidal pass along a magnetic field line. Interaction of these populations produces bootstrap current. Particle "trapping" has the same physical mechanism as discussed for the magnetic mirror above (conservation of energy plus first adiabatic invariant).

Presently, the tokamak is the most advanced concept for magnetic plasma confinement, which is the result of very intensive research during the last six decades. The temperatures required for fusion are achieved in present-day tokamaks, fusion confinement is one step away and is expected to be achieved in the next-generation experiment ITER [4] (see Fig. 2.3). In this book we will focus mainly on the tokamak device, especially on the stabilization of different instabilities in present devices and perspectives for ITER. The general overview of tokamak physics is given in J. Wesson's comprehensive book "Tokamaks" [1]. The other configurations will be also examined in some parts of this book, insofar as this is required to understand and deal with tokamak problems.

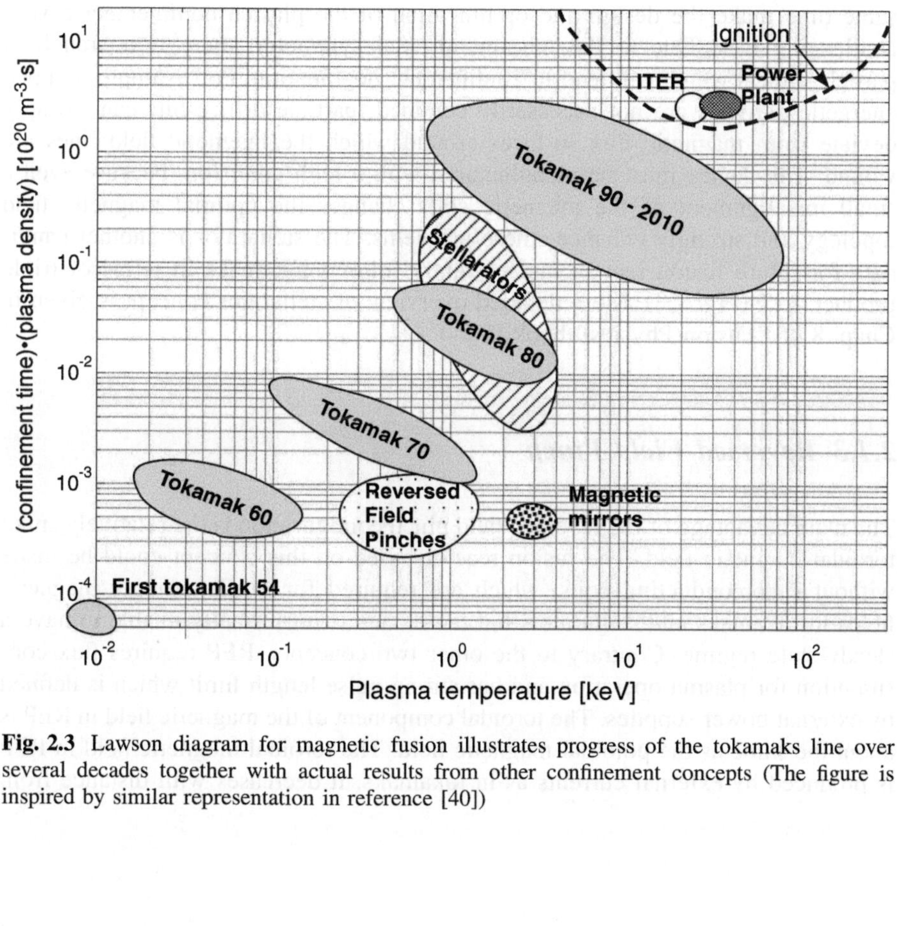

Fig. 2.3 Lawson diagram for magnetic fusion illustrates progress of the tokamaks line over several decades together with actual results from other confinement concepts (The figure is inspired by similar representation in reference [40])

2.1.2 Stellarator

The helical twist of the magnetic field lines can also be produced by additional external currents. In this case, plasma current is not required for confinement and the plasma operations are potentially steady-state. This is the so-called "stellarator" line of fusion research, an independent approach whose results are currently behind the tokamak line; however the concept is considered as long term solution.

The lack of plasma current avoids current-driven instabilities in stellarator. For example, there are no disruptions, no tearing modes and no resistive wall modes in stellarator. All these instabilities will be discussed in details in this book. In reality, some current is still produced by particle effects. The bootstrap current, which is an advantage for tokamaks, should be minimized by the magnetic design and/or external current drive techniques. The other line of the stellarator optimization relays on some plasma current for the confinement. The overall plasma behavior is calmer compared to the tokamak case.

The main advantages and disadvantages come from the stellarator's complicated geometry. The concept is fully three-dimensional, in contrast to the toroidal symmetry of tokamaks. The resulting three-dimensional plasma, magnetic field, and vessel provide wide variety of different factors for optimization and at the same time make the design and optimization of the plasma confinement a very challenging task. Due to the missing toroidal symmetry, there is a branch of particle effects which can lead to confinement degradation. For example, trapped energetic particles are not necessarily confined; particle drift orbits can strongly deviate from magnetic flux surfaces around which the magnetic field lines are wound. The device must be manufactured with a high precision, because even a small misalignment of the magnetic coils changes the optimal magnetic field topology and strongly enhances these problems. The stellarator is another candidate for future fusion reactor and currently follows tokamaks in terms of triple product, $n\tau T$ (Fig. 2.3). More detailed overview of stellarator concept is given in Chap. 8 of "Fusion Physics" book [5].

2.1.3 Reversed Field Pinch

The main advantage of the reversed field pinch configuration is the relatively small toroidal magnetic field. The fusion reactor based on this concept could be made without superconducting coils, which are required for much stronger magnetic fields in tokamaks and stellarators but the device is intrinsically unable to have a steady state regime. Contrary to the other two concepts, RFP requires flux consumption for plasma operation and has strong pulse length limit which is defined by external power supplies. The toroidal component of the magnetic field in RFP is about the same as the poloidal magnetic field. The toroidal magnetic field in RFP is produced by external currents as in tokamaks. It decreases with distance from the plasma centre and reverses direction near the plasma edge (this fact is reflected in the name of the device). The RFP configuration of the magnetic field is a result of the relaxation process in plasma to a minimal energy state accompanied by conservation of the magnetic helicity (a quantity that measures the field's degree of "twisting"). Thus, magneto-hydrodynamic instabilities (MHD instabilities) and their relaxation play a key role in the construction of RFP equilibria. The resulting RFP magnetic field is weak compared to the two previous concepts and the plasma does not satisfy the Kruskal-Shafranov stability criterion (see Sect. 2.6). This criterion says that the safety factor, defined by (2.1), has to be larger than one at the plasma boundary to keep plasma stable against large scale perturbations driven by the current gradient ($q > 1$). Thus, MHD instabilities not only create the RFP configuration but also destroy plasma confinement in a free boundary case. A conducting shell around the plasma is a way to stabilize the plasma against these perturbations. The motion of the current-carrying plasma due to global instabilities induces currents in the shell and provides a stabilizing restoring force.

In a typical RFP configuration, multiple MHD modes are unstable simultaneously. An overlap of all these perturbations produces a stochastic magnetic field where magnetic field lines wander chaotically inside the plasma. This situation is significantly different to that in tokamaks and stellarators, where the field lines normally stay on their respective flux surfaces and radial transport is suppressed. Stochasticity substantially increases energy and particle transport because particles follow the magnetic field lines going also in the radial direction (not only toroidally and poloidally!). Suppression of the MHD instabilities by reducing the free energy source (the current density gradient) is one of the possible solutions, allowing a reduction of the perturbation level and creation of nested flux surfaces inside the plasma. The other approach is to suppress all instabilities except one and construct the so-called "single helicity state". In this case, a single island dominates and forms nested magnetic flux surfaces with the island helicity inside the plasma. In order to manage these control tasks, RFPs are equipped with a large set of control coils to act on the plasma. Currents in the coils are feedback-controlled, and influence the MHD modes' behavior to control and/or suppress a particular instability. The results of such experiments are very valuable from a scientific point of view and can be used in some cases in tokamaks. In that sense, the RFP is a nice testbed for different control approaches which then can be transferred to the tokamak case. In terms of triple product required for the reactor, the concept is behind modern tokamaks and stellarators as shown in Fig. 2.3. More detailed introduction in to this concept is given in Chap. 9 of "Fusion Physics" book [5].

2.2 Fluid Description of the Plasma

The discussion in the previous section has focused on single particle motion in different magnetic field configurations. Here, we would like to identify basic stability properties of the plasma in a particular magnetic configuration. The self-consistent interaction between particles in the plasma is extremely important for this problem. Calculation of all interactions between all individual particles is impossible task and is not actually required for plasma physics problems. The macroscopic plasma behavior can be described by averaging over a sufficiently large number of particles, an approach which is also used in statistical mechanics. This formalism leads to a kinetic description of the plasma and the plasma itself is characterized by distribution functions. The kinetic models are very accurate and include a wide variety of physical phenomena which are not interesting for us in this chapter. Thus, a further simplification can be done. Integration of the equation of motion over a distribution function results in the two-fluid description of the plasma while retaining self-consistency in that the electrons and ions form two different fluids (for example see J.A. Bittencourt book "Fundamentals of Plasma Physics" [6]). This fluid model can be simplified further to a single fluid approach which is called magnetohydrodynamics (MHD) and can be also derived by combining Newton's mechanics and Maxwell's electrodynamics equations. In its simplest form, the approach treats the plasma as a conductive fluid. The MHD approximation simplifies the problems drastically and substitutes microscopic variables (particle velocities, particle trajectories, etc.) by macroscopic ones, as well as easily recognizable physical quantities (plasma density n, mass density ρ, pressure p, fluid velocity $\vec{v}$, etc.). This approach is easier to solve and it is more intuitive to understand, in comparison to the kinetic description. The equations of the single-fluid MHD model are the following:

$$\frac{\partial \rho}{\partial t} + \vec{\nabla} \cdot (\rho \vec{v}) = 0 \text{ Conservation of mass} \tag{2.2}$$

$$\rho \frac{d\vec{v}}{dt} = \vec{J} \times \vec{B} - \nabla p \text{ Force balance} \tag{2.3}$$

$$\vec{E} + \vec{v} \times \vec{B} = 0 \text{ (ideal MHD), or } \vec{E} + \vec{v} \times \vec{B} = \eta_{\parallel} \vec{J} \text{ (resistive MHD) Ohm's law} \tag{2.4}$$

$$\nabla \times \vec{E} = -\frac{\partial \vec{B}}{\partial t} \text{ Faraday's law} \tag{2.5}$$

$$\nabla \times \vec{B} = \mu_0 \vec{J} \text{ Ampere's law} \tag{2.6}$$

$$\nabla \cdot \vec{B} = 0 \text{ Absence of magnetic monopoles} \tag{2.7}$$

Here $\vec{B}$ is the magnetic field, $\vec{E}$ is the electric field, $\vec{J}$ is the plasma current density, and $\eta_{\parallel}$ is the plasma resistivity parallel to the magnetic field. It is important to note that the single fluid MHD approximation does not distinguish between electrons and ions since only their sum appears in the equations; for example, for the plasma pressure $p = p_e + p_i$. The mass density in MHD is clearly determined by the ions ($\rho \equiv m_i n$) and the momentum of the fluid is carried by the ions as well ($\vec{v} \equiv \vec{v}_i$), because they represent the dominant mass of the plasma.

The equation system is not closed. The typical closure is the adiabatic equation, which represents conservation of energy in the system: $\frac{d}{dt}\left(\frac{p}{\rho^\Gamma}\right) = 0$, where Γ is the adiabatic factor.

The equation system actually contains two different approaches depending on the assumption for the plasma resistivity in Ohm's law. In the case of quasi-static plasma configurations, dissipation rates are small, the plasma resistivity is negligible and the plasma is well described in the frame of ideal MHD ($\vec{E} + \vec{v} \times \vec{B} = 0$). Combining this variant of Ohm's law with Faraday's law results in the convection equation for the magnetic field $\frac{\partial \vec{B}}{\partial t} = \nabla \times \left(\vec{v} \times \vec{B}\right)$. Thus, evolution of the magnetic field is determined by the plasma motion. The magnetic field is "frozen into" the plasma and any topological changes of the magnetic field in ideal MHD are forbidden.

For long-term evolution of the plasma, non-linear processes with small but finite resistivity become important and the ideal MHD assumption is not valid anymore. Even a small resistivity in the resistive MHD, $\vec{E} + \vec{v} \times \vec{B} = \eta_{\parallel}\vec{J}$, allows to change the topology of the magnetic field at slow resistive time scales[2] and produces a new class of so-called "resistive" MHD instabilities as discussed later.

The typical fluid approximation is based on the assumption that the system is locally close to the thermodynamic equilibrium, which requires a certain rate of collisions and dissipation (one makes this step at the reduction of the kinetic approach to the fluid description). For that case, the mean free path λ should be short compared with the typical gradient scales, $\lambda|\nabla f| \ll f$ [7]. The mean free path in hot plasmas becomes very long, but at the same time in a magnetized plasma, for perpendicular directions, the mean free path is roughly the gyroradius, $\lambda \approx r_L$. This value is typically very small and the condition is fulfilled. Consequently, the fluid description can be applied to the plasma behavior perpendicular to the magnetic field, which is typically the case for our analysis. In the context of the fusion problem, the MHD model provides a reasonably accurate description of macroscopic equilibrium and stability [8].

2.3 Plasma Equilibrium

The equations which govern the plasma equilibrium can be easily obtained from the system of MHD equations discussed above. For an equilibrium situation, it is naturally to assume zero velocity, $\vec{v} = 0$, and constant macroscopic quantities, $\partial/\partial t = 0$. In this case, the system of equations is reduced to three equations:

---

[2] $\tau_R = \frac{\mu_0 l^2}{\eta}$, where η is the plasma resistivity and l is the characteristic width of the reconnection region (see figure 2.7b).

$$\vec{J} \times \vec{B} = \nabla p \quad (2.8)$$

$$\nabla \times \vec{B} = \mu_0 \vec{J} \quad (2.9)$$

$$\nabla \cdot \vec{B} = 0 \quad (2.10)$$

These equations describe the equilibrium properties of all magnetic configurations of our interest.[3] The main properties of these configurations can be directly obtained from these equations.

The equilibrium configurations have common properties which are widely used later and which we discuss here before the analysis of plasma stability. Figure 2.4a shows a cross-section of the tokamak plasma. The contours of constant pressure of a well-confined equilibrium form a set of nested surfaces. Two important properties of these surfaces follow directly from the force balance equation. The dot product of the MHD momentum (2.8) with $\vec{B}$ gives:

$$\vec{B} \cdot \nabla p = 0 \quad (2.11)$$

using the fact that $\vec{B} \cdot (\vec{J} \times \vec{B}) = \vec{J} \cdot (\vec{B} \times \vec{B}) = 0$. Thus, the magnetic field lines must lie in the surface of constant pressure because there is no component of $\vec{B}$ perpendicular to the surface (in the direction of the pressure gradient ∇p). The other property is obtained by forming the dot product with $\vec{J}$:

$$\vec{J} \cdot \nabla p = 0 \quad (2.12)$$

The current lines also lie on the constant pressure surfaces and there is no current component in the direction of the pressure gradient, ∇p. These surfaces of constant pressure and current are also the surfaces of constant magnetic flux, which is a typical "radial" coordinate in fusion devices with arbitrary poloidal cross-section and can be defined using different components of the magnetic field. The toroidal flux definition uses the toroidal magnetic field B_ϕ:

$$\Phi \equiv \frac{1}{2\pi} \int B_\phi dS_\Phi. \quad (2.13)$$

The poloidal magnetic flux is defined using the poloidal magnetic field component B_{pol}:

$$\Psi \equiv \frac{1}{2\pi} \int B_{pol} dS_\Psi. \quad (2.14)$$

[3] The case of an equilibrium with constant flow does not dominate in fusion configurations and requires a different treatment.

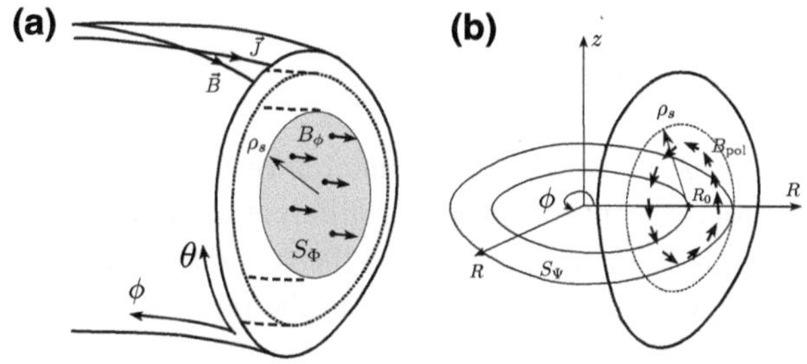

Fig. 2.4 **a** General direction of the $\vec{B}$ and $\vec{J}$ lines on the constant pressure surfaces. The toroidal flux at the position ρ_s is defined using the toroidal magnetic field B_ϕ and the area S_Φ. **b** The definition of the poloidal magnetic flux uses the poloidal magnetic field B_{pol} and area S_Ψ. The magnetic axis is R_0

The respective surfaces are indicated by the shaded areas in Fig. 2.4. Here the normalization factor $(1/2\pi)$ is introduced to simplify the relation between the flux and the poloidal magnetic field:

$$\vec{B}_{pol} = \frac{1}{R}\vec{e}_\phi \times \nabla\Psi \tag{2.15}$$

The safety factor definition (2.1) can be also written using these flux functions:

$$q(\Psi) \equiv \frac{d\Phi}{d\Psi} \tag{2.16}$$

It is convenient to define so-called "radial coordinates". These values are called poloidal (ρ_{pol}) and toroidal (ρ_{tor}) coordinates depending on the flux used in the definition:

$$\rho_{pol} = \sqrt{\frac{\Psi - \Psi_0}{\Psi_s - \Psi_0}}, \tag{2.17}$$

$$\rho_{tor} = \sqrt{\frac{\Phi - \Phi_0}{\Phi_s - \Phi_0}}, \tag{2.18}$$

where the index s refers to the separatrix and index 0 to the magnetic axis. These coordinates are scaled such that $\rho = 0$ on the magnetic axis and $\rho = 1$ at the plasma boundary.

The equilibrium equations can be also represented conveniently in terms of the flux functions, which leads to so-called Grad-Shafranov equation for the poloidal flux $\Psi = \Psi(R, z)$:

$$R\frac{\partial}{\partial R}\left(\frac{1}{R}\frac{\partial \Psi}{\partial R}\right)+\frac{\partial^2 \Psi}{\partial z^2}=-\mu_0^2 II'-\mu_0 R^2 p'\left[=\mu_0 R J_\phi\right] \tag{2.19}$$

Here, $I \equiv RB_\phi = I(\Psi)$ is the stream function of the poloidal current, which is also a flux function. The prime indicates differentiation with respect to Ψ. The right side of the equation is proportional to the toroidal current density (J_ϕ). The functions $\Psi(R,z)$ and $I(R,z)$ are connected to the plasma current density (J) and the magnetic field as follows:

$$B_R=-\frac{1}{R}\frac{\partial \Psi}{\partial z}\; B_z=\frac{1}{R}\frac{\partial \Psi}{\partial R}\; J_R=\frac{1}{R}\frac{\partial I}{\partial z}\; J_z=-\frac{1}{R}\frac{\partial I}{\partial R} \tag{2.20}$$

The Grad-Shafranov equation can be solved numerically for any plasma cross-section and provides the constant flux (and pressure) surfaces in real tokamak geometry. Detailed derivation of the equation can be found, for example, in the book by J. P. Goedbloed "Advanced magneto-hydrodynamics" [9].

Before we come to the stability analysis, it is important to highlight the two main stabilizing actions of the magnetic field. As was shown before, the conservation of the momentum equation is reduced in an equilibrium situation to a simple force balance between the magnetic force from the current in the plasma, $\vec{F}_{Lorentz} = \vec{J} \times \vec{B}$, and the pressure force, $\vec{F}_{pressure} = -\nabla p$, from the kinetic plasma pressure. This force balance has to be satisfied at each point of the plasma. Two general properties of the magnetic field lines, magnetic pressure and magnetic field line tension, play an important role in this balance (see for example book by Harra and Mason [10]). We formulate the pressure balance in the direction perpendicular to the magnetic field using the force balance (2.8), Ampere's law for the current $\vec{J}$, and the vector identity

$$\frac{1}{2}\nabla(B^2)=\frac{1}{2}\nabla(\vec{B}\cdot\vec{B})=\vec{B}\times(\nabla\times\vec{B})+(\vec{B}\cdot\nabla)\vec{B}. \tag{2.21}$$

Then the Lorentz force is

$$\vec{F}_{Lorentz}=\vec{J}\times\vec{B}=\frac{1}{\mu_0}(\nabla\times\vec{B})\times\vec{B}=\frac{1}{\mu_0}\left(\frac{\nabla(B^2)}{2}-(\vec{B}\cdot\nabla)\vec{B}\right) \tag{2.22}$$

$$F_{Lorentz\perp B}=\underbrace{\nabla_\perp\left(\frac{B^2}{2\mu_0}\right)}_{\substack{mag.field\\ pressure}}-\underbrace{\frac{B^2}{\mu_0 R_c}}_{\substack{mag.field\\ tension\perp B}}. \tag{2.23}$$

The parallel component provides tension only along the magnetic field and is not able to compensate the plasma pressure. The final pressure balance, perpendicular to the magnetic field, can be written as follows:

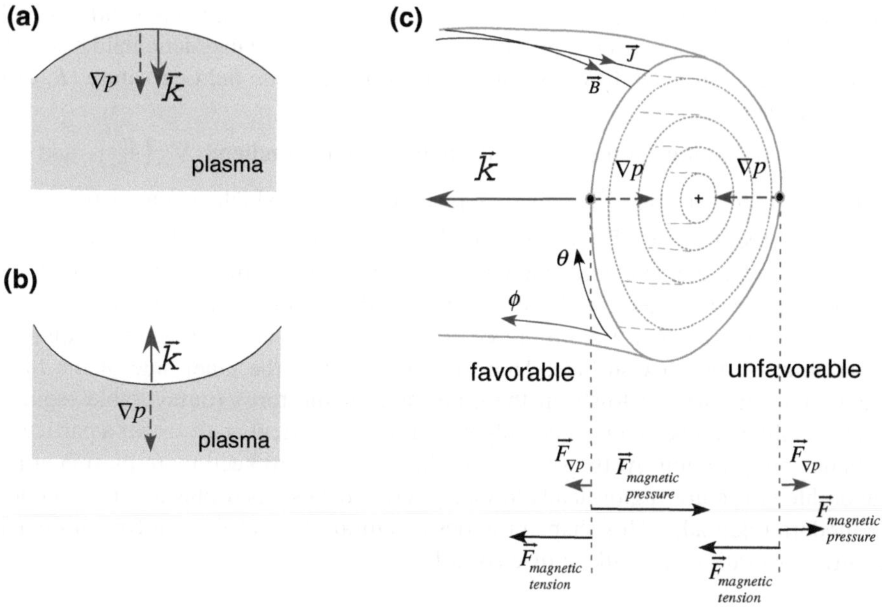

Fig. 2.5 **a** Unfavorable curvature of the magnetic field. **b** Favorable curvature of the magnetic field. **c** The tokamak case has favorable curvature (at the inner side of the torus) and unfavorable curvature (at the outer side of the torus). The resultant equilibrium forces are shown for *low* and *high* field side

$$\nabla_{\perp}\left(\underbrace{\frac{B^2}{2\mu_0}}_{\substack{mag.field\\ pressure}} + \underbrace{p}_{\substack{plasma\\ pressure}}\right) - \underbrace{\frac{B^2}{\mu_0 R_c}}_{\substack{mag.field\\ tension \perp B}} = 0 \tag{2.24}$$

where R_c is the curvature radius. The magnetic field has a property similar to plasma pressure, p, which is the magnetic field pressure, $\frac{B^2}{2\mu_0}$, and the magnetic tension, $\frac{B^2}{\mu_0 R_c}$. The magnetic field line is stretched on a particular flux surface and tries to keep the minimal length due to this tension. In the case where the curvature vector is parallel to the ∇p vector, the curvature is unfavorable and acts as a destabilizer (Fig. 2.5a). As one moves radially outward into the region of larger curvature in Fig. 2.5a, the field tends to become smaller because of the larger radius of curvature. $\langle B^2 \rangle$ decreases indicating instability. The situation is opposite for Fig. 2.5b, where perturbations in the direction of the vacuum region lead to an increase of $\langle B^2 \rangle$ and thus are stabilized. In this case, the curvature vector is antiparallel to the pressure gradient vector, and the system has so-called "favorable curvature". This result can be obtained also from changes of the system energy due to a perturbation in these two cases (see 2.35). The real tokamak situation is more complicated and has both types of the curvatures (Fig. 2.5c). In this case, decay of the magnetic field has to be taken into account ($B \sim 1/R$) together with local magnetic field curvature (R_c) for radial pressure balance.

The combination of the magnetic field pressure gradient, $\nabla_{\perp}\left(\frac{B^2}{2\mu_0}\right)$, and the magnetic field tension, $\frac{B^2}{\mu_0 R_c}$, gives a resultant force, which counteracts plasma pressure force, $F_{\nabla p} = -\nabla_{\perp} p$, and provides radial force balance (Fig. 2.5c). At the low field side, the magnetic field tension provides plasma stability, while the gradient of the magnetic field pressure makes the plasma stable on the high field side. Thus, the resultant Lorentz force counteracts the plasma pressure gradient at each point on the flux surface. This force is larger at the inner side of the torus (favorable region) and lower at the outer side of the torus (unfavorable region). The net effect can be determined only by careful numerical analysis for a particular plasma configuration. At the same time, the basic idea to keep more plasma at the favorable curvature region leads to elongated and D-shaped plasma cross-section in modern tokamaks. This shape provides a natural way for X-point formation and divertor operations as will be discussed later.

2.4 Plasma Stability

The main question in the MHD stability theory is to consider an MHD equilibrium and predict its stability. Plasma stability with respect to different perturbations can be analyzed in several different ways. Here, we restrict our consideration to the simplest form of the energy principle within the ideal MHD approximation. The perturbations are assumed to be small in comparison to equilibrium quantities, for example pressure perturbations are much less than the pressure value, $p_1 \ll p_0$. In the following, equilibrium quantities and perturbed quantities are denoted by subscript 0 and 1, respectively. This assumption, together with the assumption of the stationary equilibrium state $\left(\vec{J}_0 \times \vec{B}_0 = \nabla p_0,\ \vec{v}_0 = 0\right)$ lead to a system of linearized MHD equations:

$$\frac{\partial \rho_1}{\partial t} + \rho_0 \vec{\nabla} \cdot \vec{v}_1 = 0 \tag{2.25}$$

$$\rho_0 \frac{d\vec{v}_1}{dt} = -\nabla p_1 + \vec{J}_0 \times \vec{B}_1 + \vec{J}_1 \times \vec{B}_0 \tag{2.26}$$

$$\frac{\partial p_1}{\partial t} = -\vec{v}_1 \cdot \nabla p_0 - \Gamma p_0 \nabla \cdot \vec{v}_1 \tag{2.27}$$

$$\frac{\partial \vec{B}_1}{\partial t} = \nabla \times (\vec{v}_1 \times \vec{B}_0) \tag{2.28}$$

$$\nabla \times \vec{B}_1 = \mu_0 \vec{J}_1 \tag{2.29}$$

$$\nabla \times \vec{B}_0 = \mu_0 \vec{J}_0 \tag{2.30}$$

We use these equations to identify energy changes in the system resulting from small initial perturbations. The energy principle is based on the idea that an equilibrium is unstable if any perturbation reduces the potential energy of the system. Changes in the potential energy can be calculated for an arbitrary displacement, $\vec{\xi}$, using the system of MHD equations. The displacement is directly related to the plasma velocity, $\vec{v}_1 = \frac{d\vec{\xi}}{dt}$. The system of linearized MHD equations can be rewritten in the form of the wave equation:

$$\vec{F}\left(\vec{\xi}\right) \equiv \rho_0 \frac{\partial^2 \vec{\xi}}{\partial t^2} = J_0 \times \left[\nabla \times \left(\vec{\xi} \times \vec{B}_0\right)\right] + \frac{1}{\mu_0}\left\{\nabla \times \left[\nabla \times \left(\vec{\xi} \times \vec{B}_0\right)\right]\right\} \times \vec{B}_0 + \nabla\left(\vec{\xi} \cdot \nabla p_0\right) + \Gamma \nabla\left(p_0 \nabla \cdot \vec{\xi}\right) \tag{2.31}$$

This equation is the ideal MHD wave equation, which defines the ideal MHD force operator $\vec{F}\left(\vec{\xi}\right)$. The same conclusion can be obtained starting from the force balance (2.8). If the plasma is unstable, the force balance is not fulfilled and the resulting force is non-zero:

$$\vec{F}_{result} = \vec{J} \times \vec{B} - \nabla p \neq 0 \tag{2.32}$$

We can substitute this force into Newton's second law, where the second derivative of the displacement gives the plasma acceleration, $\vec{a} = \frac{\partial \vec{v}}{\partial t}$. The resulting equation is identical to the previous result if one substitutes total values of the current, pressure and magnetic field and makes the linear MHD assumption:

$$\rho \frac{\partial^2 \vec{\xi}}{\partial t^2} = \vec{F}_{result}\left(\vec{\xi}\right) = \vec{J} \times \vec{B} - \nabla p = \vec{F}\left(\vec{\xi}\right) \tag{2.33}$$

This resulting force operator can be used directly to calculate changes in the potential energy of the plasma as an integral over the whole system volume (including the plasma region, the vacuum region and the conducting wall):

$$\delta W = \int \vec{\xi} \cdot \vec{F}\left(\vec{\xi}\right) dV \tag{2.34}$$

If the energy change is positive, the plasma is linearly stable with respect to the perturbation $\vec{\xi}$ ($\delta W \geq 0$), otherwise the plasma is unstable ($\delta W < 0$). It is convenient to write this equation in the so-called "intuitive form" [8, 11].

$$\delta W = \delta W_{plasma} + \delta W_{vacuum} + \delta W_{surface} \tag{2.35}$$

$$\delta W_{plasma} = \frac{1}{2}\int\limits_{plasma} \left\{ \frac{|\vec{B}_{1,\perp}|^2}{\mu_0} + \frac{B_0^2}{\mu_0}\left|\nabla\cdot\vec{\xi}_\perp + 2\vec{\xi}_\perp\cdot\vec{\kappa}\right|^2 + \Gamma p_0\left|\nabla\cdot\vec{\xi}\right|^2 - 2\left(\vec{\xi}_\perp\cdot\nabla p_0\right)\left(\vec{\kappa}\cdot\vec{\xi}_\perp^*\right) - J_\parallel\left(\vec{\xi}_\perp^*\times\vec{b}\right)\cdot\vec{B}_{1,\perp}\right\} dV$$

$$\delta W_{vacuum} = \frac{1}{2}\int\limits_{vacuum} \frac{(B_1)^2}{\mu_0} dV$$

$$\delta W_{surface} = \frac{1}{2}\int\limits_{surface} \left|\vec{n}\cdot\vec{\xi}_\perp\right|^2 \vec{n}\cdot\left\|\nabla\left(p + \frac{B^2}{2\mu_0}\right)\right\| dS$$

$$\vec{B}_1 = \vec{B}_{1,\perp} + B_{1,\parallel}\hat{\vec{b}},\ B_{1,\parallel} = -B_0\left(\nabla\cdot\vec{\xi}_\perp + 2\vec{\xi}_\perp\cdot\vec{\kappa}\right) + \frac{\mu_0}{B}\vec{\xi}_\perp\cdot\nabla p_0,\ \vec{J} = \vec{J}_\perp + J_\parallel\hat{\vec{b}},$$

$\vec{\xi} = \vec{\xi}_\perp + \xi_\parallel\hat{\vec{b}}$, $\vec{\kappa} = \hat{\vec{b}}\cdot\nabla\hat{\vec{b}}$, where the indexes $\|$and $\perp$ refer to the direction parallel and perpendicular the equilibrium magnetic field B_0, perturbed magnetic field is expressed as $\vec{B}_1 = \nabla\times\left(\vec{\xi}\times\vec{B}_0\right)$, and $\|T\|$ denoting the jump in T from vacuum to plasma.

The squares of all components are always positive and provide only stabilizing effect on the plasma:

(1) $\frac{|\vec{B}_{1,\perp}|^2}{\mu_0} > 0$ *is stabilizing* because the magnetic field tries to prevent its bending;

(2) $\frac{B_0^2}{\mu_0}\left|\nabla\cdot\vec{\xi}_\perp + 2\vec{\xi}_\perp\cdot\vec{\kappa}\right|^2 > 0$ *is stabilizing* because the magnetic field tries to prevent its compression;

(3) $\Gamma p_0\left|\nabla\cdot\vec{\xi}\right|^2 > 0$ *is stabilizing* because the plasma itself counteracts compression (Γ is the adiabatic factor)

(4) $2\left(\vec{\xi}_\perp\cdot\nabla p_0\right)\left(\vec{\kappa}\cdot\vec{\xi}_\perp^*\right)$ *can be* either positive (stable) or negative (*unstable*). The term depends on the pressure gradient and can lead to the so-called "pressure-driven" instabilities. As discussed before, this term is typically more unstable at the low field side of the tokamak.

(5) $J_\parallel\left(\vec{\xi}_\perp^*\times\vec{b}\right)\cdot\vec{B}_{1,\perp}$ *can be* also either stable or *unstable*. It gives rise to the so-called "current-driven" instabilities if it is negative. Instabilities are driven by the parallel component of the current.

There is an important factor which can influence the MHD stability and is not considered in the MHD approximation. MHD instabilities can interact directly with some of the plasma particles. This process has a kinetic nature and requires more complicated approaches than MHD for a correct description. These particle-wave interactions have either a "resonant" or a "non-resonant" character. MHD instability is a perturbation of the plasma which is characterized by plasma displacement and also by perturbed electric and magnetic fields associated with this displacement. Thus, the instability can be considered as an electromagnetic wave with a particular phase velocity. The particles with similar velocities will be either accelerated or decelerated. Depending on the initial slope of the particle distribution function and its spatial gradient the net result could either make the mode unstable (energy transfer from particle to the instability) or make it more stable (energy flows from the instability into particle motion). This is an example of resonant interaction. If the particles are much faster than the frequency of the wave, they enclose a certain magnetic flux within their orbit. Conservation of this flux tries to prevent any external distortions (any displacement due to MHD instability) and stabilize the instability. This is an example of the "non-resonant" interaction. Results of all these kinetic processes are mentioned in this chapter as "particle-driven" instabilities.

The safety factor value, introduced in (2.1) and (2.16), is a relation between toroidal and poloidal flux changes which defines the inclination of the magnetic field lines at a particular flux surface, or how fast the magnetic field line winds around the torus. Neighboring flux surfaces have different inclination of the magnetic field lines and thus different safety factor values. Simple analysis shows that two situations are possible for the field line behavior at a flux surface depending on its safety factor value:

(1) The magnetic field line winds around the corresponding flux surface and never ends. In this case it covers densely the whole flux surface and the surface is "non-resonant" (Fig. 2.6a).
(2) The field line closes after a few windings around the torus if the safety factor at this flux surface can be represented as relation between two co-prime integers: $q = m/n$. In contrast to the previous situation, the field line does not densely cover the flux surface and perturbations are easily excited here. These surfaces are called "resonant surfaces". A perturbation with the same helicity as the resonant surface can occur here, with the poloidal mode number m and toroidal mode number n.

There is a wide variety of MHD instabilities which can be unstable inside the plasma or at the plasma boundary. These modes limit maximal achievable plasma parameters and the main properties of these modes have to be studied carefully. We start this study with a classification of the MHD instabilities in the next section.

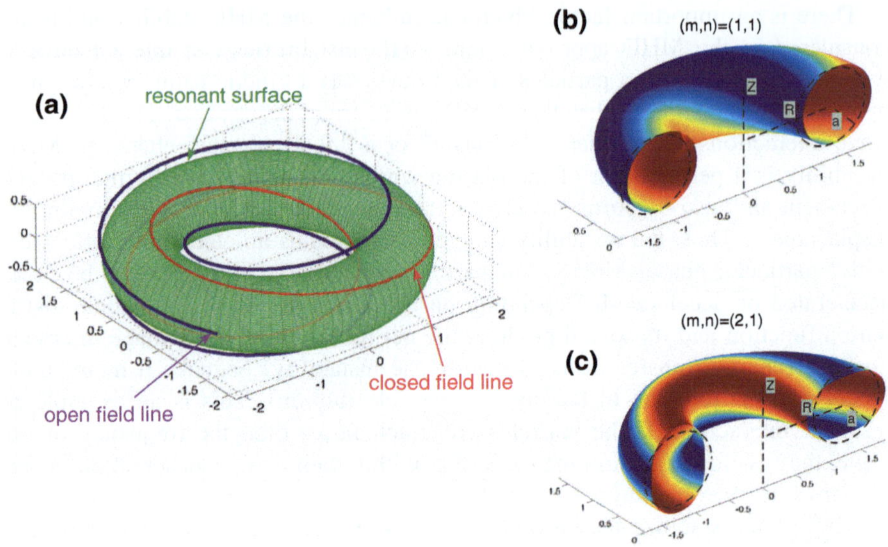

Fig. 2.6 **a** Resonant surface (*green*) and corresponding closed field line (*red*). The neighboring flux surface is non-resonant and the field line is open (*blue*). **b** Displacement of the flux surface due to an $(m, n) = (1, 1)$ perturbation. The *dashed line* represents the undisturbed contour. **c** Displacement of the flux surface due to an $(m, n) = (2, 1)$ perturbation. The *dashed line* represents the undisturbed contour

2.5 Basic Classifications of MHD Instabilities

There are several basic values which can be measured in the experiment and characterize MHD instability in tokamaks:

- growth rate of the mode (γ),
- mode numbers (m, n),
- mode frequency in the laboratory frame (ω),
- radial structure of the eigenfunction $\left(\hat{\xi}_r(\rho)\right)$, where ρ is the radial coordinate.

The growth of the mode amplitude is typically characterized by the growth rate (γ) assuming the exponential growth of the perturbation with time $\left(b(t) = \hat{b}_r e^{\gamma t}\right)$.[4] The instability with mode numbers (m, n) has the same helicity as the resonant surface where it is located. Thus, the safety factor value of the resonant surface follows directly from the helical structure of the instability $(q = m/n)$. If the mode

---

[4] The exponential growth is assumed in linearized MHD (2.25)–(2.30). It is called "linear growth" phase. This description is applicable only in some special cases, for example for resistive wall modes dynamic in RFPs. In practice, the modes are often in non-linear regime and this assumption is not valid anymore, for example for tearing modes, edge localized modes, etc.

rotates with respect to the laboratory frame, its rotation is characterized by the mode frequency (ω). The structure of the radial displacement component $\left(\hat{\xi}_r(\rho)\right)$ represents the shift of the equilibrium flux surfaces due to the perturbation at different radial locations. Identification of all these values, in particular the radial structure of the displacement, is a challenging task which is described in Chap. 3. If all these values are known, one can describe the displacement of the equilibrium due to a stationary mode in each point of the plasma:

$$\xi(\rho,\theta,\phi,t) = \hat{\xi}_r(\rho)\cdot\cos(m\theta - n\phi + \omega t)\cdot e^{\gamma t} \tag{2.36}$$

Examples of the distorted surfaces due to perturbations with $(m,n) = (1,1)$ and $(m,n) = (2,1)$ helicities are shown in Fig. 2.6b, c respectively.

Up to this point, only the structure of the instability has been discussed. It is an important ingredient for basic classification of the MHD instabilities based on its physical properties. In order to classify an instability, one has to answer three main questions:

(1) What is the dominant drive for the instability? Typical possibilities are: "current-driven", "pressure-driven" and "particle-driven" instabilities.
(2) Is the instability "resistive" or "ideal"? In case of an ideal instability, the flux surfaces in the plasma are preserved, the instability changes only the shape of the surfaces (Fig. 2.7a). It grows with the Alfvenic time ($\gamma \sim \frac{1}{\tau_A}$, $\tau_A = \frac{a}{v_A}$, $v_A = \sqrt{\frac{B^2}{\mu_0 m_i n}}$, where n is the plasma density and a is the characteristic length of the instability). This time is very short and ideal instability poses ultimate limits for plasma confinement. These instabilities are called "kink" instabilities because the plasma displacement due to such an instability tilts and kinks the plasma. Ideal MHD, which implies zero resistivity, preserves the plasma topology and is sufficient to describe the process. The so-called "resistive instabilities" requires changes of the field line topology. These changes appear only in a small part of the plasma volume, in the X-points (Fig. 2.7b). In these regions the plasma resistivity cannot be neglected and leads to topological changes. These small regions have a dramatic influence on the global equilibrium. The plasma gets access to a new degree of freedom and tries to find the minimal energy state without topological limitations. The time scale for the resistive instability is a mixture between slow resistive diffusion time ($\tau_R = \frac{\mu_0 l^2}{\eta}$, where η is the plasma resistivity and l is the characteristic width of the reconnection region, $l \ll a$) which reflects the topological changes, and fast dynamics of the plasma (τ_A). The typical manifestations of such modes in a tokamak are tearing modes (TMs) and neoclassical tearing mode (NTMs). The name implies tearing and reconnection of the magnetic field lines.
(3) Where is the dominant location of the instability? Typically, each instability develops predominantly either inside the plasma (internal mode, $\xi(a) = 0$) or

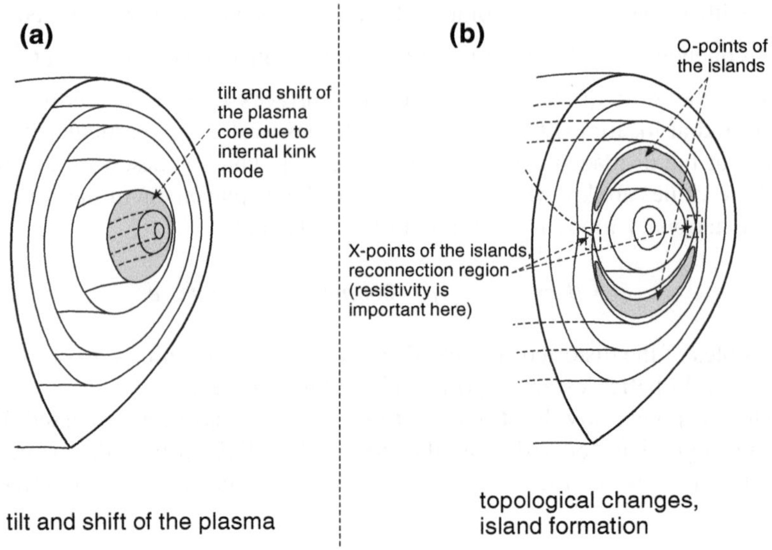

Fig. 2.7 **a** Ideal MHD instability. Internal kink mode is shown **b** Resistive MHD instability. Tearing mode is shown

at the plasma boundary (external mode/free boundary mode, $\xi(a) \neq 0$). In tokamak plasma, different modes with the same toroidal mode number n are coupled. These secondary modes, with mode numbers $(m \pm 1, n)$, have components at the plasma surface even if the dominant mode (m, n) is purely internal. In some cases, this can be neglected and one assumes the mode to be internal (Plasma shaping gives additional coupling to other poloidal mode numbers).

The presented classification is shown in Fig. 2.8. It is important to note that this classification is very basic. The toroidal geometry of tokamaks, non-circular cross-section and other effects lead to a mixture of different modes which are unstable simultaneously, for example coupling between external and internal modes. Situations with combined drives or with changes of the dominant drive are also possible, for example a current-driven classical tearing mode TM converts into a pressure-driven neoclassical tearing mode NTM when the size of the mode becomes sufficiently large and the plasma pressure is sufficient. All these complicated cases have to be studied carefully both experimentally and numerically. Nevertheless, the presented simple classification allows to get a basic idea about the problem and provides a preliminary case characterization. It is important to note that the zoo of MHD instabilities in the plasma is extremely large. All instabilities have particular names, many of them are connected to the historical background, not to the physical properties. We have considered keeping the number of names at a minimal possible level and forwarding the readers to other books with detailed descriptions, for example: A.B. Mikhailovskii "Instabilities in a confined plasmas" [12].

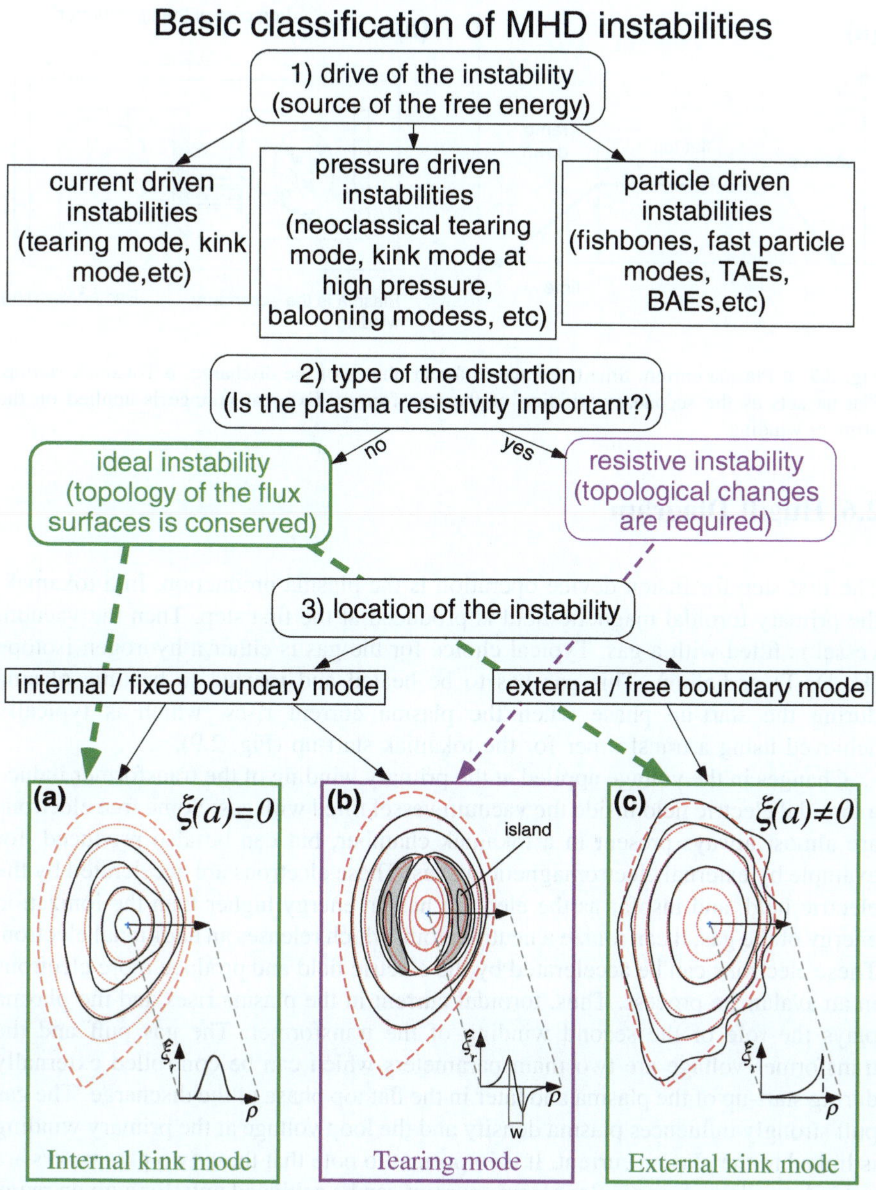

Fig. 2.8 The basic classification of MHD instabilities

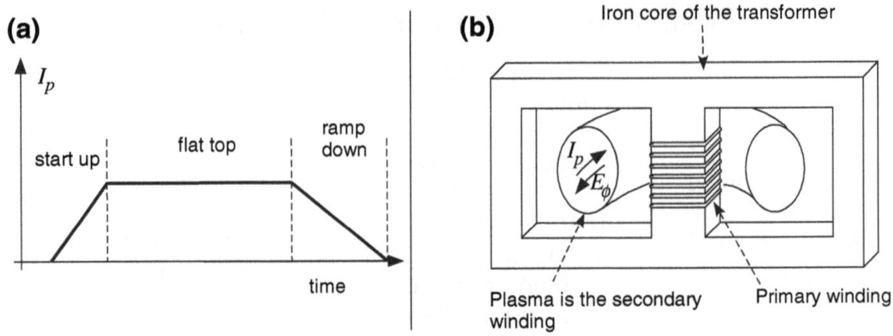

Fig. 2.9 **a** Plasma current time trace and different phases of the discharge. **b** Tokamak startup. Plasma acts as the secondary winding of the transformer. External voltage is applied on the primary winding

2.6 Hugill Diagram

The first step for fusion device operation is the plasma production. In a tokamak, the primary toroidal magnetic field is produced at the first step. Then the vacuum vessel is filled with a gas. Typical choice for the gas is either a hydrogen isotope (H, D, T) or helium. This gas has to be heated and ionized to become plasma during the start-up phase when the plasma current rises, which is typically achieved using a transformer for the tokamak start-up (Fig. 2.9).

Changes in the voltage applied at the primary winding of the transformer induce a toroidal electric field inside the vacuum vessel filled with gas. Some free electrons are almost always present in a tokamak chamber, but can be also produced, for example by external electromagnetic waves. These electrons are accelerated by the electric field, and insofar as the electron has an energy higher than the ionization energy of the gas, it can ionize a neutral atom, which releases an additional electron. These electrons can be accelerated by the electric field and produce more electrons in an avalanche process. Thus, toroidal current in the plasma rises and the plasma plays the role of the second winding of the transformer. The gas puff and the transformer voltage are two main parameters which can be controlled externally during start-up of the plasma and later in the flat top phase of the discharge. The gas puff strongly influences plasma density and the loop voltage at the primary winding is linked to the plasma current. It is important to note that these two parameters are not independent. A particular plasma current can be achieved only for a given range of the plasma densities and stable operation is possible only in a restricted area in the (I_p, n_e) space, shown in the so-called Hugill diagram (Fig. 2.10). If the stability boundary is crossed, the discharge is either terminated abruptly (plasma disruption, hard limit), or the plasma confinement degrades over a longer time (soft limit).

There are three main stability boundaries on this diagram: (i) runaway limit, (ii) current limit, and (iii) Greenwald limit. All together they define the stable operation space in terms of the plasma current and average plasma density. This

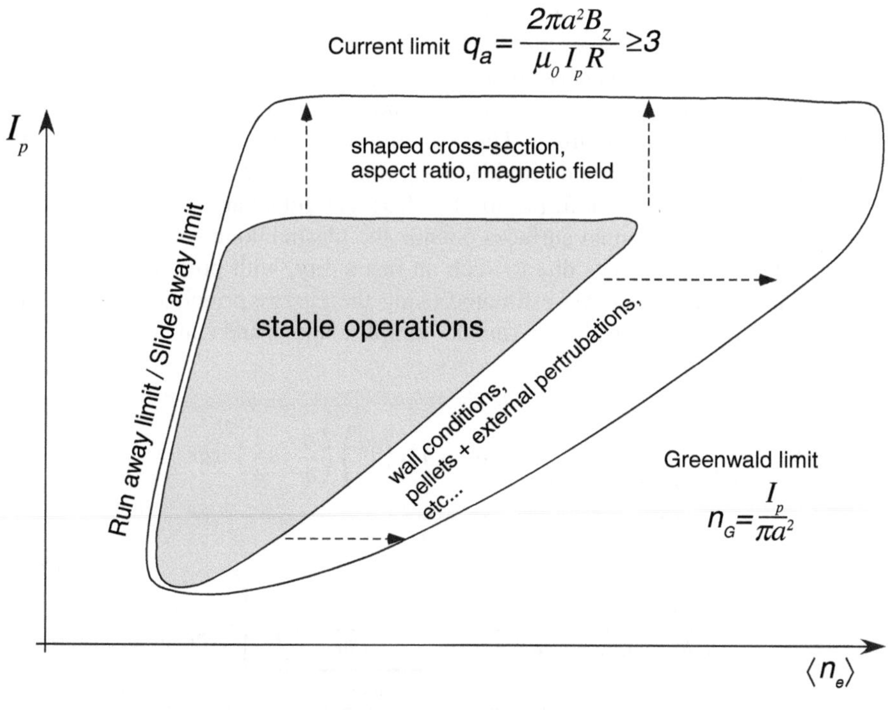

Fig. 2.10 The Hugill diagram and the main limits for plasma operations

diagram does not present all limitations for the plasma operations, but provides a general idea regarding possible parameters.

The *runaway* (or *slideaway* limit) is near the left border of the graph at low plasma density. As discussed before, plasma electrons are continuously accelerated by the toroidal electric field. If an electron gains more energy per toroidal turn than it loses by collisions in low density plasma, it does not transfer the energy to the plasma ions anymore. Operation close to this limit is not very interesting for two reasons:

- the fusion reactor requires higher density operation for higher fusion performance, which is not the case in this regime;
- the lost electrons may cause damage of the first wall.

The other two limits are more important and we discuss them in detail.

2.6.1 Current Limit

The maximum current limit of a tokamak depends on many different factors and has to be calculated numerically, but main ideas can be derived from simple MHD models. The simplest MHD model for a tokamak is the "straight tokamak" approximation, a column of plasma having a circular cross-section and length $L = 2\pi R_0$. The fields have to satisfy the tokamak inverse ratio expansion $B_\theta/B_Z \sim \varepsilon = a/R_0 \ll 1$, and periodic boundary conditions. In this case, Z is the "toroidal" coordinate along the plasma cylinder instead of ϕ and r is the radial coordinate of the plasma column. These assumptions exclude any influence of the toroidicity on the problem and provide an estimation for the current-driven modes. The most dangerous instabilities in the "straight tokamak" are external kink modes, which have resonant surfaces outside the plasma boundary, in the vacuum region. The energy changes due to such an instability, with poloidal and toroidal mode numbers (m,n), can be estimated using the energy principle (2.35), which takes into account the "straight tokamak" simplifications and can be written in the following form [8, 13, 14]:

$$\delta W = \underbrace{\frac{2\pi^2 B_z^2}{\mu_0 R_0} \int_0^a \left(\left(r\frac{d\xi_r}{dr} \right)^2 + (m^2-1)|\xi_r|^2 \right) \left(\frac{n}{m} - \frac{1}{q} \right)^2 r dr}_{plasma} \\ + \underbrace{\frac{2\pi^2 B_z^2}{\mu_0 R_0} \left(\frac{2}{q} \left(\frac{n}{m} - \frac{1}{q} \right) + \underbrace{(1+m\lambda)\left(\frac{n}{m} - \frac{1}{q} \right)^2}_{wall} \right) r^2 |\xi_r|^2 \Big|_{r=a}}_{vacuum} \tag{2.37}$$

where $\lambda = \frac{1+(a/b)^{2m}}{1-(a/b)^{2m}}$ represents the influence of the wall at radius b.

The first part of the expression is the integral from the plasma center, $r = 0$, to the plasma boundary $r = a$. All terms under the integral are positive. Thus, if the wall is at the plasma boundary, the system is stable.[5] The second part describes the vacuum region and influence of the conducting wall, $a < r < b$. The effect of the wall is represented by the second term in the vacuum part. This contribution is also stabilizing, because $\lambda > 1$ and this term is always positive. The only potentially unstable contribution comes from the first term in the vacuum part. It gives a destabilizing contribution if the following relation holds:

$$\left(\frac{n}{m} - \frac{1}{q} \right) < 0 \tag{2.38}$$

Thus, the necessary condition for the mode to be unstable is $q_a < {}^{m}/_{n}$, and the resonant surface is in the vacuum region in this case. Potentially, the most unstable external kink mode has poloidal and toroidal mode numbers $(m,n) = (1,1)$, and

---

[5] The plasma part of the system is stable only in the ε^2 order. In case of the $(m,n) = (1,1)$ internal kink mode, the plasma integral is zero and higher order expansion has to be taken into account, see for example Bussac et al., Phys. Rev. Let. Vol 35, p. 1638.

the stability condition for the safety factor at the plasma boundary is $q_a > 1$. This condition is called the Kruskal-Shafranov limit for the plasma current because the edge safety factor is directly related to the plasma current:

$$q_a = \frac{2\pi a^2 B_Z}{\mu_0 I_p R} \quad \text{(``straight tokamak'' model)} \tag{2.39}$$

In reality, stable operations require higher values of the edge safety factor ($q_a \geq 3$) and the maximum plasma current is strongly limited. Special, non-circular plasma cross-sections can improve the situation. The D-shaped cross-section, which is used in all modern tokamaks, utilizes the combined effect of the plasma elongation and triangularity. This shape naturally allows higher plasma currents for a given aspect ratio and magnetic field. Stability of the kink modes in real tokamaks is investigated numerically taking into account real plasma geometry, the current and the pressure profiles. If the kink mode is unstable, it leads to a strong and fast deformation of the plasma boundary and the plasma disrupts (hard limit).

2.6.2 Greenwald Limit

To the right of the operation space we encounter the density limit, which determines maximum achievable plasma density for a given plasma current. This is the most interesting corner of the operation space, since the fusion reaction rate scales with n^2. The plasma current has to be also maximised to reach this goal as seen from Fig. 2.10. In the majority of cases, the density limit shows up as a disruptive event in which the plasma thermal energy is quenched in about 1 ms, while the energy stored in the plasma current is dissipated on a time scale of the order of 10 ms [15]. The most common relation for the density limit is the empirical scaling, so-called Greenwald limit:

$$n_G = \frac{I_p}{\pi a^2} \tag{2.40}$$

where n_G is the line average density in units of 10^{20} m^{-3}, I_p is the plasma current in MA, and a is the minor radius in m. The presented formulation describes relatively well the limits for circular and shaped plasma cross-sections (Fig. 2.11) [16, 17].

Unlike the current limit, which depends on MHD physics alone, the density limit physics includes transport and atomic processes. The general picture of the density limit involves edge cooling followed by the current profile shrinkage. The current profile becomes unstable with respect to tearing and/or kink modes and the plasma becomes stochastic and cold. Typical manifestations of this process is multifaceted asymmetric radiation from the edge (MARFE), which creates a toroidally symmetric zone of high radiation either at the high field side (limiter) or

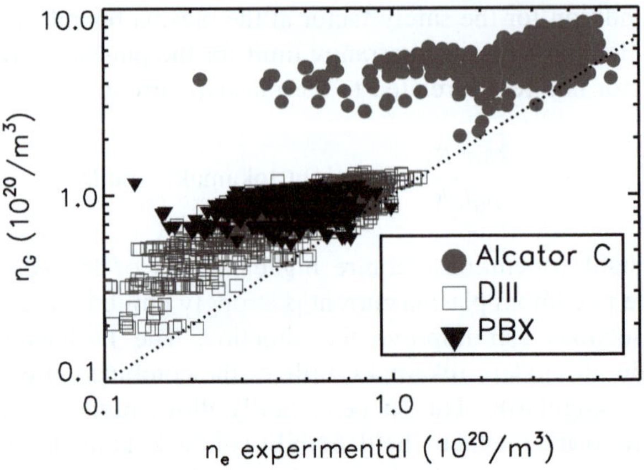

Fig. 2.11 Recorded densities versus Greenwald predictions in circular and shaped tokamaks. (The figure is from [16]. All rights reserved.)

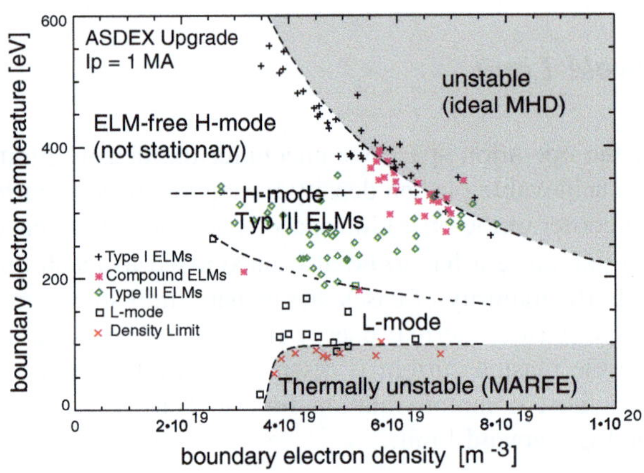

Fig. 2.12 Operation space of the ASDEX Upgrade tokamak is represented in terms of the edge plasma parameters. The density limit occurs when the edge temperature falls below a threshold and plasma goes into a thermally unstable regime (MARFE regime) (The figure is from [19]. All rights reserved.)

around the separatrix X-point (divertor) [15, 18]. The origin of the density limit is not yet fully understood, but the main findings show strong connection to the edge plasma physics. This idea is supported by the fact that the operation regimes can be defined in terms of edge temperature and density (Fig. 2.12) [19]. Here, the density limit occurs when the edge temperature falls below a threshold and plasma goes into a thermally unstable regime (MARFE regime). The electron temperature in this region is very low and the electron density becomes very high. The line radiation due to ionization and charge exchange of the incoming neutral particles provide the largest part of the energy losses.

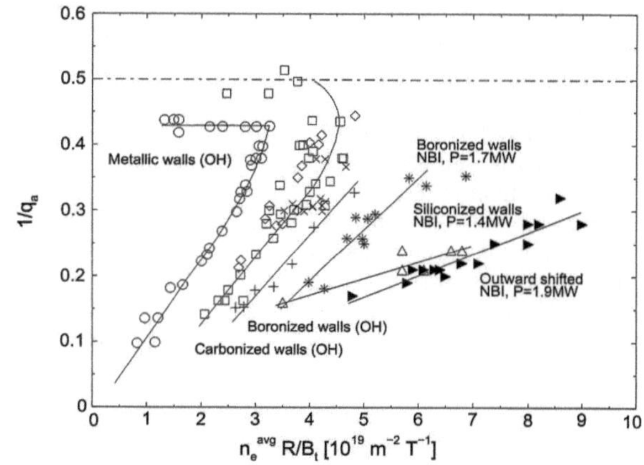

Fig. 2.13 Hugill diagram for the TEXTOR tokamak with different wall conditioning and different heating scenarios. Advanced wall coating methods extend the operation space (The figure is from [20])

There are several ways to extend the density limit. The achievable density is clearly reduced if the plasma is contaminated by impurities. Wall conditioning is a standard technique to avoid this situation. An example is presented in Fig. 2.13 for the TEXTOR tokamak. Different groups of data points in Fig. 2.13 show that the limit has increased due to the application of advanced wall coating methods: carbonized wall, boronized wall, and siliconized wall [20].

The basic idea is that the impurities released at the first wall can dilute the plasma and cause strong line radiation. The disruption becomes unavoidable when the total radiated power exceeds the heating power. Any action against this scenario helps to keep the plasma healthy and improves the situation. For example, additional heating permits higher radiation losses and enhances the density limit. Unfortunately, the situation is more complicated and increase in power not always helps. Generally, the density limit is not found to increase strongly with input power [16]. This is in clear contradiction to the simple power balance model.

The other way to increase the density well above the Greenwald limit in present experiments is pellet fueling (Fig. 2.14). Results from the ASDEX Upgrade tokamak show that pellet fueling in combination with magnetic fields from active in-vessel saddle coils allow to sustain very attractive plasma scenarios. In this case, core densities up to $1.6n_G$ have been reached while maintaining mitigation of dangerous big edge localized modes (ELMs) [21].

There are also other factors which influence the density limits; for example, plasma triangularity has been found to be important in determining the density at which confinement begins to drop. This is probably connected to the pedestal stability, which depends on the plasma shape. There are several ideas which try to explain the mechanism for the onset of the density limit. For example, the theory of D.A. Gates and L. Delgado-Aparicio assumes that the onset of the density limit is the same as the onset criterion for radiation-driven islands. One can reproduce the density scaling using this assumption [22]. This theory provides a nice explanation for cases with radiation from the island, but it has difficulties to explain cases when radiation from the islands is not observed.

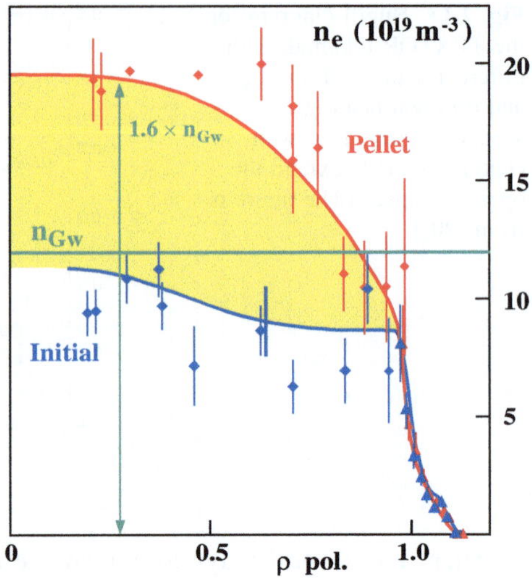

Fig. 2.14 Density profiles in ASDEX Upgrade with and without pellet fueling are shown. Greenwald density limit is marked by green line. The density is well above Greenwald limit in the plasma core but is less than this limit at the plasma edge (The figure is from [21]. © IAEA. All rights reserved)

Operations in other fusion devices are also limited by the density limit. In reversed field pinches, the density limit is the same as in tokamaks. Situation in stellarators is completely different. This device has intrinsically no current which makes the plasma more calm. In order to apply the Greenwald formula the helical winding of the field lines in stellarator should be converted into equivalent plasma current (2.39). The result data points from present stellarators clearly exceed the Greenwald limit and show no indications for "absolute" limits. Operation limit is typically set by radiative/thermal instabilities. Density limit for stellarators approximately obeys Sudo scaling $n_c \sim (PB/V)^{0.5}$, where P is the heating power, B is the magnetic field and V is the plasma volume [23], with densities up to 5 times the Greenwald limit. Exceeding the density limit in stellarator leads to degradation of the plasma confinement on a transport time scales, which is much longer compared to fast disruptive MHD instabilities in tokamaks. Thus, the two main advantages of the stellarator concept are higher density limit and absence of dangerous disruptive instabilities. The reason for two different density scalings is not clear. It could be either a different physics in the machines with and without plasma current or wrong interpolation for the effective plasma current in stellarator case. In spite of the long history of the density limit, there is no widely accepted first principle theory which can explain it. It has to be noted that the density and current limitations of Hugill diagram give basic limitations for plasma operation. Other limits, which are not covered by a simple $(I_p, \langle n \rangle)$ diagram, will be discussed in the next subsections.

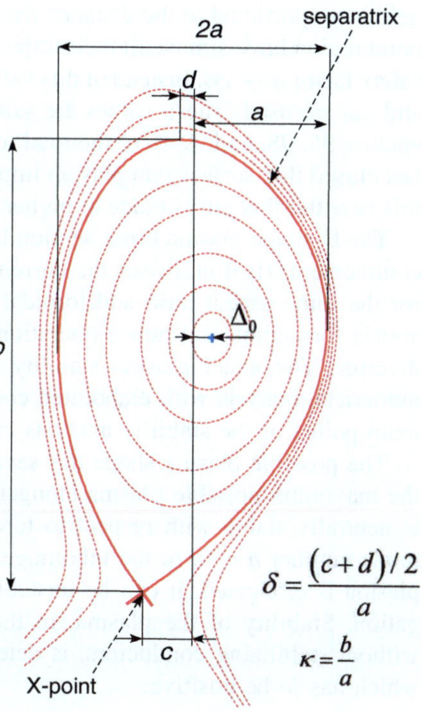

Fig. 2.15 Plasma shape of a typical ASDEX Upgrade discharge is shown. Main characteristics are: elongation (δ), triangularity (κ) and Shafranov shift (Δ)

2.7 Restriction Due to Plasma Shaping

Modern tokamaks have strongly non-circular, D-shaped plasma cross-sections in the poloidal plane (Fig. 2.15).

This shape is characterized by two main dimensionless quantities:

(1) plasma elongation

$$\kappa = \frac{b}{a} \tag{2.41}$$

(2) plasma triangularity[6]

$$\delta = \frac{(c+d)}{2a} \tag{2.42}$$

The flux surfaces are shaped and shifted due to the Shafranov shift, Δ_0, which is the result of the force balance and comes directly from the solution of the Grad-Shafranov (2.19). The other important feature is the X-point. The poloidal magnetic field, B_θ, is zero at the X-point. In its vicinity, the magnitude of the poloidal magnetic

[6] In praxis, also upper triangularity, $\delta_u = d/a$, and lower triangularity, $\delta_l = c/a$, are used.

field is proportional to the distance from the X-point. Thus, the field lines in the X-point region have almost all their trajectories close to the X-point and the separatrix safety factor $q \to \infty$. Instead of this value, for practical purposes, the values q_{95}, q_{98}, and q_{99} are used. These values are safety factor values at the flux surfaces, which enclose 95, 98, and 99 % of the total magnetic flux respectively.[7] The shape of the last closed flux surface can play an important role for free boundary instabilities; in this case the flux cut is made at higher values, for example at 99,98 %.

The D-shape plasma cross-section has two main advantages in terms of plasma confinement: (i) it increases the current limit, which allows higher plasma current for the same aspect ratio and toroidal magnetic field; (ii) it leads to natural separatrix formation, and hence operation in high confinement mode (H-mode) with divertor. The exact plasma stability boundaries in such configurations require numerical analysis with elaborated code packages (for example [24–26]), but the main points of the stability analysis can be shown on a simple basis.

The possible plasma shape has several basic limitations. The first limitation is the maximum possible plasma elongation. The plasma with circular cross-section is neutrally stable with respect to toroidally symmetric vertical motion, toroidal mode number $n = 0$, in the inhomogeneous magnetic field of the tokamak. If the plasma is elongated, it can be unstable to a motion in the direction of the elongation. Stability of the plasma, in the simple case of large aspect-ratio plasma without stabilizing conductors, is determined by the magnetic field decay index, which has to be positive:

$$n_v = -\frac{R}{B_Z}\frac{dB_Z}{dR} > 0, \tag{2.43}$$

where B_Z is the externally applied vertical magnetic field (discussed in the next section). R is the major radial coordinate. A conducting wall and internal conducting structures inside the device change stability properties and numerical calculations are required for practical cases.

If this stability criterion is violated, then in the absence of any wall or other conducting structure surrounding the plasma, instability occurs on a fast inertial time scale. However with surrounding conducting structures then there is a range of $n_v < 0$ in which the growth is slowed to the resistive decay time of the surrounding structures. The vertical instability can be controlled by active feedback stabilization system up to an elongation $\kappa = 2.2$ [27]. At the same time, the practical limit for modern tokamaks with relatively close conducting shells is about $\kappa \approx 1.8$ [28].

The second limitation comes from ballooning and kink modes. It was shown that proper choice of the plasma elongation and triangularity brings the plasma into a stable operation window even at high pressure as discussed in the next section.

---

[7] This is different to the limiter plasma where the safety factor at the plasma edge, q_a, is well-defined.

2.8 Beta Limit

The plasma in fusion devices is confined by a magnetic field. The strength of the toroidal magnetic field should be sufficient to stabilize the external kink mode for a given current (Hugill diagram). At the same time, the poloidal magnetic field should provide the pressure balance in the fusion device. To characterize this balance, a new variable is introduced. The beta value is the ratio of the average plasma pressure and magnetic field pressure:

$$\beta = \frac{\langle p \rangle}{\langle B^2 \rangle / 2\mu_0}. \tag{2.44}$$

Based on this definition, other variants of beta can be defined. For example, poloidal beta $\left(\beta_{pol} = 2\mu_0 \langle p \rangle / \langle B_\theta^2 \rangle\right)$ is defined relative to the poloidal magnetic field (B_θ). In this part we discuss two main beta limits for tokamak plasmas and their consequences for the operation space of the tokamak. The first limit arises from equilibrium force balance. In the tokamak configuration, the plasma is bent to avoid the end losses along the magnetic field, but this bending generates three forces—all directed outward along the major radius. These forces are [28]:

- The "hoop force", which is the same as the outward expansion force for a current in a circular loop of wire. The poloidal magnetic field, produced by the plasma current, is higher at the inner side of the torus. Thus, the resulting Lorentz force is directed outward, which leads to expansion of the current ring (see Fig. 2.16a).
- The "1/R force" is associated with the 1/R decay of the toroidal magnetic field from the inner side of the torus to the outer side. The plasma is usually diamagnetic $\left(\beta_{pol} > 1\right)$, producing the field inside the plasma that partially cancels the applied toroidal magnetic field. In this situation, the forces on the high field side and on the lower field side are different, which provides the net outward force. In the opposite, paramagnetic situation $\left(\beta_{pol} < 1\right)$, the net force is directed inward (see Fig. 2.16b).
- The "tire tube force" has a hydraulic nature and appears due to the difference in the inner and outer torus surfaces for the same pressure in the tube (see Fig. 2.16c).

The sum of all these forces has to be balanced in the device. In the presence of a perfectly conducting wall, the outward shift of the plasma compresses the flux at the low field side of the device and the increased magnetic field provides a restoring force in the inward direction as shown in Fig. 2.16d. This restoring force is not enough to completely suppress the outward force due to resistive decay, and this is typically done by an applied vertical magnetic field. Superposition of this vertical field on the poloidal magnetic field leads to X-point formation on the high field side of the tokamak. With increase of the plasma pressure, a higher vertical field is required to balance the forces and the X-point moves in the direction of the

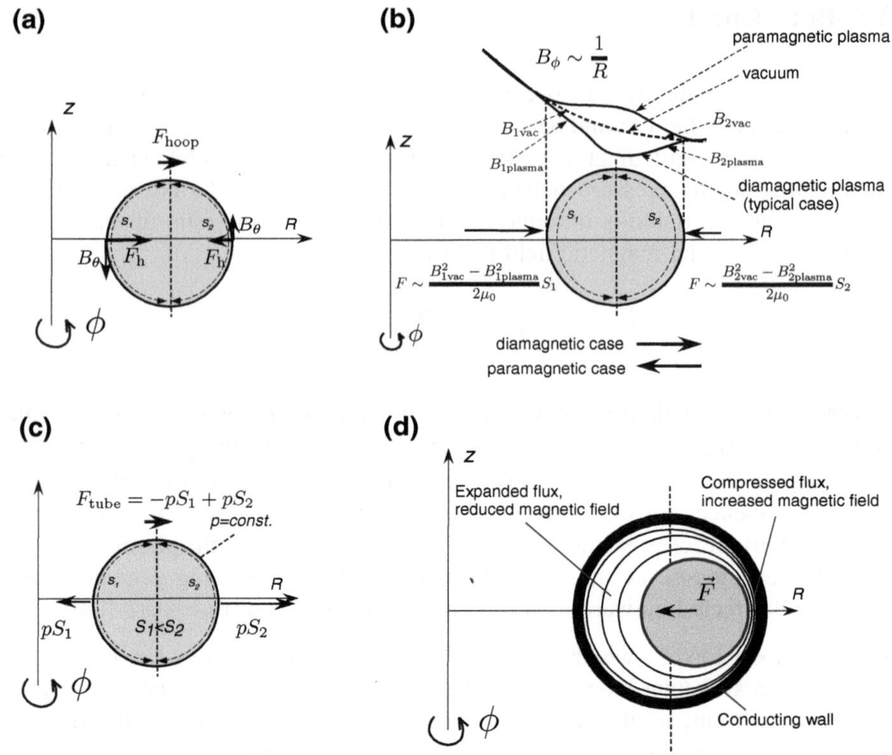

Fig. 2.16 Different forces acting on the plasma are shown schematically for a circular cross-section tokamak. **a** The "hoop force" is the same as the outward expansion force for a current in a circular loop of wire. Assuming only the surface current for a torus, one gets variation of the result poloidal magnetic field from the plasma current, I_p. This poloidal magnetic field is higher at the inner side of the torus and the result magnetic force has outward direction. **b** The "1/R force" is associated with the decay of the toroidal magnetic field, $B_\phi \sim 1/R$, from the inner side of the torus to the outer side. One obtains the force from the difference between the vacuum and plasma magnetic field taking into account the field decay. **c** The tire tube force is due to the difference in the inner and outer torus surfaces for the same pressure in the tube. The inner surface is smaller than the outer surface, $s_1 < s_2$. **d** Outward plasma movement leads to flux compression at the *low field side*. A higher magnetic field at this point produces an inward restoring force

plasma boundary. The plasma has reached its equilibrium β limit when the X-point touches the plasma surface (Fig. 2.17). Further attempts to increase β require cutting off and eliminating the inside portion of the plasma. For a fixed geometry, the actual limit is a limit on β/I_p^2. For tokamaks with circular cross-section this limit can be written in the following form [28]:

$$\left(\frac{2\pi a^2 B_0}{\mu_0 R_0 I_p}\right)^2 \frac{\beta}{\varepsilon} \leq \frac{\pi}{16} \approx 0.62 \tag{2.45}$$

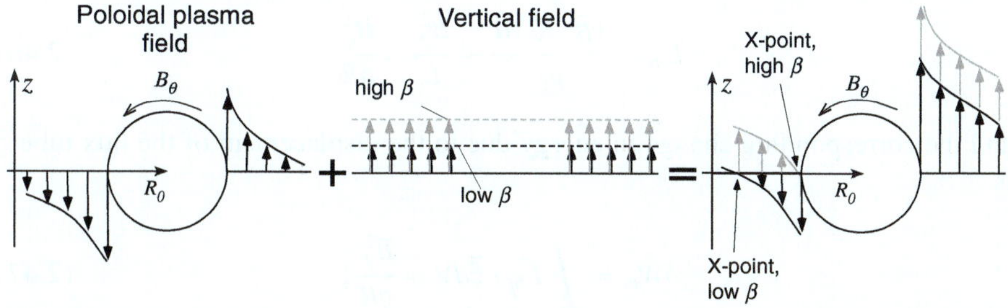

Fig. 2.17 Combination of the poloidal magnetic field (from plasma current) and external vertical magnetic field forms X-point at the *high field side* (inner side of the torus). With increase of the plasma pressure, a higher vertical field is required to balance the forces and the X-point moves in the direction of the plasma boundary. The plasma has reached its equilibrium β limit when the X-point touches the plasma surface

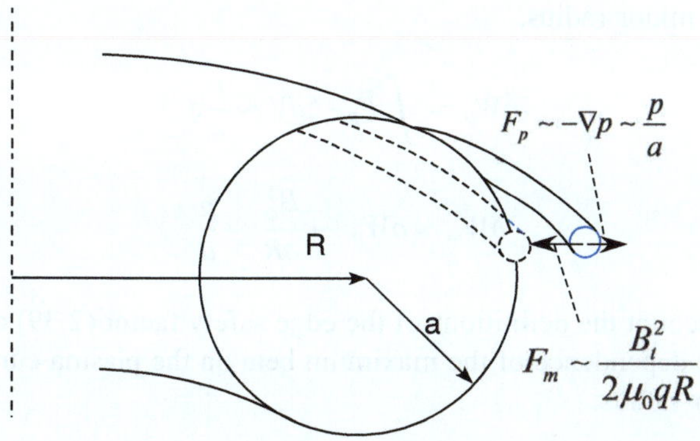

Fig. 2.18 A flux tube with length $L \sim qR$ is shifted outwards by plasma pressure. The situation is stable only if the magnetic force is sufficient to counteract the pressure force

The second β limit is connected to the final achievable pressure with respect to ballooning modes. These modes result from the toroidal geometry of the tokamak magnetic field, which implies a stronger field at the inner part of the torus ($B_\phi \sim 1/R$, where R is the major radius). Thus, the outer side of the torus has a lower magnetic field and is more sensitive to the perturbations. The physical nature of the ballooning beta limit comes from the energy balance for a particular flux tube exploring the destabilizing unfavorable curvature on the low field side (Fig. 2.18). This is shown schematically in the Fig. 2.18.

In tokamak geometry, the length of the flux tube at the low field side is proportional to the major radius and safety factor: $L \sim qR$ [1]. The stabilizing energy contribution comes from the magnetic tension force (2.22). This force can be estimated as follows:

$$F_m = \frac{(\vec{B}\cdot\nabla)\vec{B}}{\mu_0} \sim \frac{B_\phi^2}{L} = \frac{B_\phi^2}{qR}, \tag{2.46}$$

and the corresponding change in energy due to the displacement of the flux tube ξ is

$$\delta W_m = \int \vec{F}_m \cdot \vec{\xi} dV \sim \frac{B_\phi^2}{qR}\xi. \tag{2.47}$$

This energy should be sufficient to compensate the destabilizing pressure contribution. The pressure force is approximately

$$F_p \sim -\nabla p \sim \frac{p}{a}, \tag{2.48}$$

where a is the minor radius.

$$\delta W_p = \int \vec{F}_p \cdot \vec{\xi} dV \sim \frac{p}{a}\xi \tag{2.49}$$

$$\delta W_m \sim \delta W_p \rightarrow \frac{B_\phi^2}{qR} \sim \frac{p}{a} \tag{2.50}$$

Taking into account the definitions of the edge safety factor (2.39) and beta (2.44) one obtains the dependence of the maximum beta on the plasma current, magnetic field and minor radius:

$$\beta_{crit} \sim \frac{I_p}{B_\phi a}. \tag{2.51}$$

Historically, this relation was first formulated by Troyon based on ideal MHD stability simulations [29]. In these simulations Troyon and co-authors investigated the stability of the plasma with respect to all ideal instabilities: external kink modes, ballooning-kink modes, internal kink modes, etc. It was shown that the typical limit is set by kink modes with the same dependence as above:

$$\beta_{crit} = \beta_N \frac{I_p}{B_\phi a}, \tag{2.52}$$

where β_N varies depending on the plasma profile, plasma shape and other factors. This value is typically around 3×10^{-2} but in the fusion community it is typical to write this value as a percentage, omitting the percent indices (in our case this will be $\beta_N = 3$ instead of 3 % or 3×10^{-2}). The maximum beta defined by Troyon includes only the plasma part. An external conducting structure can stabilize instabilities and lead to other normalized values of beta. This will be discussed in details in the Resistive Wall Mode Chap. 6, where a normalized beta will be defined for two situations (with and without ideal wall) for the same plasma.

Numerical investigation of the ballooning stability reveals that there are two stability regions for plasma operations [30]. It is expected that at low radial pressure gradient values the pressure-driven ideal ballooning modes are stabilized by field line bending. This leads to the first stability window at low pressure gradient $\left(\alpha = -\frac{2\mu_0 R q^2}{B^2}\frac{dp}{dr}\right)$ and large shear value $\left(s = \frac{r}{q}\frac{dq}{dr}\right)$. The local value of the magnetic shear s shows how strong are the changes in helicity between neighboring flux surfaces. It is also a characteristic of the current profile gradient, because the safety factor is inversely proportional to the integrated current inside the flux surface. High shear is in general favorable for the plasma stability. The more surprising result is the existence of a second stability region at low magnetic shear and high pressure gradient. This is connected to the local stability of the ballooning modes in the negative shear case, which corresponds to a flat or reversed q-profile. Increase of the pressure in the tokamak builds up the pressure gradient and increases the Shafranov shift. These two modifications of the equilibrium extend the negative local shear into the favorable curvature region, which provide global stability in the second region [31].

The first and second stability regions are typically separated in the ideal MHD approximation (Fig. 2.19, dashed curves). The plasma shape, current profile and pressure profile determine the boundary of the stable regions in ideal MHD. The situation in a real plasma is more complex and effects beyond the fluid

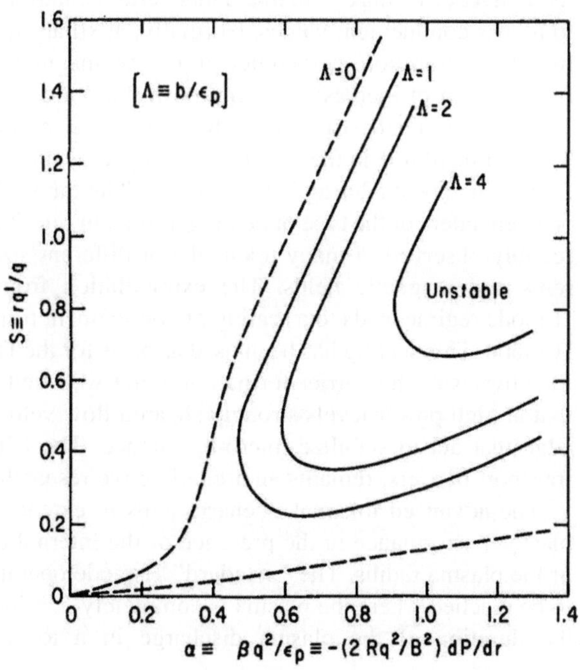

Fig. 2.19 Stability boundaries for ballooning modes as a function of the shear parameter and the pressure parameter at various values of finite gyroradius parameter $\Lambda = b/\varepsilon_p$ (The figure is from [32].)

approximation have to be taken into account. For example, finite Larmor radius, r_L, effects influence plasma stability and provide stronger stabilization for higher values of the finite gyroradius parameter $\Lambda = b/\varepsilon_p$, where $b \equiv \frac{n^2 q^2 r_L^2}{r^2 2}$ is a measure of the stabilizing gyroradius contributions and $\frac{1}{\varepsilon_p} \equiv -R\frac{d(\ln p)}{dr}$ is a measure of the destabilizing pressure-driven forces [32].

In stellarators, the plasma behavior at the beta limit is similar to its behavior at the density limit. There are no disruptive events and the beta limit manifests itself as a soft limit. Moreover, the archived values of beta clearly exceed the predicted linear stability boundaries for pressure-driven MHD modes [33]. Global MHD modes, which were found in stellarators, are consistent with predictions based on the linear stability theory. However, in most cases the observed modes saturate on a harmless level, which allows further increase of beta. Therefore, the linear stability threshold significantly underestimates the achievable beta [34]. The highest values archived in stellarators are $\langle\beta\rangle \approx 3.4\,\%$ (W7AS, Germany) and $\langle\beta\rangle \approx 5\,\%$ (LHD, Japan) [35].

2.9 Different Plasma Scenarios and their Limits

A remarkable finding in fusion plasmas is the existence of different confinement regimes: low confinement mode (L-mode), high confinement mode (H-mode) and advanced operation regime with internal transport barrier (ITB). It was first found at the ASDEX tokamak [36], that under certain conditions there is an abrupt transition to higher confinement values. Normally, auxiliary heating of the plasma enhances the radial transport perpendicular to the magnetic field lines (L-mode), but a combination of increased heating power and a divertor configuration leads to the formation of a barrier region with reduced transport at the plasma edge (H-mode). The dominant role in the transition is played by the shear of the radial electric field, E_r, which appears during L-H transition. The threshold of the transition to H-mode is independent of the type of heating as seen in Fig. 2.20 [37]. The H-mode has been reliably observed in many tokamaks of different sizes, aspect ratios, plasma currents and magnetic fields. The extrapolation from different tokamaks for the H-mode regime leads to a scaling of confinement time for the so-called "standard" H-mode. This scaling has been used as basis for the ITER design. At the same time, the physics of the barrier formation is not well understood. The dominant idea is that at high power levels strongly sheared flow velocities develop near the plasma edge that act to stabilize micro-turbulence. The L-H transition, as well as other transport barriers, remains an area of active research.

The advanced tokamak scenario aims to extend the H-mode regime to higher plasma performance in the presence of the internal transport barrier at the middle of the plasma radius. The "standard" H-mode operation does not allow conditions to be reached where the plasma is completely non-inductively driven, which limits the duration of the plasma discharge in a tokamak. The advanced scenario

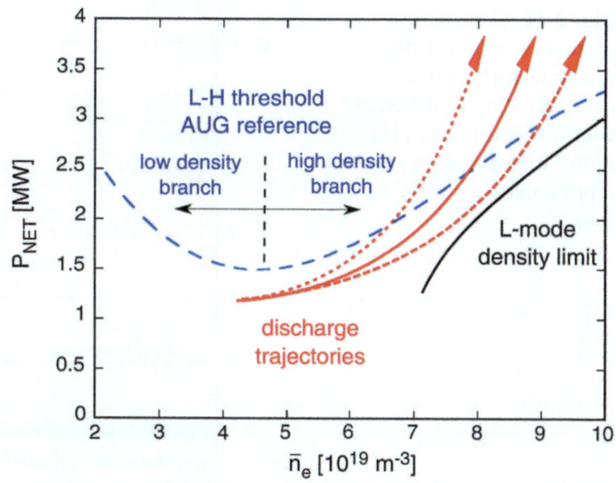

Fig. 2.20 Plasma density and heating power must be increased simultaneously, remaining above density limit, for transition into H-mode (The figure is from [37]. © IAEA. All rights reserved.)

promises to achieve the steady state operation substituting the inductively driven current by bootstrap current [3]. This scenario is more disruptive and not as robust as the H-mode.

The current density profile, and hence the safety factor profile, plays an important role in plasma stability and confinement. The slope of the current profile is characterized by the magnetic shear $s = (r/q)dq/dr$ (where r is the minor radius), which is widely used in stability analysis. The "standard" H-mode profile has a safety factor value at the center, q_0, just below 1, while at the plasma boundary, q_{95}, is 3 or higher, and the shear is always positive. The situation is completely different in the advanced scenario. It uses a range of non-monotonic q-profiles as shown in Fig. 2.21. Depending on the bootstrap current fraction, q_0 varies from 2 to ∞. The later situation appears in the so-called "current hole" variants [38], when the plasma current density is zero at the magnetic axis. Tailoring and control of the current density profile is clearly the key for the development of these scenarios. It is possible to imagine a continuum of regimes between the non-inductive and inductive scenarios in which the bootstrap current is only a fraction of the total current. This situation is typical for improved H-mode' discharges, which are characterized by a broad region in the plasma core of $\sim$ zero shear and $q_0 \geq 1$ [39]. This scenario is also often called a "hybrid" scenario. It opens the way to either longer plasma pulse duration at reduced plasma current, or to improved performance at full plasma current. Hybrid scenarios typically require moderate additional heating during the plasma current ramp-up to allow the formation of a low central shear profile. The "improved" H-mode is a promising operation option for ITER which combines the current advantages of the standard H-mode, as well as the advanced scenario.

The basic equilibrium limits for plasma confinement have been discussed in the previous sections. MHD limitations are stricter. The ideal kink instability limits the maximum plasma current, ballooning and kink modes limit maximum achievable β_N values, etc. These are limits for the economic operation of future

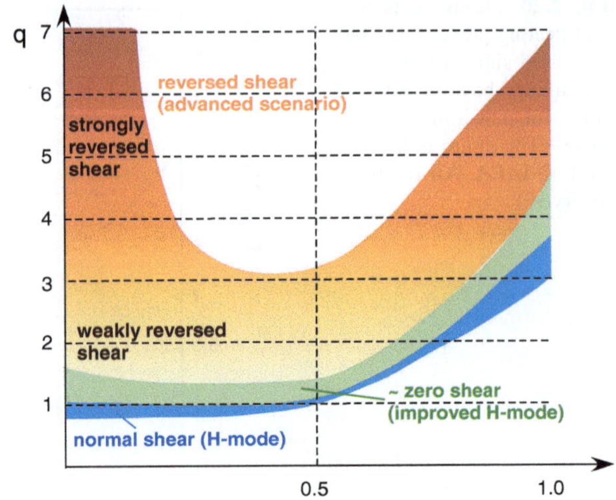

Fig. 2.21 Schematic representation of different safety factor profiles corresponding to different confinement modes (The figure is inspired by similar representation by A.C.C. Sips.)

fusion power plants. However, the other instabilities set the limits well below the ideal limits in real plasmas. In the following we briefly discuss the main limiting instabilities for different plasma scenarios. All these instabilities have to be either avoided or controlled for successful operation of future fusion reactor. A special chapter is dedicated for each of them in this book. Here, we only briefly describe where the instability becomes important.

The safety factor profile of the conventional tokamak scenario contains the $q = 1$ resonant surface. The *sawtooth* oscillations, discussed in Chap. 4, develop at this resonant surface. They are associated with abrupt changes in central plasma confinement due to the growth of an $(m, n) = (1, 1)$ mode, where m and n are poloidal and toroidal mode numbers. Whilst the plasma usually survives the drop in the core temperature and density due to this instability, the triggering of other, more dangerous instabilities is the main concern. It is often observed that larger sawtooth crashes trigger a resistive instability: the *neoclassical tearing mode* (NTM). The maximum achievable beta limit in conventional scenarios is usually set by this mode (see Chap. 8). Thus, both instabilities have to be studied and controlled.

The safety factor profile in the advanced tokamak scenario is more elevated with respect to the conventional scenario. This allows to avoid sawteeth $((1,1) \rightarrow q = 1)$ and NTM $((3,2) \rightarrow q = 1.5)$. For even higher central safety factor values, the most dangerous $q = 2$ resonant surface of the (2, 1) NTM can also be avoided. This is a positive consequence of the elevated safety factor profile. Advanced tokamak requires high bootstrap current fraction, which can only be achieved with high pressure gradients. These gradients drive the external kink instability. In the presence of a resistive wall, this kink instability is converted into the slower-growing *resistive wall mode* (RWM), as described in Chap. 6. This mode has to be stabilized to achieve the reactor relevant beta in this scenario. The same type of mode poses serious problems for RFP operation and has to be stabilized. Due to its high reproducibility, RWMs in RFPs are ideal for testing different mode control algorithms as will be discussed later.

There are also several important issues which appear in both types of scenarios. The transition from L-mode to H-mode creates high pressure and current gradients at the plasma boundary. These gradients drive the so-called *edge localized modes* (ELMs). ELMs lead to a fast loss of energy and particles from the plasma edge. These particle and energy losses provide a high heat flux at the divertor. In particular, the peak values are crucial for the internal components. This instability has to be mitigated or suppressed to reduce the peak flux and is the subject of the Chap. 5.

Future burning plasmas will contain fast particles resulting from fusion reaction. Satisfactory confinement of the energetic fusion products for sufficiently long time that the fuel ions are heated by them to a level where they start to fuse and a self-sustained burn process is critical for future reactor. Interaction of these fast particles with MHD instabilities may destroy this heating mechanism. The main interaction is expected with Alfvenic types of modes, characterized by high mode frequency, which could effectively interact with the fast particles. This is the subject of Chap. 9 which is dedicated to *fast particle instabilities*.

Plasma confinement in a tokamak can be lost in a fast catastrophic event called *disruption*. This event produces large power and force loads on the structures that surround the plasma. In a reactor, these loads have the potential to produce unacceptable damage to the machine itself. Disruptions must be avoided or done in controllable manner (to avoid severe damage). Also, there can be different reasons for disruptions, and all related points are discussed in a special disruption Chap. 7 of this book.

Important instabilities in the stellarator concept differ compared to the tokamak. Some instabilities either do not exist (current driven modes) or are much less dangerous (pressure driven modes). At the same time, the fast particle modes (Chap. 9) are probably more important for stellarators. The 3D nature of magnetic field produces an even larger number of potentially unstable modes in stellarators, compared to 2D tokamaks. Fast particle confinement is the subject of current research and is one of the key points to investigate in ITER.

2.10 Further Reading

Overview of the operation space in this chapter is an attempt to provide concise description of the main operation limits. It is not possible to discuss all details within this format. There are several books which nicely cover first part of this chapter and can be recommended for further reading.

- J.P. Freidberg, "Plasma Physics and Control Fusion", Cambrige University Press, 2007 (This book gives nice description of main concepts of magnetic fusion).
- More advanced description can be found in "Fusion Physics", Ed. M. Kikuchi et.al., IAEA VIENNA, 2012, (this book is freely available online http://www-pub.iaea.org/books/iaeabooks/8879/Fusion-Physics).

- Tokamaks are discussed in the famous book of J. Wesson, “Tokamaks”, 4th Edition, Clarendon Press, Oxford, 2011.

There are several books which discuss magneto-hydrodynamic theory and MHD instabilities with different level of required background:

- J.P. Freidberg, “Ideal Magnetohydrodynamics”, Plenum Press, New York and London, 1987 (This is one of the best standard textbooks for MHD).
- Hartmut Zohm “Magnetohydrodynamic Stability of Tokamaks”,Wiley-VCH, Weinheim, Germany, 2014 (The book gives nice description of the MHD instabilities and their connection to the operation limits in tokamak).
- D.D. Schnack, Lectures in Magnetohydrodynamics: With an Appendix on Extended MHD, Lect. Notes Phys. 780 (Springer, Berlin Heidelberg 2009) (This book provides detail derivation of main MHD concepts and will be extremely useful for students).
- J.A. Bittencourt, “Fundamentals of Plasma Physics”, 3ed Edition, Springer 2004 (Connection of the kinetic and the fluid descriptions is nicely shown in this book).
- Johan Peter Goedbloed and Stefaan Poedts, “Principles of Magnetohydrodynamics: With Applications to Laboratory and Astrophysical Plasmas”, Cambridge University Press, 2004 and J.P. Goedbloed, Rony Keppens, Stefaan Poedts, “Advanced Magnetohydrodynamics: With Applications to Laboratory and Astrophysical Plasmas”, Cambridge University Press, 2010 (These two volumes give extremely detailed description of MHD theory starting from basic assumptions).
- A B Mikhailovskii, “Instabilities in a Confined Plasma”, IOP Publishing, 1998, (This book is focused on MHD experts and discusses systematically theory of MHD instabilities).

Different aspects of the operation limits in tokamaks are described in the following books and papers:

- L. C. Woods, “Theory of Tokamak Transport. New Aspects for Nuclear Fusion Reactor Design”, Wiley-VHC Verlag GmbH Weinheim, 2006 (Some aspects of the MHD and thermal limits can be found in this book).
- R. Koslowski, “Operation Limits and Limiting Instabilities in Tokamak Machines”, Transaction of Fusion Science and Technology, Feb. 2006, Vol 49, p. 147. (The paper gives nice and very concise introduction into the subject).
- D. Frigione, “High-density operation in tokamaks”, Rivista del Nuovo Cimento Vol. 22, N8, 1999, p. 1 (Mainly density and radiation limits are discussed).
- M Greenwald, “Density limits in toroidal plasmas”, Plasma Phys. Control. Fusion 44 (2002) R27–R80 (This is a detailed review of the density limit).

References

1. J. Wesson, "*Tokamaks*", 3rd edn. (Clarendon Press Oxford, Oxford, 2004)
2. N.J. Fisch, Rev. Mod. Phys. **59**, 59175 (1987)
3. A.G. Peeters, Plasma Phys. Control. Fusion **42**, B231 (2000)
4. M. Shimada et al., Nucl. Fusion **47**, S1 (2007)
5. E.M Kikuchi et al., Fusion Physics, IAEA (2012), http://www-pub.iaea.org/books/IAEABooks/8879/Fusion-Physics
6. J.A. Bittencourt, *Fundamentals of Plasma Physics*, 3rd edn. (Springer, Berlin, 2004)
7. D. Biscamp, *Nonlinear MHD* (Cambridge University Press, Cambridge, 1993)
8. J. Freiberg, *Ideal MHD* (Plenum Press, New York, 1987)
9. P. Goedbloed, *Advanced Magneto-Hydrodynamics* (Cambrige University Press, Cambrige, 2010)
10. L. Harra, K. Mason (eds.), *Space Physics* (Imperial College Press, London, 2004)
11. D.D. Schnack, *Lectures in Magnetohydrodynamics: With an Appendix on Extended MHD*, Lecture Notes Physics (Springer, Berlin, 2009), p. 780
12. A.B. Mikhailovskii, *Instabilities in a Confined Plasmas* (IOP Publishing, Bristol, 1998
13. J. Wesson, Nucl. Fusion **18**, 1 (1978)
14. H.J. de Blank, Trans. Fusion Sci. Technol. **49**, 111–118 (2006)
15. D. Frigione, Rivista del Nuovo Cimento **22**, 1 (1999)
16. M. Greenwald, Plasma Phys. Control. Fusion **44**, R27–R80 (2002)
17. M. Greenwald, Nucl. Fusion **28**, 2199 (1988)
18. W.M. Stacey, Fusion Sci. Technolo. **52**, 29 (2007)
19. W. Suttrop et al., Plasma Phys. Control. Fusion **39**, 2051 (1997)
20. J. Rapp et al., Scaling of density limits with respect to global and edge parameters in TEXTOR-94, in P*roceedings of 26th International Conference on Control Fusion and Plasma Physics,* Maastricht, vol. 23. European Physical Society, Geneva (1999), p. 665
21. P.T. Lang, Nucl. Fusion **52**, 023017 (2012)
22. D.A.Gates et al., Phys. Rev. Let. **108**, 165004 (2012)
23. S. Sudo, Nucl. Fusion **30**, 11 (1990)
24. R. Gruber et al., Comput. Phys. Commun. **21**, 323 (1981)
25. L.C. Bernard et al., Comp. Phys. Commun. **24**, 377 (1981)
26. W. Kerner et al., J. Comput. Phys. **142**, 271 (1998)
27. F. Hofmann et al., Nucl. Fusion **40**, 767 (2000)
28. J. Freiberg, *Plasma Physics and Fusion Energy* (Cambridge University Press, Cambridge, 2007)
29. F. Troyon et al., Plasma Phys. Control. Fusion **26**, 209 (1984)
30. A. Sykes et al., Plasma Phys. Control. Fusion **29**, 719 (1987)
31. K. Miyamoto, *Plasma Physics and Controlled Nuclear Fusion* (Springer, Berlin, 2005)
32. W.M. Tang et al., Nucl. Fusion **21**, 891 (1981)
33. K.Y. Watanabe et al., Nucl. Fusion **45**, 1247–1254 (2005)
34. A. Weller et al., Fusion Sci. Technol. **50**, 158 (2006)
35. A. Komori et al., Nucl. Fusion **49**, 104015 (2009)
36. F. Wagner et al., Phys. Rev. Lett. **49**, 1408 (1982)
37. F. Ryter et al., Nucl. Fusion **52**, 114014 (2012)
38. T. Fujita, Nucl. Fusion **50**, 113001 (2010)
39. J. Stober et al., Nucl. Fusion **47**, 728 (2007)
40. D. Meade, Nucl. Fusion **50**, 014004 (2010)

Chapter 3
Identification of MHD Instabilities in Experiments

Anja Gude

Abstract There is a wide range of MHD instabilities which can develop in tokamak plasmas. These instabilities can have ideal or resistive (tearing) character, can be driven by gradients of pressure, current or energetic particles distributions (in position and momentum space), and can reside inside the plasma or affect the plasma boundary significantly. Investigation of their characteristics is the first step towards their identification. The poloidal and toroidal mode number, the mode frequency, the mode amplitude and spatial structure are important features to describe an instability. Although such a plain description does not directly make any statement on stability, driving forces and growth rate or saturation of modes, it is basic information for such analysis and for validation of stability codes. In the following we describe experimental possibilities to determine the above mentioned mode characteristics, mode coupling and how to distinguish between ideal and tearing modes.

In the following chapters of this book it will be shown that there is a wide range of MHD instabilities in tokamak plasmas. In Chap. 2 it has been introduced that instabilities can have ideal or resistive (tearing) character, can be driven by gradients of pressure, current or energetic particles distributions (in position and momentum space), and that they can reside inside the plasma or affect the plasma boundary significantly. Investigation of their characteristics is the first step towards their identification. The poloidal and toroidal mode number, the mode frequency, the mode amplitude and spatial structure are important features to describe an instability. Although such a plain description does not directly make any statement on stability, driving forces and growth rate or saturation of modes, it is basic information for such analysis and for validation of stability codes. In the following we describe experimental possibilities to determine the above mentioned mode characteristics, mode coupling and how to distinguish between ideal and tearing modes.

A. Gude (✉)
Max Planck Institute for Plasma Physics, Boltzmannstr. 2, 85748 Garching, Germany
e-mail: Anja.Gude@ipp.mpg.de

V. Igochine (ed.), *Active Control of Magneto-hydrodynamic Instabilities in Hot Plasmas*, Springer Series on Atomic, Optical, and Plasma Physics 83,
DOI 10.1007/978-3-662-44222-7_3

For feedback control of instabilities, an additional requirement is that the data are available in real-time and the subsequent analysis can be performed on a time scale that is of the order of the control system time constant (for ITER ≈ 1 ms). An even larger difficulty is the automated reliability control. While in off-line analysis, the operator can decide whether a result is trustable based on experience, and in case of doubt can just repeat it with slightly different parameters, in real-time analysis, the decision has to be made based on quantitative results. Avoiding this discussion completely, the focus of this chapter is on general requirements and techniques for mode analysis, regardless of whether they can be performed in real-time.

3.1 Mode Numbers, Frequency and Mode Coupling

3.1.1 Mode Numbers in an Axisymmetric Tokamak

Coherent modes in tokamak plasmas develop dominantly close to rational surfaces, where the safety factor q can be described by $q = m/n$ with m and n being small coprime integers: $m(n)$ is the number of toroidal(poloidal) turns that are necessary to close the field line (or a perturbation with the same helicity). For an instability at this surface, m and n are called the poloidal and toroidal mode number, respectively, because $m(n)$ is the symmetry number in a poloidal(toroidal) plane. This is shown for ($m = 1$, $n = 1$) and ($m = 2$, $n = 1$) in Chap. 2, Fig. 2.6.

In a periodic cylindrical geometry ("straight tokamak") the perturbation of an individual instability, $\tilde{A}_c(\theta,\ \phi)$, with mode numbers m, n, located at a flux surface with $q = m/n$ can be described by

$$\tilde{A}_c(\theta, \phi) = \hat{A}_c \cos(m\theta - n\phi), \tag{3.1}$$

with constant perturbation amplitude, $\hat{A}_c$. In an axisymmetric tokamak, the symmetry in toroidal direction is unchanged, but generally the perturbation amplitude varies in θ with a maximum period of 2π. Any such amplitude variation can be represented by a spatial Fourier expansion (see also Sect. 3.2.1) in θ:

$$\hat{A}(\theta) = \hat{A} + \sum_{k=1}^{\infty} \hat{A}_k cos(k\theta + \varphi_k) \tag{3.2}$$

with amplitudes, $\hat{A}$, $\hat{A}_k$, and phases, φ_k. The $cos(\theta)$ term represents the toroidicity effect, the $cos(2\theta)$ term the ellipticity of the plasma and higher k terms correspond to higher shaping moments. Taking for example only the toroidicity into account, the mode amplitude varies as $\hat{A}(\theta) = \hat{A} + \hat{A}_1 cos(\theta + \varphi_1)$, so the perturbation is represented by:

$$\begin{aligned}\tilde{A}(\theta,\phi) &= \hat{A}(\theta)cos(m\theta - n\phi)\\ &= \hat{A}cos(m\theta - n\phi)\\ &+ \frac{1}{2}\hat{A}_1(cos((m+1)\theta - n\phi + \varphi_1) + cos((m-1)\theta - n\phi - \varphi_1))\end{aligned} \tag{3.3}$$

In the spatial Fourier decomposition, this poloidally varying amplitude of the (*m*,*n*) mode produces sidebands with mode numbers ($m - 1$, n) and ($m + 1$, n). Correspondingly, an ellipticity induced amplitude variation leads to sidebands with ($m - 2$, n) and ($m + 2$, n). These shaping induced additional helicities (($m + k$)/n and ($m - k$)/n) can couple to the respective flux surfaces. The mode stability therefore has to be examined for the whole ensemble of coupled poloidal mode numbers (but still for a single toroidal mode number). The poloidal coupling is also influenced by the difference in plasma rotation angular frequencies between the flux surfaces. Strong differential rotation can counteract mode coupling and therefore also influence mode stability.

In real tokamaks, also the symmetry in ϕ is not perfect. First, the toroidal field varies between the toroidal field coils, causing a field ripple with the symmetry number defined by the number of coils. Second, components in the vessel (e.g., ICRH antennas, vacuum ports) and outside lead to non-axisymmetric modulations of the magnetic field. Last but not least, in many present-day tokamaks the axial symmetry is broken on purpose by perturbation field coils. The produced **R**esonant or non-resonant **M**agnetic **P**erturbations (RMPs or non-RMPs) are applied for mitigation of ELMs (Chap. 5) and control of NTMs (Chap. 8). The tools to deal with this increased complexity are being developed. In order to keep the complexity low and because the degree of asymmetry in toroidal direction is much smaller than in poloidal direction, here we only consider the perfectly axisymmetric tokamak, where perturbations have a well defined toroidal mode number, but consist of a mixture of poloidal mode numbers. Moreover, in the following we talk about single mode numbers also in *m*, because the mode coupling is often such, that one poloidal mode number is clearly dominant.

3.1.2 Mode Rotation

The instabilities discussed here are in most cases only a small deviation from the unperturbed plasma. Thus, a direct observation of the perturbation from absolute measurements is not impossible in general (see Sect. 3.9) but often difficult. Therefore one takes advantage of the fact that most instabilities move in the laboratory frame on timescales much faster than changes in the equilibrium quantities.

The observation systems are fixed in the laboratory frame. A rotation of a mode within this frame transfers the spatial variation due to the mode to a temporal oscillation of the observed quantities. When neither the mode structure nor the

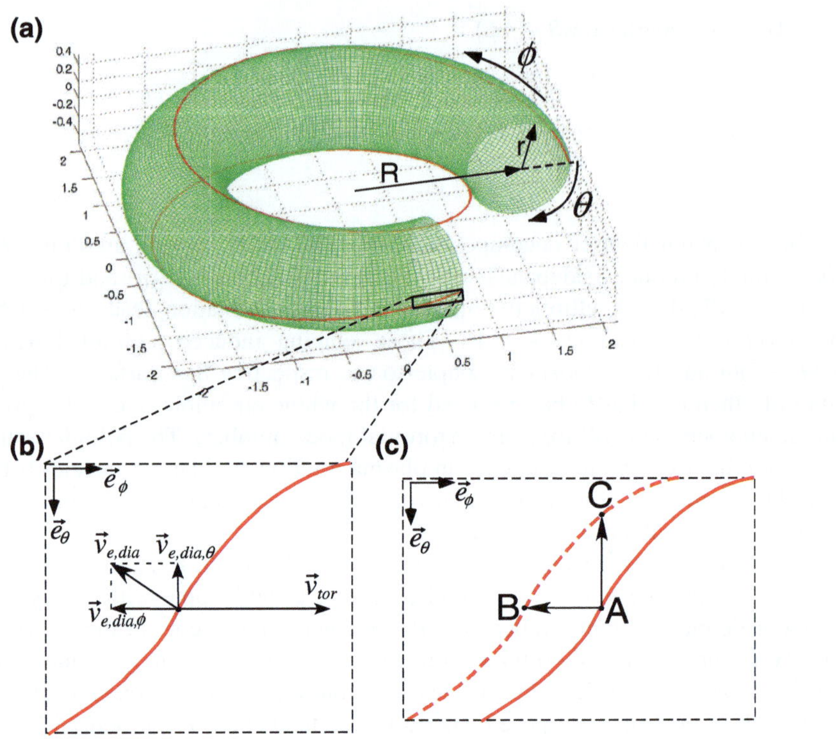

Fig. 3.1 **a** Torus with major radius, R, minor radius, r, toroidal angle, ϕ, poloidal angle, θ, and a closed ($m = -2, n = 1$) *field line* (*red*), **b** zoom: example for velocity components contributing to the mode rotation, **c** equivalence of poloidal and toroidal shift

rotation frequency vary significantly during the considered time window, the oscillation can be regarded as periodic. This allows the observation of the perturbation, e.g., using spectral analysis techniques, and is therefore an important feature for mode observation. In the following, the mode frequency, $f_{(m,n)}$, always refers to the frequency of the oscillation observed in the laboratory frame.

$f_{(m,n)}$ has an effect on which modes can be observed in a specific measurement: When a mode changes significantly over few (or even one) periods of its rotation, its observation and characterisation is more difficult. From this aspect, a high $f_{(m,n)}$ is favourable. On the other hand, high $f_{(m,n)}$ requires faster data sampling in the observation system, which restricts the possible observation systems for mode analysis.

When instabilities rotate in the plasma, its direction and—connected with this—the coordinate system have to be discussed. The standard convention for a toroidal coordinate system (r, θ, ϕ) is defined as shown in Fig. 3.1, where (r, θ, ϕ) is a right-handed coordinate system. In this system a field line on a flux surface with

$q = m/n$ can be described by $m\theta - n\phi = const$, such that a right-hand helix corresponds to positive q.[1]

Any velocity along a flux surface can be described as the sum of its toroidal and poloidal component: $\vec{v} = \vec{v}_\phi + \vec{v}_\theta = v_\phi \cdot \vec{e}_\phi + v_\theta \cdot \vec{e}_\theta$ with $\vec{e}_\phi$ and $\vec{e}_\theta$ the unit vectors in ϕ and θ direction and v_ϕ and v_θ the corresponding scalar velocities. In a straight tokamak, these can be translated into angular frequencies by $\omega_\phi = v_\phi/R$ and $\omega_\theta = v_\theta/r$ with R being the major radius[2] and r the minor radius of the flux surface. For a shaped tokamak, similar relations apply, but corrections according to the flux coordinates have to be considered.

A rotating instability is described by $(m\theta - n\phi + \omega_{(m,n)}t) = const$ with time t, $\omega_{(m,n)} = 2\pi f_{(m,n)}$ and $f_{(m,n)}$ the mode frequency in the laboratory frame. Due to the helical structure of the instabilities, toroidal (ω_ϕ) and poloidal (ω_θ) rotation components contribute to $f_{(m,n)}$ according to

$$2\pi f_{(m,n)} = n\omega_\phi - m\omega_\theta, \tag{3.4}$$

where the signs of n, m and of the angular frequency components have to be considered. Figure 3.1b shows an example for a mode rotation due to a toroidal plasma rotation velocity, v_{tor}, and an electron diamagnetic drift velocity, $v_{e,dia}$, which is always perpendicular to the field line. The resulting mode frequency is $f_{(m,n)} = (n(v_{tor}/R + v_{e,dia,\phi}/R) - mv_{e,dia,\theta}/r)/(2\pi)$ with $v_{e,dia,\phi}$ and $v_{e,dia,\theta}$ both being negative.

The toroidal and poloidal frequency components cannot be distinguished from any diagnostic that observes the mode. This is similar to the movement of a screw (as long as the screw ends are not visible). A rotation around the screw axis (corresponding to a poloidal rotation of the helical field line) results in the same picture as a shift along the axis (in toroidal direction). Figure 3.1c demonstrates this: the two movements shift point A to points C and B, respectively, but B and C cannot be distinguished because they lie on the same field line and thus mode phase.

The mode frequency in the laboratory frame is determined by the plasma rotation and the mode movement in the plasma rest frame. Since contributions to the mode frequency depend on its nature, the mode frequency can give hints to the mode character. Some modes, e.g., TAEs (see e.g., [1]), have a significant intrinsic velocity within the plasma frame in the order of several hundred kHz. The ($m = 1$, $n = 1$) sawtooth precursor or tearing modes have comparably slow rotation within the plasma frame, which is mainly induced by diamagnetic drifts ([2] and references therein) and **N**eoclassical **T**oroidal **V**iscosity (NTV) effects ([3] and references therein). Diamagnetic drifts are induced by pressure gradients, so they

[1] There are also left-handed systems in use and the field line is sometimes represented by $m\theta + n\phi = const$.

[2] In a straight tokamak, $2\pi R_0$ is the length of the periodic cylinder, such that $R = R_0$ everywhere in the plasma.

vanish in regions with flat pressure, which can occur for example due to large magnetic islands. The direction of ion and electron diamagnetic drifts is always opposite. Therefore rotation is often termed as being in ion or in electron direction, which is unambiguous and independent from the coordinate system. The mode frequency can help to localize the mode: when mode numbers, plasma rotation profile and all other contributions to the mode frequency are known, the radial position of the mode can be extracted.

In plasmas with significant torque input (mainly by neutral beam injection) and thus fast toroidal plasma rotation in the order of several kHz, the frequency of tearing modes and the (1,1) mode is as a first approximation determined by their toroidal mode number and the toroidal plasma rotation. Poloidal plasma rotation is strongly damped in the core plasma. Neoclassical calculations show that poloidal rotation only becomes significant in regions with high gradient, like close to the plasma edge in H-mode (see Sect. 2.9) scenarios.

It should be kept in mind that coupled modes keep a constant phase relation, thus they are always observed at only one common frequency in the laboratory frame. The mode coupling causes another force that influences the mode rotation.

3.2 Signal Analysis

Important methods for mode analysis first perform a spectral analysis of the individual signals (e.g., using Fourier transform or wavelet analysis) and subsequently correlate the spectral information from different locations. Another approach is singular value decomposition (SVD), where the spatial and temporal information from different signals are treated simultaneously. All these methods can be used for individual characterisation of various modes occurring simultaneously and with low amplitude compared to the absolute measured quantities. We cannot give a comprehensive description but present only some important features.

3.2.1 Fourier Transform

Fourier transform is a very common tool to analyse periodic signals. The macroscopic modes considered here are periodic in space (in poloidal and toroidal direction), which is taken into account by the discrete poloidal and toroidal mode numbers. With uniform plasma and mode rotation, the recorded time traces of mode modulated signals are also periodic in time. The Fourier transform projects the signals from the time (or space) to the temporal (or spatial) frequency domain. The result is also called Fourier transform or frequency spectrum and consists of amplitude and phase as a function of frequency.

3.2.1.1 Harmonics

For a single infinite and perfectly sinusoidal signal, the complex Fourier transform is non-zero only at the frequency of the sine function. The phase is simply the phase of the initial function with respect to the reference time point chosen. For a perfectly periodic but not sinusoidal function, the Fourier transform is a set of discrete values at the fundamental frequency f (corresponding to the period, P, of the signal by $f = 1/P$) and at multiples of it, called harmonics: $f_h = hf$ with $h \in 2, 3, \ldots$ For the mode numbers n, m and frequency applies

$$f_h/f = n_h/n = m_h/m \tag{3.5}$$

where the subscript h indicates the hth harmonic. For example the second harmonic of a (1,1) with $f = 8$ kHz is a (2,2) with $f_2 = 16$ kHz, the third harmonic is a (3,3) with $f_3 = 24$ kHz. When f_2 is selected, the methods described below indeed give the mode numbers $m = 2$ and $n = 2$. However, one should be aware that this set of mode numbers is different from the (e.g. toroidal) coupling of modes. Toroidally coupled modes have the same n but different m. They have different helicities and thus they extend over a set of flux surfaces. For harmonics, m/n is preserved, so they have a common helicity and belong all to the same flux surface.

Higher harmonics can result from the magnetic mode structure directly (non-sinusoidal deformation of the flux surfaces), but they often strongly depend on the measured quantity and observation method. Figure 3.2 shows spectra of a

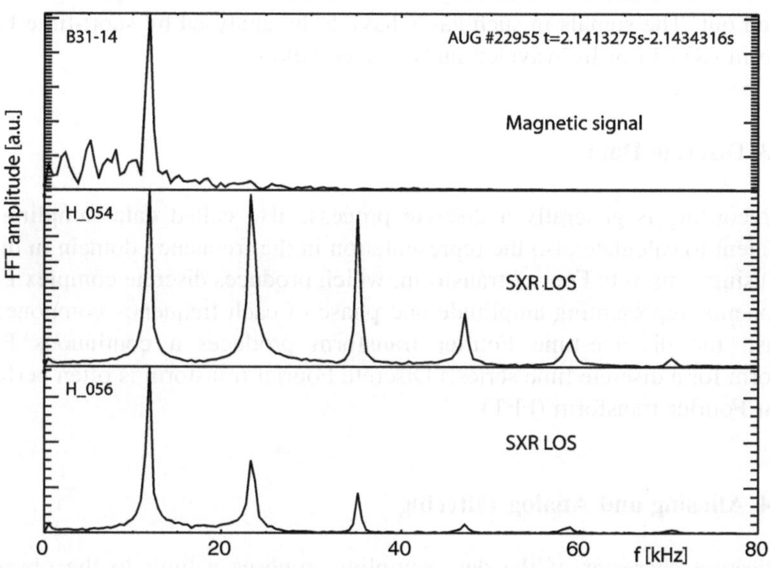

Fig. 3.2 Fourier amplitude of a magnetic pick-up coil signal and 2 soft X-ray lines of sight for a tungsten snake

magnetic pick-up signal and a soft X-ray signal (see Sect. 3.3 for diagnostics description) during a so-called tungsten snake, a (1,1) instability with tungsten accumulation inside the magnetic (1,1) island [4]. The soft X-ray signals show strong harmonics of the fundamental frequency due to the very localized radiation enhancement in the island, which do not show up in the magnetic structure.

Since the Fourier transform is reversible, the original signal can be reconstructed from it. Reconstruction of the signal using only specific Fourier components allows analysis of individual modes that occur simultaneously (but with different frequency) and to clean the signals from noise. These reconstructed time traces can be used instead of the original ones to study a single mode. For a full reconstruction of the mode structure, also the higher harmonics of the fundamental frequency have to be considered. Note that coupled modes have a common frequency and are not separated by Fourier transform.

3.2.1.2 Periodicity Limit

In reality, the observed signals are never perfectly periodic. The modes evolve in time and so one is interested in Fourier analysis of a finite wave. Technically, this is performed by multiplying the signal with a window function, that differs from zero only in a finite time window. The simplest such window function is a rectangle, but other functions can be used according to the purpose. When the signal is periodic with a large number of oscillations in this time window, deviations from perfectly periodic results are small. When either the mode changes fast or its rotation is slow, this condition may not be fulfilled and the frequency peaks get smeared out. The signals in such cases have to be analysed by short-time Fourier transform (STFT) or by wavelet analysis (see below).

3.2.1.3 Discrete Data

Data recording is generally a discrete process, also called data sampling. It is convenient to calculate also the representation in the frequency domain in discrete form, using a discrete Fourier transform, which produces discrete complex Fourier components, representing amplitude and phase of each frequency component. (In contrast, the discrete-time Fourier transform produces a continuous Fourier transform for a discrete time series.) Discrete Fourier transform is often performed as Fast Fourier transform (FFT).

3.2.1.4 Aliasing and Analog Filtering

The discrete character of the data sampling imposes a limit to the observable frequencies. The highest unambiguously detectable frequency, the Nyquist frequency, is $f_{Ny} = 0.5\, f_s$, where f_s is the sampling frequency ($f_s = 1/dt$ with dt the time between two adjacent samples). If the signal contains frequencies, f, above f_{Ny}, they will be represented in the frequency spectrum by a frequency below f_{Ny} according to $f \rightarrow |f - Nf_s|$ with an integer number N. This corresponds to a mirroring at f_{Ny} (and again at $f = 0$ and so on until the frequency lies within $[0, f_{Ny}]$).

The aliased frequencies appear in the FFT with the same amplitude as the original frequencies would have had with faster sampling. To avoid confusion with true and aliased frequencies, often analog antialiasing filters are applied before the data are sampled (digitized). These are usually low-pass filters, which ideally damp frequencies above the filter cutoff frequency. In reality, this damping is a continuous function of frequency, thus already lower frequencies are damped and higher still occur in the signals. Moreover, the filter induces a time delay which manifests in a phase shift of the filtered signals with respect to the original signals, which increases with frequency. Since relative phases of different signals play a large role in MHD analysis, care has to be taken that all considered signals have the same phase delay or that it is known and can be corrected.

3.2.2 Spectral Analysis for Non-stationary Signals

3.2.2.1 Spectrogram

Applying a Fourier transform to successive time windows, each of them long enough to produce a satisfactorily resolved frequency spectrum, the temporal development of an amplitude spectrum can be obtained by joining the spectra in the third dimension, the time. Figure 3.3 shows this in a so-called spectrogram. The ordinate is frequency, abscissa is time and the oscillation amplitude is colour coded. An alternative representation can be any three-dimensional graph with the amplitude as third dimension. Spectrograms allow a good overview on occurring modes and their development in frequency and amplitude. Furthermore, parasitic frequencies, that are artefacts of the observation system itself or are picked up from the environment, can be recognized by a characteristic frequency evolution.

The time window length has to be adjusted to the studied signal: there is always a competition between resolution in time and in frequency. A long time window allows to observe slow (low frequency) oscillations but reduces the time resolution of the spectrogram. A short time window allows high temporal resolution, but is only suited to resolve large differences in frequency, so it is only suited for high frequencies. This is a fundamental limitation which is expressed in the uncertainty relation between energy (or frequency) and time.

Instead of choosing time windows that only touch their neighbours, they can also more or less overlap. This is especially useful for window functions that approach zero at the edge, because the information of the time points at the edge is lost otherwise.

When the signals change rapidly, special care has to be taken to extract correct spectral information. When only few oscillations can be regarded as stationary,

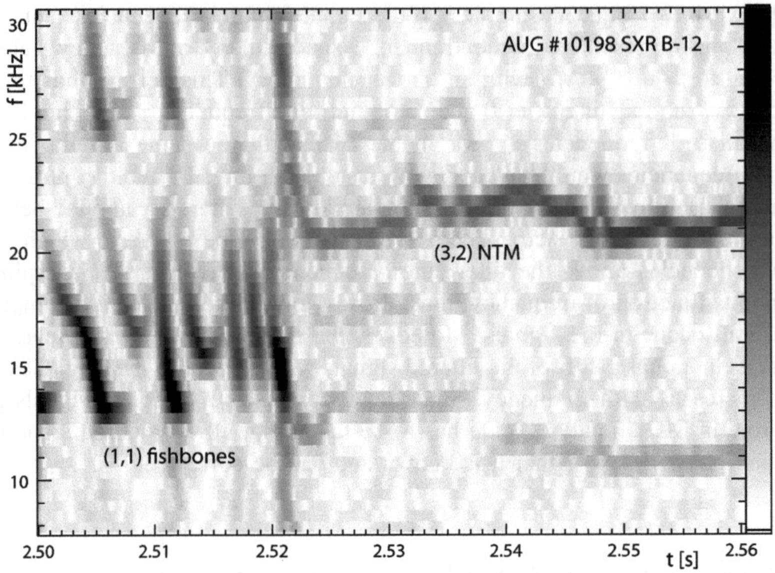

Fig. 3.3 FFT amplitude spectrogram of a soft X-ray signal showing fishbone instabilities (a-fast particle driven instability, described e.g., in [15], Chap. 7) that are terminated by a sawtooth crash (at 2.52 s), which triggers a (3, 2) NTM. The sample rate is 125 kHz, the FFT was performed over 256 samples (2,048 μs) with a rectangular window and is shifted by 16 samples (128 μs) for the subsequent FFT window

Short Time Fourier Transform has to be applied. In STFT the transform is calculated by moving a short—usually bell-shaped—window function, over the time region of interest. The window function, its length and the amount of overlap have to be adjusted to the signal. STFT gives a two-dimensional transform (in contrast to the one-dimensional Fourier transform) in frequency and time. The visualization of this is a spectrogram. The fixed length of the window function has to be optimized for a specific frequency according to the frequency-time uncertainty relation, thus strongly different frequencies cannot be treated optimally simultaneously.

3.2.2.2 Wavelet Transform

When there are Fourier components at high and at low frequencies simultaneously in a signal to be analysed, it is desirable to adjust the resolution to the component: For low frequencies, the frequency resolution must be high, whereas a low temporal resolution is sufficient, since one oscillation takes a long time anyway. For high frequencies a high temporal resolution is required to detect fast changes in the signals, whereas a lower resolution in frequency still corresponds to sufficient relative resolution.

Wavelet analysis provides exactly this: the width of the time window is adjusted to the probed frequency. Wavelets are wave packages, which are localized in frequency and in time. The width of the wavelets is adjusted to its frequency, such that with changing frequency the shape is approximately preserved. Similar to Fourier transform there are wavelet transforms for continuous and for discrete signals and the result can also be continuous or discrete. The discrete wavelet transform is usually calculated as fast wavelet transform, where the wavelet width is varied in powers of 2 (so it is kept constant for a frequency band).

The result of a wavelet analysis is the amplitude and phase in time-frequency space. The visualization is called scaleogram, which is the equivalent of a spectrogram, but with varying temporal and spatial resolution for different frequencies.

3.2.3 Multi-signal Analysis

3.2.3.1 Singular Value Decomposition

Fourier and wavelet transforms are applied on individual time traces (signals) from different measurement locations. Any spatial information has to be considered in a later step. A different approach is the singular value decomposition (SVD, see e.g., [5, 6]), where spatial and temporal dependencies are considered simultaneously. SVD is a method for analysing signals of a multidimensional measurement.

Similar to FFT, the original data are decomposed into orthogonal eigenvectors, but in the case of SVD, in two dimensions, space and time. The measured data x_{ij} at the jth observation position (out of p) and the ith time point (out of N) form the $N \times p$ matrix $\mathbf{X}$, with typically $N \gg p$. SVD represents $\mathbf{X}$ by pairs of spatial and temporal eigenvectors, $\vec{a}_k$ and $\vec{b}_k$, also called topos and chronos, with an eigenvalue λ_k, indicating the significance of the eigenvector pair $\vec{a}_k$ and $\vec{b}_k$. Topos and chronos are the eigenvectors of $\mathbf{X}^T\mathbf{X}$ and $\mathbf{X}\mathbf{X}^T$, respectively, where $\mathbf{X}^T$ is the transpose of $\mathbf{X}$

$$(\mathbf{X}^T\mathbf{X}) \cdot \vec{a}_k = \lambda_k \vec{a}_k \text{ and } (\mathbf{X}\mathbf{X}^T) \cdot \vec{b}_k = \lambda_k \vec{b}_k \tag{3.6}$$

The original data matrix can then be represented by $k_{max} = \min(p, N)$ eigenvector pairs $\vec{a}_k$, $\vec{b}_k$, and the corresponding eigenvalue λ_k

$$\mathbf{X} = \sum_{k=1}^{k_{max}} \sqrt{\lambda_k} \vec{b}_k \otimes \vec{a}_k^T \quad \text{or shorter :} \quad \mathbf{X} = \mathbf{B}\mathbf{\Lambda}\mathbf{A}^T, \tag{3.7}$$

where the quadratic matrices $\mathbf{A}$ and $\mathbf{B}$ consist of the eigenvectors $\vec{a}_k$ and $\vec{b}_k$, respectively. The $p \times N$ matrix $\mathbf{\Lambda}$ consists of the diagonal $k_{max} \times k_{max}$ submatrix with $\sqrt{\lambda_k}$ in the diagonal elements, and is 0 elsewhere. Equation (3.7) is the generalized principal axis transformation for non-quadratic matrices.

Typically a subtraction of the mean value for each measurement and a normalization with the standard deviation is applied to $\mathbf{X}$. In the case of offset subtraction alone $\mathbf{X}^T\mathbf{X}$ represents the covariance matrix, a normalization with the standard deviation results in the correlation matrix of the data. Therefore the eigenvectors are constructed in a way to maximize the correlated information content within the data. They are sorted with decreasing relevance for the data (decreasing eigenvalue in (3.7) and diagonal elements in $\mathbf{\Lambda}$). Like for FFT, the original signals can be reconstructed from the SVD components and thus, SVD can be used to select specific features of a signal or simply for noise reduction. For the latter, the number of relevant eigenvectors can be identified, for example, by the Akaike Information Criterion [7, 8]. An example for noise reduction by SVD can be found in [9].

For harmonic signals, the chronos of SVD are equal to the FFT transform: only two eigenvalues, corresponding to the sine and cosine part of the complex Fourier component, are different from zero. In case of periodic but non-sinusoidal signals, these two chronos combine all harmonics of the fundamental frequency, as long as they are correlated also spatially. For multiple frequency components without phase coupling, each is represented by two pairs of eigenvalues with the corresponding topos and chronos. In general, SVD temporal eigenvectors can have any form and can therefore represent any temporal evolution of a signal, not only periodic signals. Examples for such non-periodic signals are oscillations with rapidly changing frequency or amplitude and events like sawtooth crashes (Chap. 4) or ELMs (Chap. 5).

For MHD oscillations (also with changing frequency), the spatial information like the mode number is contained in the topos. SVD has been applied for mode number determination in many tokamaks to magnetic measurements [10, 11] and soft X-ray data [11, 12] by interpreting, in particular, the topos. Care has to be taken to correctly interpret the eigenvectors, since the temporal and spatial information on the modes can be distributed in different eigenvectors, which have to be identified. In [13] another application is presented, where SVD is used to select the most relevant of the temporal and spatial Fourier components of a set of signals. This method should be able to resolve modes of different mode numbers that have the same frequency.

A proper normalization of the signals is vital for SVD, in order to avoid overestimation of parts of the signals. The best procedure depends critically on the type of data. This point is especially important when different types of data, like radiation and magnetic perturbation, are combined in SVD analysis. For a physical interpretation of the eigenvectors, the signals may have to be calculated in their original quantity using the inverse of the normalization process.

To summarise, the main advantages of SVD are its "unprejudiced" correlation in space and time, and that the technique is applicable to non-stationary and in particular to non-periodic signals.

3.2.3.2 Cross Correlation

Cross correlation of two discretely sampled signals f and g is defined as

$$(f * g)[l] = \sum_{k=-\infty}^{\infty} f^*[k] \cdot g[k+l] \tag{3.8}$$

with f^* being the complex-conjugate of f. It is a measure of the similarity of f and g as a function of the time shift between them. When the signals are sufficiently similar, the maximum of $f * g$ indicates the time delay, thereby it can be used to determine propagation velocities. This is especially interesting for signals that are not well described by a single periodic signal (e.g., short oscillation pulses or modes with rapidly changing frequency) or even fully non-periodic.

The Fourier transform of the cross correlation, the cross spectrum, amplifies frequency components that are common to both signals. The cross spectrum with a reference signal, that shows the desired mode of interest well, can be used to enforce these frequencies in other signals, where the signal-to-noise ratio is much lower and the mode cannot be well recognized from the signal alone.

The technique can be applied to signals from different diagnostics (different measured quantities). In this case, the results must be interpreted with care since there is not always a near-linear relationship between such signals. The amplitude and phase of line integrated signals, for example, cannot be directly compared with that of local data (see also in Sect. 3.5.4).

3.3 Diagnostics for MHD Mode Observation

3.3.1 Basic Requirements

Before any mode analysis can be performed, suitable data have to be recorded. The detection systems, in the following called diagnostics, must fulfil three basic requirements for MHD mode analysis:

- The measurement must be sensitive to a quantity that oscillates with the instability, i.e., the magnetic field structure, e.g., temperature, density, radiation or magnetic field components. For other than magnetic quantities it has to be considered that the oscillation amplitude is not only determined by the oscillation amplitude of the magnetic structure, but also by the spatial variation of the observed quantity perpendicular to the flux surfaces (see Sect. 3.6.1). This means that the oscillation is only visible if the observed quantity is not constant over the affected plasma area.
- The spatial distribution of observed signals from the same diagnostic (in the following called channels) must be suited for determination of the spatial mode structure.
  - The toroidal symmetry number can be obtained best from data that are taken at different toroidal positions at equal poloidal position, and vice versa for the poloidal symmetry number.

  - Structures smaller than the distance between observation positions are not detectable (spatial aliasing).
  - The radial mode structure can only be determined from data that are in some way localized in radius.

- The sampling frequency, f_s, must be high enough. Since the highest unambiguously detectable mode frequency, $f_{(m,n)}$, with discrete data is half the sampling frequency, f_s should at least exceed $2f_{(m,n)}$. In order to allow for a reconstruction of non-sinusoidal signals, the significant higher harmonic frequencies have to be resolved. Therefore, for a proper representation of the signal (and mode) structure, $f_s > 6f_{(m,n)}$ is desirable.

  Since MHD modes have frequencies in the laboratory frame in the range of several hundred Hz to several hundred kHz, only diagnostics with times between two samples well below the ms range come into consideration. Spectroscopic methods are usually too slow for MHD analysis since they have to integrate over some time to get reliable photon statistics.

  In case the mode to be analysed is stationary within a time window (does not change apart from its rotation), there is a possibility to get good temporal resolution even with a low sampling rate as long as the data are instantaneous (i.e., they represent a time range much smaller than the time between two samples): When a fast sampled reference signal exists (either from the same diagnostic or a different one), the time points can be mapped in time according to the phase of the mode as determined from the reference signal.

3.3.2 Diagnostic Geometry

We cannot present a complete overview or detailed descriptions of plasma diagnostics here (please refer to literature, e.g., [14, 15]), but give examples of diagnostics that can be used for MHD analysis. The diagnostics can be sorted into groups with respect to the location of the observed quantity, which has relevant impact on the methods applicable and the information that can be extracted from the data. There are 3 geometric groups of diagnostics:

- Local quantity in the plasma. Examples are **E**lectron **C**yclotron **E**mission Spectroscopy (ECE) for electron temperature and microwave reflectometry for electron density and its fluctuation.
- Line integrated measurement through the plasma. Examples are Soft X-ray radiation and laser interferometry for electron density.
- Quantity outside the plasma. Examples are magnetic pick-up coils for the temporal derivative of magnetic field components (B_r, B_θ) and fast particle loss detectors for energy and pitch angle distribution of fast particles escaping the plasma.

In addition there are other diagnostics whose output is used in some way for MHD analysis, although they do not give information on the fast time scale directly. Such are the diagnostics used for equilibrium reconstruction, diagnostics for plasma rotation or diagnostics with lower temporal resolution that are used to "calibrate" faster sampled data. In the following we introduce main features of the 3 diagnostics mainly covered in this chapter.

3.3.3 Magnetic Pick-Up Coils

Magnetic coils, like Mirnov coils, provide a direct observation of the perturbed magnetic field, since they measure the currents induced in the coils by the variation of magnetic field components. The poloidal and radial magnetic fields, B_θ and B_r can be decomposed in a time-independent equilibrium component ($B_{X,0}$, where X stands for θ or r) and a perturbation ($\tilde{B}_X$): $B_X = B_{X,0} + \tilde{B}_X$. Since the induced current in the coils is proportional to the temporal derivative of the magnetic field, $\dot{B}_X = dB_X/dt$, the Fourier amplitude of a coil signal depends on the frequency, f. The true amplitude, $\hat{B}_X$, of a specific oscillation, $\tilde{B}_X = \hat{B}_X \cdot exp(2\pi ift + \varphi(f))$, can be obtained from the integrated signal or—much easier—by dividing the measured amplitude $\hat{\dot{B}}_X$ by the oscillation frequency: $\hat{B}_X \propto \hat{\dot{B}}_X/f$. This is important, e.g., when the temporal development of a mode with varying frequency is analysed.

The measurement is well localized in poloidal and toroidal position, but it is influenced not only by the region of interest (inside the plasma) but also by currents in the vessel components, that can modify amplitude and phase of the coil signals. Measuring the magnetic field outside the plasma implies that modes closer to the magnetic axis are superposed by modes further out and that there is no radial information contained. Due to the fall-off of the magnetic field component B_x from the resonant surface at r_{res} to the coil position, r_{coil}, the signals from distant regions (central modes) are weakened, especially for high poloidal mode numbers, m. In vacuum approximation the fall-off follows

$$\tilde{B}_X \propto (\frac{r_{res}}{r_{coil}})^{m+1} \tag{3.9}$$

The plasma modifies the perturbation amplitude decay. The vacuum solution is a good approximation close to the plasma edge and gets worse towards the plasma centre. For a correct description, the tearing mode equation ((8.3) in Sect. 8.2) has to be solved. The dependency of the measured amplitude on the distance between coil and resonant surface is an additional source of poloidal amplitude variation in the magnetic signals, which has to be considered when methods depend on the relative amplitude from different coils (not only the phase), e.g., for SVD analysis (Sect. 3.2.3).

The advantages of magnetic pick-up coils are that the measurement is quite simple (simple calibration, no large dynamic range, except for the strong signals before disruptions, and sufficient signal-to-noise ratio), available for all plasma configurations and machine parameters and that there are usually plenty of coils all around the torus, which are needed for equilibrium reconstruction and plasma control, anyway.

3.3.4 Soft X-Ray Measurements

Soft X-ray (SXR) measurements are performed in many tokamaks to dominantly view the central (hot) part of the plasma along many lines of sight (LOS).

The observed quantity is the radiation power density of the plasma, the so-called emissivity. The radiated power, $d\epsilon(E)$, per unit volume and energy interval, dE, is a complicated function of electron temperature and electron and impurity densities, where various atomic processes play a role (see for example in [14]). This can complicate the interpretation of emissivity oscillations, but contains also additional information if most quantities (like electron temperature and density) are known from other diagnostics.

Detectors have a specific energy dependent sensitivity, $f_{det}(E)$. The detector specific emissivity is then

$$\epsilon_{det} = \int_0^\infty f_{det}(E) \cdot \frac{d\epsilon(E)}{dE} dE \tag{3.10}$$

Thus, one has to distinguish between total emissivity (with $f_{det} \equiv 1$) and the diagnostic specific emissivity, which depends on the detector system. By selecting the photon energy sensitivity, one can influence which part of the plasma is dominantly observed. SXR detectors cut off the low-energy photons ($\lesssim 1keV$) coming especially from the colder (edge) plasma regions. SXR detectors are therefore more sensitive to core radiation and its fluctuations. With lower cut-off energies, a similar detector can observe a larger, or the whole, plasma region, thus getting more sensitive to instabilities close to the plasma edge.

In the presence of medium to high-Z impurities, the local SXR radiation can vary significantly on flux surfaces. e.g. through centrifugal forces in rotating plasmas (see Sect. 3.8.2). Therefore the *SXR emissivity is not necessarily a flux surface constant*, which has to be considered in the interpretation.

The individual detectors record the absorbed radiation power. This is proportional to the integrated SXR emissivity along the LOS, the so-called intensity, I:

Fig. 3.4 Geometry of lines of sight of SXR diagnostics at ASDEX Upgrade (*coloured lines*), and example for ECE measurement positions (*black circles*) projected to the same poloidal plane. The bold *blue line* is the plasma separatrix

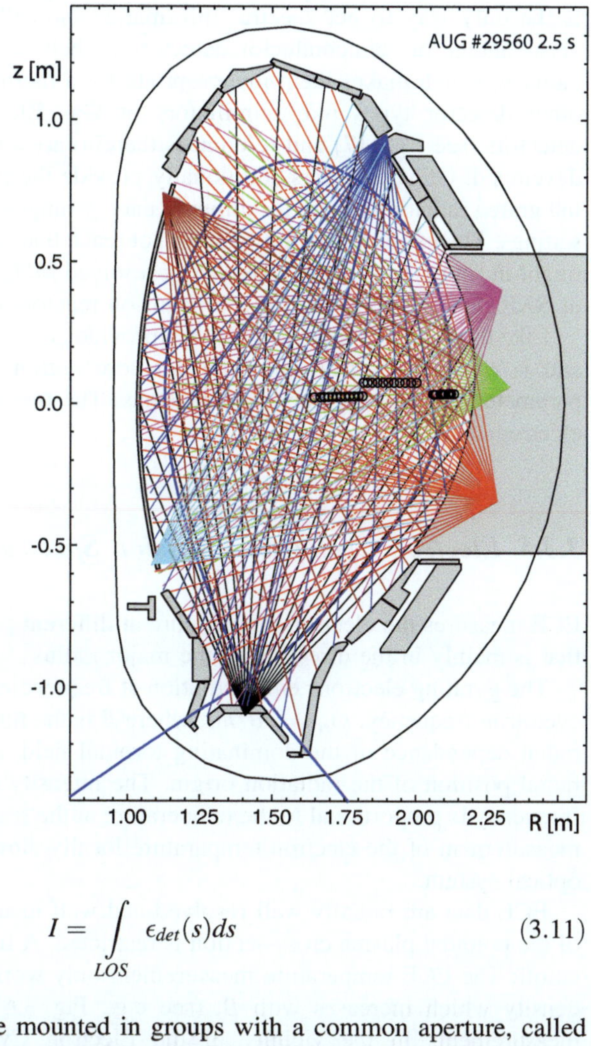

$$I = \int_{LOS} \epsilon_{det}(s)ds \tag{3.11}$$

Usually, the detectors are mounted in groups with a common aperture, called cameras. A standard geometrical design is a number of SXR cameras positioned at different poloidal angles in one poloidal plane with all LOS lying in the same poloidal plane. Figure 3.4 shows an example for ASDEX Upgrade. The spatial (poloidal and radial) information must be extracted from the line integrals, e.g. by tomographic reconstruction. Another type of SXR diagnostics have LOS with a dominant toroidal (ϕ) or tangential (along the field lines) component. Also 2D detector arrays are in use (for SXR imaging) [16]. If not mentioned otherwise, in the following we talk about the standard poloidal plane geometry.

The detectors currently used are mainly semiconductor diodes, which allow fast measurements (modes up to several hundred kHz have been observed). The low energy photons are blocked by material filters, mostly Beryllium, sometimes Aluminium. Applying different filter thicknesses at the same observation position is the only way to get spectral information (so-called multi energy-SXR). The disadvantage of semiconductor detectors is their sensitivity to neutron radiation damage, which makes them inappropriate for a thermonuclear reactor. Therefore, other detector types like scintillators or **G**as **E**lectron **M**ultiplication (GEM) detectors (see e.g. [17] and references therein) are already used and further being developed. GEM detectors additionally provide the possibility to measure energy integrated radiation power for different energy ranges simultaneously by electronic settings. The temporal resolution does not reach that of the conventional diodes but might in future be sufficient for energy resolved MHD analysis. A good discussion of SXR detector concepts for use in fusion reactors is given in [18].

Like the magnetic signals, SXR can provide good coverage of the whole plasma and is especially sensitive to the hot core region. SXR works for all plasma parameters, except for very cold plasmas. The tiny signal currents require good electronics.

3.3.5 Electron Cyclotron Emission Spectroscopy

ECE measures the electron temperature at different positions along a line-of-sight that is mainly in the direction of the major radius.

The gyrating electrons emit radiation at frequencies that depend on the electron cyclotron frequency, $\omega_{ec} = eB/m_e$, where B is the total magnetic field. Due to the radial dependence of the dominating toroidal field, ω_{ec} is directly related to the radial position of the radiation origin. The intensity of the radiation at a specific frequency is proportional to the temperature at the respective position. This allows measurement of the electron temperature locally along a line-of sight of the ECE optical system.

ECE data are radially well resolved and well localized, although the coverage of the poloidal plasma cross-section is restricted. A further limitation is due to the cutoff: The ECE temperature measurement only works correctly below the cutoff density which increases with B, (see e.g., Fig. 3.6 in [9]). To allow for ECE measurements in the vicinity of the Electron Cyclotron Resonance Heating (ECRH) deposition position, a special protection filter (notch filter), which suppresses frequencies in a narrow range, is required.

An interesting extension of the standard ECE is the ECE imaging (ECEI) or 2D-ECE diagnostics (see e.g., [9]). For ECEI, the LOS is split in vertical direction such that each new LOS is evaluated like the 1D ECE (frequency selective intensity). This results in a two-dimensional array. The diagnostics combines the advantage of local measurement and radial resolution of ECE with an additional poloidal resolution.

3.4 Toroidal Mode Number Determination

In toroidally symmetric plasmas, the toroidal mode number, n, is a well defined quantity: Since the equilibrium quantities do not depend on the toroidal angle, all n numbers are decoupled (see Sect. 3.1.1). Generally, n can be inferred from any signals that are sampled at different toroidal positions but at equal poloidal position and with sufficient temporal resolution. The main diagnostics in most tokamaks for this purpose are toroidal arrays of Mirnov coils measuring $\dot{B}_\theta$ or magnetic coils for $\dot{B}_r$.

Another method to determine n is possible with line integrated measurements with a mainly toroidal direction analogous to the m determination in Sect. 3.5.4, where it is described.

3.4.1 Phase Fitting

The relation between the signal phase, φ_{mode}, and the toroidal angle of the measurement position, ϕ_{coils}, is linear, thus

$$n\Delta\phi_{coils} = -\Delta\varphi_{mode}, \tag{3.12}$$

where $\Delta\phi_{coils}$ is the toroidal angle between 2 coils at the same poloidal position and $\Delta\varphi_{mode}$ the phase difference of the mode signals measured with these coils (their cross-phase) in the same time interval. This relation is used to determine n by fitting a straight line through the points of equal phase in the time traces as in Fig. 3.5. This can be done either with the signals directly (for a strong single mode) or using the signal reconstructed from the mode's fundamental Fourier component.

An alternative way is to fit the signal phase for the fundamental Fourier component of the mode as function of the toroidal angle of the coils (see Fig. 3.6). The inclination defines the toroidal mode number.

The example in Fig. 3.5 shows coils which are $\pi/4$ (or multiples thereof) apart. The observed oscillation can be fitted with either $n = 2$ or $n = -6$. This ambiguity is due to the spatial aliasing, as will be explained in the following.

3.4.2 Spatial Aliasing

The angular distance between measurement positions determines the highest unambiguously detectable mode number. For example, with two coils $\pi/2$ apart, the determination of n numbers up to 4 is possible, provided the rotation direction is known and $n \neq 0$. Higher n numbers show the same cross-phase as one from 1 to 4, e.g., $n = 5$ is not distinguished from $n = 1$. This effect is known as spatial

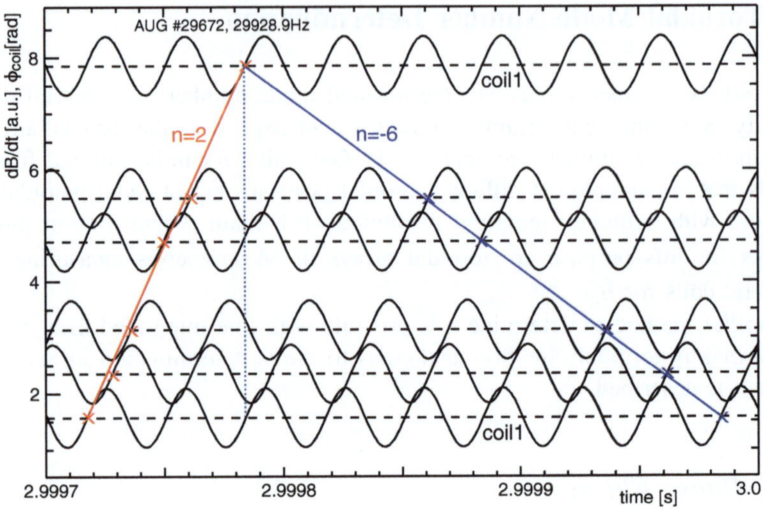

Fig. 3.5 Mirnov coil signals reconstructed from the fundamental frequency centred at their toroidal position with amplitude 1. The angular distance between the coil positions is $\pi/4$ or multiples of it. Time points with the same phase are connected in both directions giving mode numbers $n = 2$ and $n = -6$ respectively

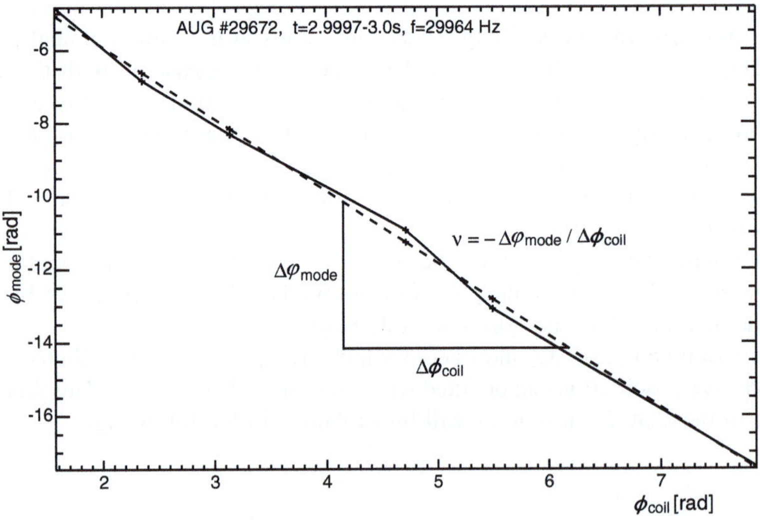

Fig. 3.6 Signal phase for fundamental mode frequency for a toroidal Mirnov coil array (same poloidal position) versus the toroidal angle of the coil position. The inclination of the linear fit (*dashed line*) determines n

aliasing. If the rotation direction of the mode is unknown, the aliasing has to be extended to the opposite sign, such that $n = 3$ (in "normal" direction) gives the same signals as $n = -1$ (an $n = 1$ mode rotating in the opposite direction).

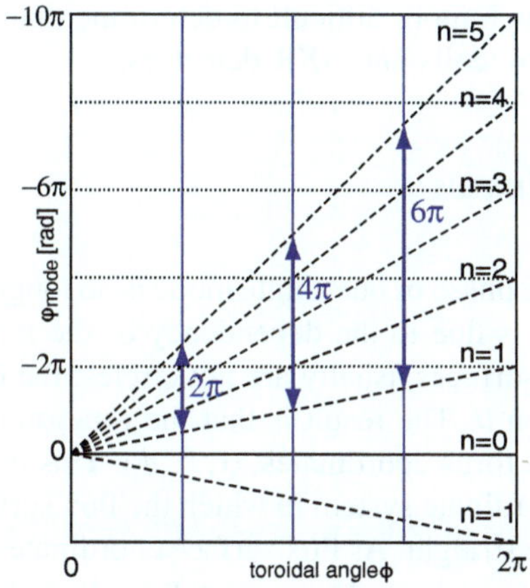

Fig. 3.7 Mode phase φ versus toroidal angle ϕ for different n. Coils are at the position marked by *vertical lines* (0, $\pi/4$, $\pi/2$, $3\pi/4$). Since φ is periodic in 2π, at all measurement positions modes with n and $n \pm 4$ give the same signal phase

The general rule is: when $\Delta\phi_{coils}/(2\pi) = a/b$ with a and b being coprime numbers, the distinguishable toroidal mode numbers represent a residue system modulo b. In other words: the modes cannot be distinguished if the difference in mode numbers is a multiple of b. Consider one coil at $\phi_1 = 0$ and the second at $\phi_2 = a/b \cdot 2\pi$. Two mode numbers, n_1 and n_2 cannot be distinguished if $n_1 \cdot \phi_2 = n_2 \cdot \phi_2 + k \cdot 2\pi$ *with* $k \in \mathbb{Z}_0$. This is equivalent to $(n_1 - n_2) = k/a \cdot b$ or—since the right hand side must be an integer value—to $(n_1 - n_2) = k' \cdot b$ *with* $k' \in \mathbb{Z}_0$.

For perfect signals (and perfect toroidal symmetry), the closest possible position of measurements would allow the highest unambiguous n number detection. However, due to imperfections and noise, the optimal distance varies with signal strength and toroidal mode number n. This applies for any pair of detection points. Therefore, a set of non-equally spaced positions is preferable. The optimum depends on signal-to-noise ratio, the highest mode number to be detected and on the envisaged number of coils. A higher number of coils with equal distance does not increase the resolution, as shown in Fig. 3.7, but it can help if one coil fails or to reduce uncertainty due to noisy information. A sophisticated discussion of this problem is found in [19].

3.5 Poloidal Mode Number Determination

In a straight cylindrical tokamak (see Sect. 2.6) the poloidal mode number, m, can be inferred similar to the n determination described above, using a set of signals in a poloidal plane. However, the toroidicity and the plasma shaping induce several effects that make m much more difficult to determine than n. Main diagnostics for this purpose are Mirnov coils and SXR detectors.

3.5.1 Flux Coordinates

In tokamak plasmas the phase of one single mode is no longer a linear function of the poloidal angle θ. This is due to the dependency of the field line inclination on θ. Furthermore, the flux surfaces usually are not circles: the distance to the magnetic axis, r, also depends on θ. The result is that the rotation of an instability is not a uniform rotation in the torus coordinates, (r, θ, ϕ). This is taken into account by a transformation to a coordinate system in which the flux surfaces are circular and the magnetic field lines are straight. As flux surface coordinate one can choose $\rho = \rho_{pol}$ or $\rho = \rho_{tor}$, as long as the same is always used. For tokamaks, a coordinate system in which the ignorable coordinate, the toroidal angle ϕ, is kept constant is preferable: (ρ, θ^*, ϕ). In [20] the transformation from θ to the straight field line angle, θ^*, is described. A general treatment of flux coordinates can be found in [21]. In the simple case of circular plasmas, the Merezhkin formula [22] accounts for the correction of θ for the toroidicity effect alone, whereas in shaped divertor plasmas generally an equilibrium reconstruction [23] is necessary for determining θ^*. Figure 3.8 shows an example for a shaped divertor plasma. In the (ρ, θ^*, ϕ) system, the mode rotation is uniform, thus the mode phase becomes a linear function of ϕ and θ^*:

$$\Delta\varphi_{mode} = -n\Delta\phi + m\Delta\theta^* \tag{3.13}$$

3.5.2 Toroidal Mode Coupling

In toroidal geometry, poloidal mode numbers are more or less accompanied by sideband mode numbers and toroidal coupling of modes with same n but different m can be induced, as explained in Sect. 3.1.1. The poloidal mode number is usually not a single number but a mixture of several m. Coupled modes have the same frequency in the laboratory frame, so they are also not directly distinguished by spectral analysis nor by SVD. The different rational surfaces (radial locations) of coupled modes can allow for distinguishing different m components when the radial resolution is sufficient. One should keep in mind that there are also modes with the same n number which are not coupled and rotate with different frequencies. In such cases, individual analysis of the modes is straight forward.

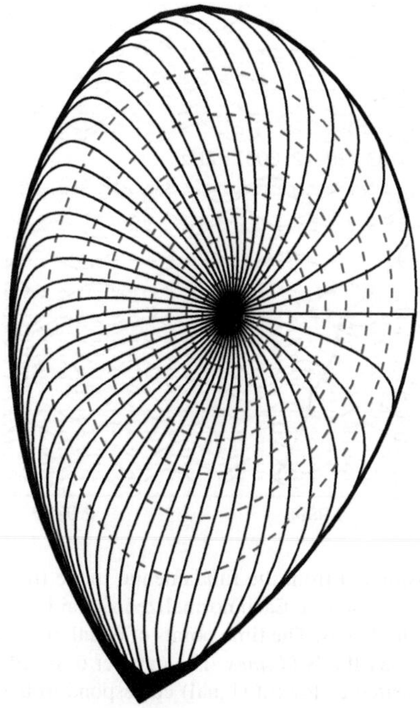

Fig. 3.8 Poloidal cross-section in a shaped divertor plasma with lines of constant ρ (flux surfaces, dashed) and *lines* of constant θ^* (*solid*)

3.5.3 m from Local Phase

3.5.3.1 Magnetic Pick-Up Coils

The main diagnostics for determination of *m* are currently magnetic pick-up coils, that measure the magnetic field perturbation outside the plasma. This means that coupled modes produce a common signal phase and cannot be easily separated. The signal amplitude at the coil position decays with distance to the resonant surface (e.g., (3.9) in vacuum approximation). Furthermore, Mirnov signals are affected in phase and amplitude by vessel components through induced mirror currents, which are difficult to describe correctly. In toroidal direction, the distances between coils and resonant surface are usually independent from ϕ and most mirror currents have only small variation in ϕ, while for the poloidal mode number, *m*, these effects can play a large role. Therefore, the θ^* correction is often only a minor improvement. Sophisticated codes can perform a forward modelling of the Mirnov signals (phase and amplitude) assuming a mixture of various *m* numbers for the (formerly determined) *n* number. Each mode is represented by a helical perturbation current distribution located at (or around) its resonant surface. Examples are described in [20] in vacuum approximation and in [24] including

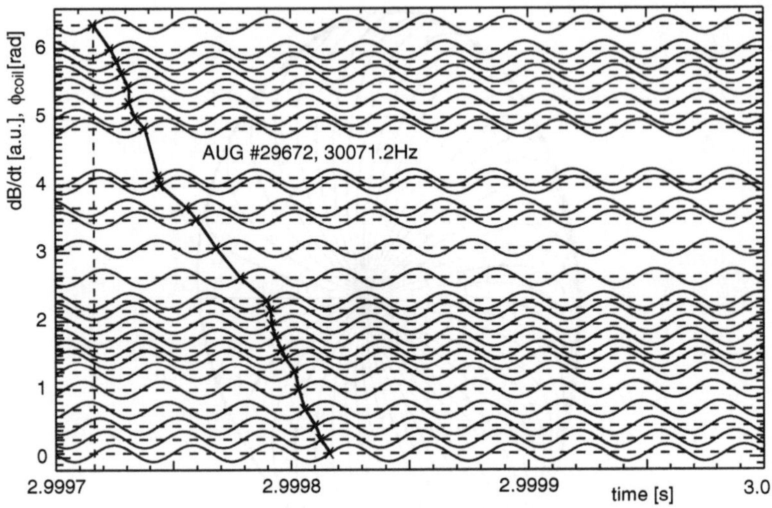

Fig. 3.9 Time traces reconstructed from the fundamental mode frequency of a poloidal Mirnov coil array showing an $m = 3$ oscillation. Each normalized signal is centred at its poloidal position marked by the horizontal *dashed lines*. The time points of equal mode phase are connected around the poloidal array (2π in θ_{coil}) by the *bold lines*. The number of oscillations that are passed in the uppermost coil (which is repeated as lowest signal) correspond to the poloidal mode number, m

plasma response. In [10] a comparison of this method with m determination using SVD is presented.

In spite of the difficulties for magnetic pick-up coils, the dominant poloidal mode number can often be inferred in a much simpler way from a full array of Mirnov coils in one poloidal plane. Like for the toroidal mode number, the mode phase can be followed, as in Fig. 3.9 using either the original signals or the Fourier filtered reconstructed time traces. This phase relation can also be analysed in the FFT phase of the mode frequency plotted versus the coil position (see Fig. 3.10).

However, the mode phase is not always a monotonic function of the coil position. This often becomes obvious especially in case of the internal (1,1) kink, where the phase versus coil position often shows a region in which $\partial\phi_{mode}/\partial\theta_{coils}$ changes sign (see Fig. 3.11). This phase reversal has been described in [25] as a result of strong poloidal harmonics of the mode (in case of the (1,1) by (m,1) components with $m > 1$).

3.5.3.2 SXR

Since the SXR signals do not represent local quantities, there is no location to directly assign to them. To overcome this, the SXR LOS can be mapped to the flux parameter, ρ_{tan}, of the innermost flux surface along the LOS, i.e., the flux surface, to which the LOS is tangent. The idea behind this is that the contribution (length)

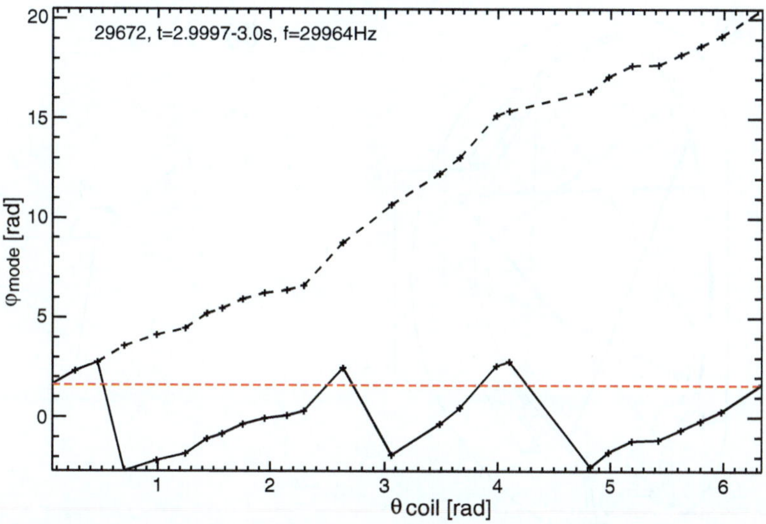

Fig. 3.10 Mode phase, φ_{mode}, versus poloidal angle of coil position, θ_{coil}, for the same oscillation as in Fig. 3.9 with the phase as calculated in the interval $[-\pi/2, \pi/2]$ (*solid line*) and with monotonically increasing phase (*black dashed line*). The first coil is repeated as the last. The number of 2π sweeps of the phase when going around the poloidal plane is generally equal to the dominant mode number. The mean inclination of the *black dashed line* also corresponds to the poloidal mode number

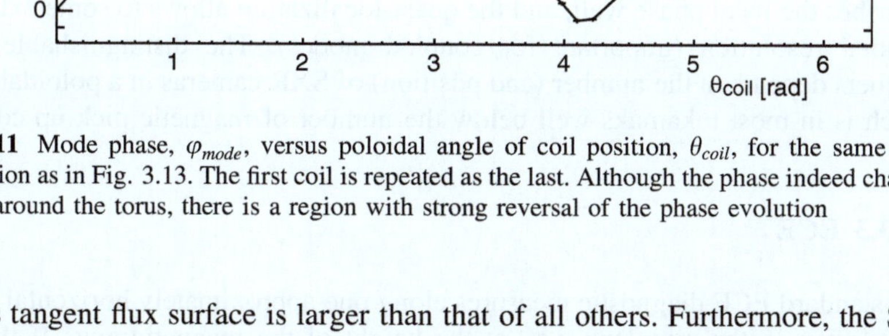

Fig. 3.11 Mode phase, φ_{mode}, versus poloidal angle of coil position, θ_{coil}, for the same (1,1) oscillation as in Fig. 3.13. The first coil is repeated as the last. Although the phase indeed changes by 2π around the torus, there is a region with strong reversal of the phase evolution

of this tangent flux surface is larger than that of all others. Furthermore, the SXR emissivity usually increases towards the magnetic axis, such that the contribution from the innermost flux surface is further increased. To distinguish between the two sides of the camera with respect to the magnetic axis, the LOS that pass on one side of the magnetic axis can be assigned $-\rho_{tan}$.

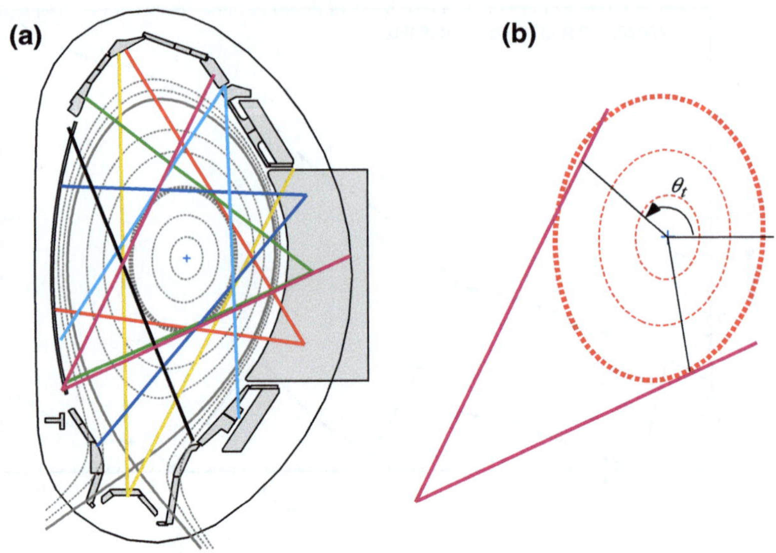

Fig. 3.12 **a** SXR LOS at ASDEX Upgrade that are nearly tangent to the selected flux surface (*bold dashed grey line*). Different cameras have different colour. **b** Two of such LOS with the tangent point and the corresponding angle, θ_t, indicated

This quasi-localization of LOS data can allow for m determination according to (3.13). The SXR LOS (from different cameras in one poloidal plane) that are tangent to the rational surface of an instability show mainly the fluctuation of their tangent points. Plotting their phases against the angle θ (or better: θ^*) of the connection between magnetic axis and tangent point (see Fig. 3.12), m can be determined from the slope.

This method has been used for a (1,1) mode in [26]. The advantages compared with magnetic data are that the vessel currents do not affect the SXR phase, so θ^* describes the local phase well, and the quasi-localization allows to some extent for a radial resolution (important for coupled modes). The distinguishable mode numbers depend on the number (and position) of SXR cameras in a poloidal plane, which is in most tokamaks well below the number of magnetic pick-up coils.

3.5.3.3 ECE

The standard ECE diagnostic measures along one approximately horizontal line of sight, often in the midplane, i.e., at the height of the magnetic axis. If the data extend to both sides of the magnetic axis, this corresponds to a spatial phase difference of $\approx 180°$ in θ, allowing at most to distinguish the poloidal parity (odd or even m) for all modes and mode components where measurements on HFS and LFS exist simultaneously.

ECEI has an additional resolution in poloidal direction, thus the poloidal mode number can be determined from it similar to an incomplete array of Mirnov coils. An example is found in [9]. In contrast to Mirnov coils, ECEI provide local data inside the plasma. Therefore, knowing θ^*—from an equilibrium reconstruction—the global m can be determined from the observed local wavelength in the core plasma.[3] Furthermore, coupled modes that are sufficiently separated in radius can be identified. Providing local and spatially well resolved measurements, ECEI data are a very valuable extension of diagnostic capabilities for MHD analysis.

3.5.4 *m* from Line Integrated Signals

The most common of line integrated diagnostics used for MHD analysis is Soft X-ray measurements, on which we concentrate here. SXR has the advantage of viewing the core plasma directly, so that core modes are often also visible in case of coupled modes, although the line integration can make the interpretation more difficult. On the other hand, line integration has the advantage that already with a single poloidal observation position (one pinhole camera with an array of LOS in one poloidal plane) the determination of poloidal mode numbers is possible. This procedure is very different from the one described above and requires some more explanation.

3.5.4.1 SXR Camera and Mode Profile

Often SXR data are shown for each camera individually. A straightforward parameter for the individual channels is the angle of the LOS with respect to a reference direction, e.g., the $\theta = 0°$ direction. Another possible parameter is the ρ_{tan} introduced in Sect. 3.5.3. Figure 3.13 shows an example of a SXR camera profile together with the LOS geometry in a poloidal plane and some SXR signals.

The time traces of the central LOS in Fig. 3.13 reveal a clear oscillation. The FFT spectrum (Fig. 3.14a) of one central time trace gives the oscillation frequency. At this frequency, the amplitude and phase of all LOS are determined. The plot of these over the LOS angle or ρ_{tan} are the mode profiles (Fig. 3.14b).

3.5.4.2 Phase Jumps and Amplitude Minima

When a single mode is analysed (no significant coupling to other poloidal mode numbers), the line integration of the SXR signals results in amplitude minima and

[3] Towards the plasma edge of a shaped divertor plasma, the radial variation of θ^* (or rather of $d\theta^*/d\theta$, see Fig. 3.8) gets so large that the evaluation becomes ambiguous.

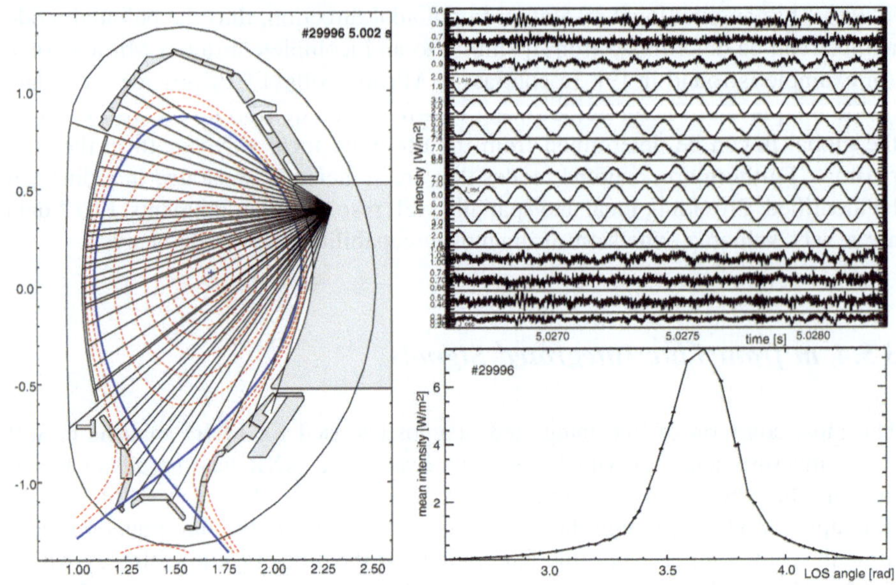

Fig. 3.13 Geometry of all LOS of a SXR camera at ASDEX Upgrade (*left*), subset of SXR time traces with suppressed zero (*upper right*) and background SXR camera profile (averaged over the depicted time range)

phase jumps by π, whose number is equal to m. To demonstrate this, let us consider a single perturbation with poloidal mode number m in a cylindrical geometry, located in a thin layer around its rational surface only. Figure 3.15 shows the geometry for a single LOS. Since the mode rotates in the laboratory frame, the LOS detects a modulated signal, $I = I_0 + \tilde{I}$. The modulated part, $\tilde{I}$, is the sum of the two contributions where the LOS crosses the resonant surface only—all other regions have constant emissivity—and can be written as

$$\begin{aligned} \tilde{I} &\propto (exp(i(m\theta_1 - \omega t)) + exp(i(m\theta_2 - \omega t)))/\sin(\theta_d) \\ &= exp(i(m\theta_m - \omega t) \cdot 2\cos(m\theta_d))/\sin(\theta_d) \end{aligned} \tag{3.14}$$

with $\theta_m = \frac{1}{2}(\theta_1 + \theta_2)$, $\theta_d = \frac{1}{2}|\theta_1 - \theta_2|$ and $\theta_{1,2,m} \in [0, 2\pi[$. The dependency of the amplitude on the LOS geometry ($\sin(\theta_d)$) is caused by the integration length inside the resonant layer. The modulated signal amplitude vanishes for a LOS which fulfils

$$m\theta_d = \pi/2 + k\pi \quad (k \in \mathbb{N}_0) \tag{3.15}$$

Since θ_d is within [0, π], there are exactly m roots in the amplitude profile. At each of them, the cosine changes sign, such that the phase is reversed. So the phase profiles show m jumps by π.

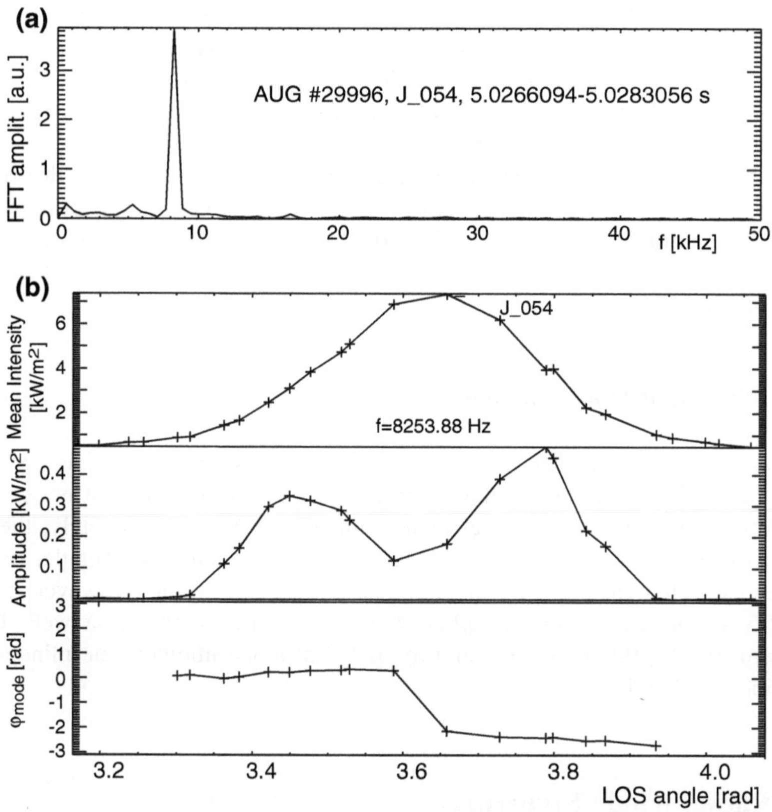

Fig. 3.14 **a** FFT spectrum of a central SXR time trace for the example in Fig. 3.13, **b** SXR mean, amplitude and phase profiles at the peak frequency in (**a**)

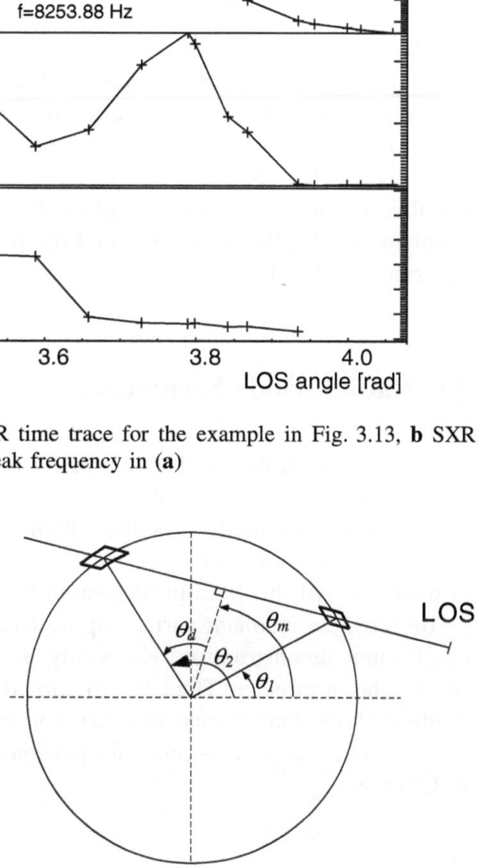

Fig. 3.15 Schematic geometry of a line-of-sight

For parallel LOS (a camera infinitely far away), θ_m is constant for all LOS, resulting in constant phase of the signals between the jumps. For a realistic pinhole geometry (finite angle between neighbouring LOS), the change in θ_m leads to an inclination of the phase profile proportional to m. In realistic tokamak geometry the phase slope is distorted by toroidicity and plasma shaping, but the qualitative results of this consideration are still valid: The SXR amplitude profiles of a mode with mode number m show m minima which correlate to m jumps by $\approx \pi$ in the phase profiles. Figure 3.14 shows an example for a (1,1) mode.

Noise and finite spatial resolution limit the m numbers that can be determined. However, mode coupling can complicate or prevent already the determination of low mode numbers. The phase profiles of coupled modes depend on the relative position and amplitude of the modes as well as their radial structure, making an interpretation difficult.

3.5.4.3 Toroidal Mode Number

Similar to m, the toroidal mode number, n, can also be determined with this method. The LOS have to be mainly in toroidal direction and must not cross the magnetic axis. Since the toroidal plane cannot be covered completely, the evaluation of phase jumps and phase slope must be calculated per angular distance between selected channels, not for the flux surface as a whole. However, because the flux surfaces in a toroidal plane are perfect circles, this is a well defined problem. In [27] this procedure and the poloidal mode number determination are described in detail.

3.6 Radial Mode Structure

So far we have considered the modes to be located at their resonant surface only. Realistically the perturbations also have a radial structure. Ideally, we want to determine the full mode structure which contains information on resonant surface position, mode character (e.g., ideal or tearing mode), size and coupling to other m numbers. All this is important input for the identification of the instability.

Besides, the resonant surface up to here has been regarded as known from the equilibrium reconstruction. Especially in the plasma core, this reconstruction is often rather imprecise. Thus, information on the position of resonant surfaces from mode analysis can be used as restriction for improved equilibrium reconstruction or—when gained in real-time—for plasma control tools (see the NTM stabilization in Chap. 8).

3.6.1 Radial Displacement of Ideal Modes

A single ideal mode in the plasma can be described as a displacement of the equilibrium magnetic field lines with the radial component of the displacement

$$\xi_\rho = \hat{\xi}_\rho(\rho)\cos(m\theta^* - n\phi + \omega t) \tag{3.16}$$

where ω is the angular frequency of the mode rotation ($\omega = 2\pi f$). The poloidal and toroidal components of $\vec{\xi}$ are to be determined from additional constraints like incompressibility of the plasma ($\vec{\nabla} \cdot \vec{\xi} = 0$) and the slip motion condition ($\vec{\nabla} \times (\vec{\xi} \times \vec{B}_\phi) = 0$), which means that the instability "slips" through the toroidal field, $\vec{B}_\phi$, without changing it [28]. In the simplified case of a straight tokamak, these conditions become equivalent.

To extract the radial mode structure (or only its resonant surface position), diagnostics with radial resolution are required. Such measurements (of e.g., temperature, density or radiation) can be used to extract the radial mode structure. In first approximation, the Fourier amplitude, $\hat{S}$, of a measured quantity, for a harmonic perturbation with displacement amplitude $\hat{\xi}_\rho(\rho)$ is

$$\hat{S}(\rho) = \hat{\xi}_\rho(\rho)\frac{\partial S}{\partial \rho} \tag{3.17}$$

where S is the unperturbed quantity. Figure 3.16 illustrates this relation. One important consequence is that in regions with constant measured quantity (e.g., flat temperature profile after a sawtooth crash), no instability is visible. Generally, a quantity with a small background gradient in the mode region will give a smaller signal amplitude than a quantity with a large gradient. Since the gradient can also

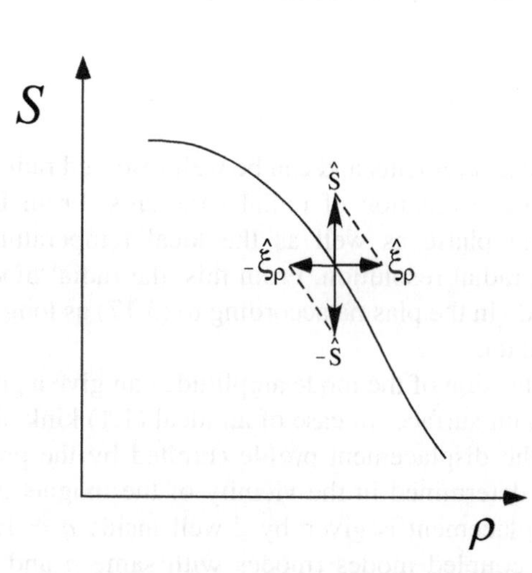

Fig. 3.16 The radial displacement amplitude $\hat{\xi}_\rho(\rho)$ shifts the original flux surface to smaller and larger ρ (*dashed lines*). This results in a fluctuation of the measured quantity at the (fixed) measurement position (*vertical arrows*). The relation between the observed Fourier amplitude, $\hat{S}(\rho)$, and $\hat{\xi}_\rho(\rho)$ is determined by the local gradient of S

change within an amplitude profile for one mode, the signal amplitudes should be corrected for the background gradient, if possible.

Note that in case of reversed gradient, also the signal is reversed, i.e., its phase is shifted by π. This is important for so-called hollow profiles, where the quantity (e.g., radiation or temperature) has a local minimum inside the plasma. The local mode phase undergoes a phase jump at the position where the gradient is reversed, which is not related to the structure of the instability but only to the background plasma.

Care has to be taken which background gradient is used for the correction and what is represented by the resulting mode amplitude profile. This depends on the specific instability. For small ideal perturbations of the equilibrium, where the background gradient is not significantly modified by the instability, it can be replaced by the (mode averaged) gradient and the resulting mode amplitude represents the above described radial displacement of the mode. The case of tearing modes is discussed in the Sect. 3.6.3.

3.6.2 Mode Localization and Radial Structure

In the following section we restrict to modes with well defined dominant mode numbers (which are assumed to be already known from previous analysis). The radial structure of Alfvén Eigenmode instabilities is a special case and will not be treated in this section. Examples for investigations on the radial structure of energetic particle modes are given in [29, 30].

3.6.2.1 Local Data

ECE data are local measurements and can be well resolved radially. Therefore they are well suited for determination of radial structures. From ECE data the local signal amplitude and phase as well as the local temperature gradient can be extracted with good radial resolution. From this, the radial mode structure can be obtained for any mode in the plasma according to (3.17) as long as the mode region is covered by ECE data.

Generally, the extension of the mode amplitude can give a good estimate for the position of the resonant surface. In case of an ideal (1,1) kink, the mode amplitude directly represents the displacement profile (limited by the precision with which the gradient can be determined in the vicinity of the magnetic axis). The magnitude of the core displacement is given by ξ well inside $q = 1$.

When toroidally coupled modes (modes with same n and frequency but different m and q) have sufficient spatial separation, determination of phase and amplitude relation between them is possible. Furthermore, the different rational surface positions can be determined simultaneously. A sufficient spatial separation of coupled modes is likely for localized modes in the core plasma, for example for coupled (2,1) and (3,1) magnetic islands, while in the edge region the rational surfaces usually lie too close together to be resolved.

In the edge region of the plasma, gradients of temperatures, densities, and also the safety factor are often large such that the modes are strongly localized. Therefore, the mode position can be well determined even if the resolution should not be sufficient to resolve its structure.

3.6.2.2 SXR

For any mode visible in the SXR data, the position of the resonant surface can be estimated from the radial extension of the mode amplitude. The restriction in precision is the rather low spatial resolution, which is also due to the line integration of SXR LOS, and the not fully determined resonant position within the radial mode profile. The usual guess is that the resonant position is in the flank of the mode amplitude profile. However, the SXR mode profile can deviate from the mode amplitude profile due to different local emissivity gradients. Since the local emissivity gradients cannot be directly inferred from the line integrals, a correction according to (3.17) is usually not possible, leading to incorrect mode localization. Furthermore, the line integration in case of coupled modes at different q makes only the outer mode position visible and completely hides the position of inner modes. Altogether this method is rather imprecise. The main advantage is that—in contrast to e.g., ECE data—SXR data cover the whole plasma (with the exception of the cold edge) and this method can always be applied.

3.6.3 Structure of Magnetic Islands

Tearing modes are most commonly presented by the helical magnetic perturbation flux, Ψ_1, with which the region outside the island separatrix can be described (see [31]). The ideal radial displacement amplitude (according to (3.16)) connected with this is given by

$$\hat{\xi}_\rho = \frac{\Psi_1}{B_{0,\theta}(1 - \frac{nq}{m})} \tag{3.18}$$

Figure 3.17 shows a schematic island displacement amplitude profile for a symmetric case (same extension to both sides of the resonant surface). In tokamaks, magnetic islands grow non-symmetrically around the resonant surface, extending significantly more to smaller ρ (see e.g., [32] and references therein). Around the rational surface $\hat{\xi}_\rho$ is not defined because it describes the tearing instability only in the region where ideal MHD is applicable and for a linearized (small) perturbation, while the island formation itself has to be treated nonlinearly with resistive MHD.

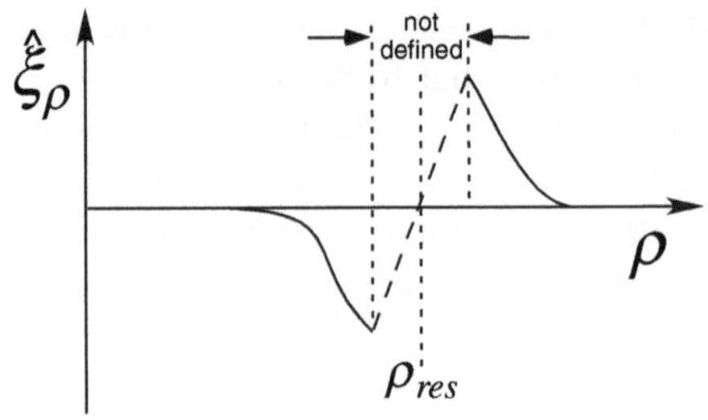

Fig. 3.17 **a** Schematic radial displacement amplitude profile for a tearing mode

A magnetic island often has constant electron temperature inside the island separatrix as long as the island is not too small [31]. Ion temperatures, electron and impurity densities and radiation are at least flattened within the island. In the O-point phase, the island separatrix extends radially to both larger and smaller ρ than the X-point position, where the latter reflects the original resonant surface position, ρ_{res}. The main features of a magnetic island to determine from experiment are its X-point position and its width. In the following, the island width, $W = W_\rho$, is given in units of the radial flux parameter, ρ. This is a convenient description because W_ρ is independent of the poloidal angle. The width in centimeter can be obtained by $W_r = W_\rho \cdot \partial r/\partial \rho$ at the poloidal angle θ. W_r varies with θ. Usually, in literature, when island widths are given in cm, this represents W_r at the LFS midplane position.

3.6.4 Determination of Resonant Surface for Magnetic Islands

With flat profiles in the island region, this causes the signals in the O-point phase to be higher (lower) than in the X-point phase for channels measuring outside (inside) ρ_{res} (see Fig. 3.19). In the FFT analysis at the mode frequency this results in a sharp phase shift of π between channels measuring inside and outside ρ_{res} (accompanied by a minimum in the amplitude at this position). This phase jump is a very precise indication of the resonant surface position and does not even require a gradient correction of signal amplitudes or precise signal calibration. However, its detection requires sufficient radial resolution and a non-vanishing radial gradient of the observed quantity.

An important application is the control of neoclassical tearing modes (see Chap. 8). Linear combination of Mirnov signals from different poloidal/toroidal positions (i.e., spatial Fourier filtering) is used to enhance specific sets of m and n numbers. Cross correlation of these filtered Mirnov coil signals with ECE channels allow tracking of the phase jump position [33], which indicates the resonant surface.

If the resonant surface can be determined at more positions, e.g. when ECE data cover both sides (HFS and LFS) of the plasma, the two individual resonant surface positions can also be used as equilibrium reconstruction constraint (independent from inaccurate calibration and small temperature gradients).

3.6.5 Determination of Island Width W

3.6.5.1 W from Magnetic Data

Magnetic pick-up coils measure magnetic fluctuations outside the plasma, so they do not allow for a direct extraction of the mode structure. The island width of a tearing mode can nevertheless be extracted from the Mirnov data using codes like those already mentioned in Sect. 3.5.3 [20, 24]. In [20] the position of the rational surface has to be taken from the equilibrium reconstruction or local measurements (e.g., the perturbation phase jump in ECE time traces), while [24] requires the whole equilibrium current profile as input.

The island width, W, is related to the integrated signal amplitude, $\hat{B}$, for constant safety factor profile as $W \propto \sqrt{\hat{B}}$. Therefore the relative changes—the temporal evolution—of the island width can be calculated from magnetic signals. A common way is to "calibrate" the Mirnov signal amplitude with an absolute island width determined for one (or few) time point(s) by any suitable method, like e.g., determination with ECE data, as described below.

3.6.5.2 W from Flat Temperature Region in the O-Point Phase

The flat region in the O-point phase ECE temperature profile reflects the island width, assuming that the temperature is flat inside the island separatrix. Since this method requires precise temperature values in the individual channels, problems can arise due to noise and imperfect calibration. The noise can be reduced by averaging over several periods of a stationary rotating mode, as was done in [32] or equivalently by reconstruction of the signal from the fundamental frequency and all significant harmonics. An imperfect relative calibration can be partly overcome by a correction based on a smoothed fitted profile in an island-free time range. Sometimes the island width is estimated roughly from the flattening of the mode-averaged temperature profile. This is easier but underestimates the island width.

For line integrated signals like SXR, a completely flat local emissivity profile does not lead to a flat region in the line integrated profile, as can be seen in Fig. 3.18a, b for a (1,1) island. Only with a local maximum in the O-point, a flat part in the SXR profiles occurs (Fig. 3.18c, d).

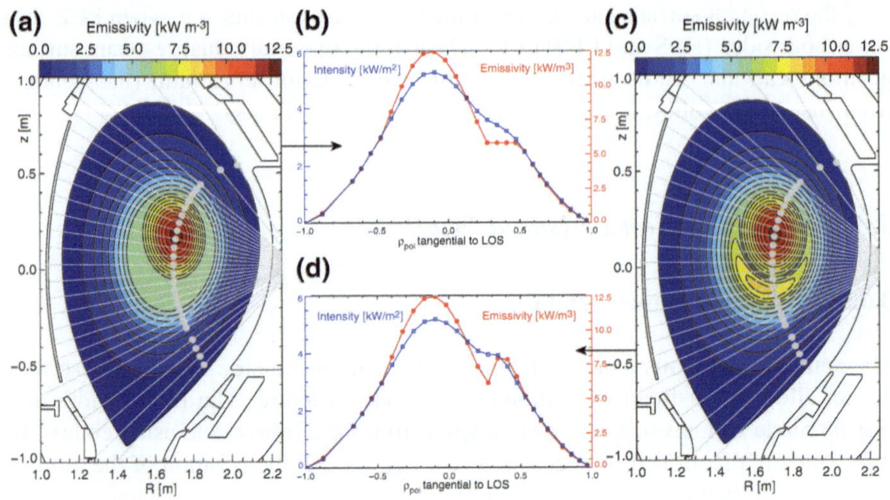

Fig. 3.18 2D emissivity distribution in a plasma with a magnetic island with constant emissivity inside the separatrix (**a**) and with a local emissivity maximum at the island O-point (**c**). Overplotted are the LOS of a horizontal SXR camera (*grey lines*) and the points at which an equilibrium flux surface is tangent to the LOS (*grey dots*). (**b** and (**d**: SXR profiles for the mode phase as shown in **a**) and **c**), respectively (*blue symbols*) together with the local emissivity in the tangent points (*red symbols*) (By courtesy of Markus Weiland)

3.6.5.3 W from Temperature Amplitude Maxima

Figure 3.19 shows the temperature oscillation for an island with flat temperature inside the separatrix. The temperature signals from a position that is crossed by the island, are non-sinusoidal. In the Fourier transform they will have significant amplitudes in the higher harmonic frequencies. Furthermore it can be seen that the magnitude of the temperature modulation increases from the O-point towards the island separatrix, and decreases outside. Although the signals are not well described by a single frequency, the amplitude of the fundamental frequency alone often provides valuable information. In a simulation of the temperature perturbation by a magnetic island [32], it was shown that the (1,1) Fourier amplitude has its maxima very close to the island separatrix, thus indicating the island width. The mode amplitude profile determined from ECE data therefore allows determination of the island width—even without higher harmonics, which might not be available due to slow data sampling—avoiding the problems mentioned for the determination of the flat region.

The amplitude maxima values also represent the island width, as can be seen in Fig. 3.19. Consider a flat electron temperature, T, inside (and at) the island separatrix. The temperatures in the points B and D are identical to that in the X-point, since they all lie on the island separatrix. For a constant temperature gradient of the X-point profile, T_X, over the region around the island $(T_A - T(X))/(\rho(A) -$

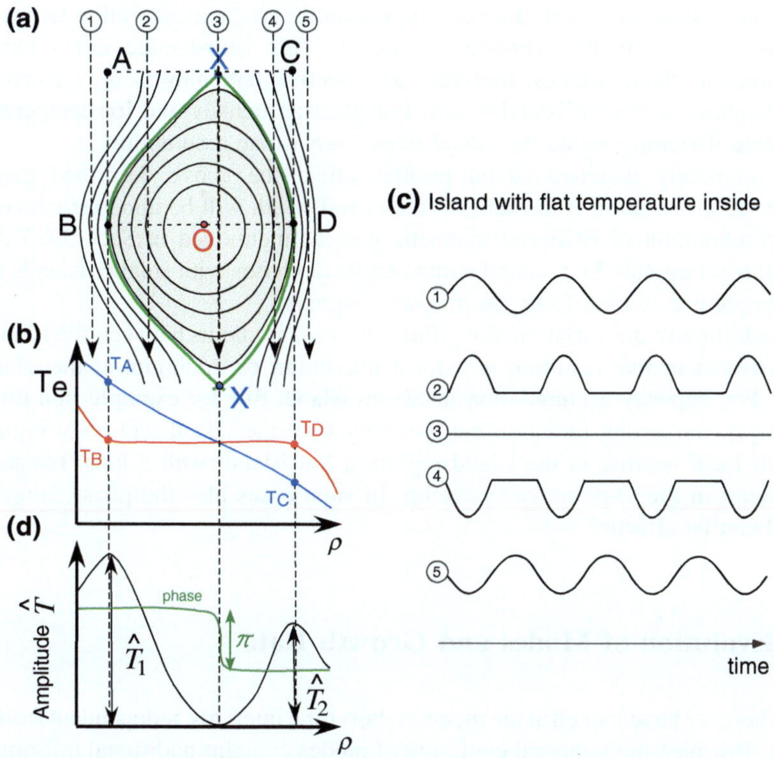

Fig. 3.19 **a** Schematic island structure, **b** Temperature profiles through X-point (*blue*) and O-point (*red*) for constant temperature inside the island ("flat" island), **c** Time traces of temperature measurement at positions *1* to *5* (as shown in **a**) and **d** Radial profile of the Fourier amplitude of the fundamental mode frequency with phase profile inserted

$\rho(X)) = \vec{\nabla}_\rho T_X$ and $(T(X) - T_C)/(\rho(X) - \rho(C)) = \vec{\nabla}_\rho T_X$ applies. The island width W is the distance between points A and C, which can be written as $W = (\rho(X) - \rho(A)) + (\rho(C) - \rho(X))$. The Fourier amplitude $\hat{T}_1$ is the half of the full width oscillation between the temperature in points A and B: $T_A - T_B = 2 \cdot \hat{T}_1$. Using $T(X) = T_B = T_D$, one gets the island width in units of ρ:

$$W = \frac{2(\hat{T}_1 + \hat{T}_2)}{|\vec{\nabla}_\rho T_X|} \tag{3.19}$$

3.6.6 *Effects of Incomplete Flattening and Local Extrema in the Island*

As derived in [31], the electron temperature inside small islands is not completely flat. Close to the separatrix the ratio of parallel to perpendicular heat conductivity is reduced, such that significant radial temperature gradients as well as temperature varitions along the flux surface can occur. This island edge effect gets less important for large islands, therefore an effective flattening is only reached for islands above a critical island width. For electron density and ion temperature a complete flattening inside the island is an even worse assumption.

Incompletely flattened island profiles affect the above described methods. While the phase jump is unchanged, the island width will be underestimated [32]. The combination of ECE and magnetic data as mentioned in Sect. 3.6.5 can be useful to extrapolate ECE island width results to phases with smaller islands taking the temporal evolution from the magnetic signals.

In addition to the variation along flux surfaces, gradients between the island flux surfaces can evolve resulting in a local maximum or minimum at the island O-point. For impurity accumulation inside an island, like for example in a tungsten snake [4], the strong radiation can possibly cool the island's O-point region. In case of local heating in the island region, a hot island, with a local temperature maximum in the O-point, can develop. In such cases also the phase jump of an island can be affected.

3.7 Evolution of Modes and Growth Rate

Up to here we have looked at the mode in short time intervals, independent from each other. However, the temporal evolution of modes contains additional information.

When a mode frequency can be tracked in time (e.g., with the help of a spectrogram), the identification of its mode numbers is not necessary for each single time point. So the temporal development of frequency and amplitude can be studied more easily.

The temporal development of the toroidal plasma rotation frequency, f_{tor}, is reflected by the mode frequency according to $\Delta f_{mode} \approx n\Delta f_{tor}$, provided the contributions to the mode frequency in the plasma frame do not change significantly. A mode rotating in the same direction as the plasma will have increasing frequency with increasing f_{tor}. When a change in the plasma rotation is correlated with an opposite change of the mode frequency (e.g., a decreasing mode frequency with increasing f_{tor}) this can have two reasons: Either the mode rotation in the laboratory frame is opposite to the plasma rotation direction (which is only possible if the plasma rotation is slower than the mode rotation within the plasma frame), or the observed mode frequency is aliased from higher frequencies (above the Nyquist frequency). Thus, the temporal development of an observed mode frequency can indicate its rotation direction and, in case of aliasing, the true mode frequency.

The mode amplitude evolution can be inferred from experiment and compared to code predictions and theoretical expectations, which can help to identify the origin of the instability. For example, the island width of an NTM is correlated with the normalized plasma pressure, β_N, which is a measure of the pressure drive (see Chap. 8). For classical tearing modes, which are current driven, this correlation is not observed.

Another application of the temporal evolution of the mode amplitude is the determination of the mode growth rate. When an instability is a small perturbation of the equilibrium quantity, A_0, which only depends on ρ, it can be described by the linear theory: $A = A_0(\rho) + A_1(\rho, \theta, \phi, t)$ with $|A_1| \ll A_0$ and the Fourier ansatz $A_1(\rho, \theta, \phi, t) = \sum_m \hat{A}_1^{(m,n)}(\rho) \cdot exp(\gamma^{(m,n)} t + i(m\theta^* - n\phi + \omega t))$ (we have assumed coupling only in the poloidal mode number and a well defined toroidal mode number of the instability). Here, ω and γ are real values and represent the mode angular rotation frequency and the linear growth rate, respectively. For negative γ the oscillation is damped, for positive γ its amplitude grows like $\hat{A}_1(t) \propto exp(\gamma t)$. Since ideal instabilities grow on the Alfvén time scale while resistive instabilities on the much slower resistive time scale, the growth rate observed in the experiment can help to reveal the nature of the instability. The linear theory is only applicable to small perturbations. For larger perturbations, the mode evolution has to be treated with non-linear models, which allow for a variation of the mode growth rate and a saturation of the mode amplitude.

3.8 Local Quantities from Line Integrated SXR Data

Having many lines of sight across one poloidal plane, the standard SXR diagnostics can be analysed using inversion algorithms to reconstruct the local emissivities from the line integrated intensities. This section deals with this topic. Information on the mode structure and localisation can be gained from such reconstructions.

3.8.1 1D Deconvolution

Assuming constant values on flux surfaces the radial distribution can be reconstructed from line integrated data. Since this method is only consistent with an axisymmetric distribution, it cannot be used for mode analysis. However, for the determination of the mode amplitude profile, the local radial gradient of the observed quantity is required. For this purpose, a mode-averaged quantity can be sufficient. The requirements are good equilibrium reconstruction on the one hand and constant quantities on flux surfaces on the other. The latter is not always fulfilled, as is shown in the following.

3.8.2 Deviation from Flux Surface Constants

Due to the high parallel transport coefficients for electron density and temperature, these quantities can be considered in most cases as flux surface constants. The same usually applies for hydrogen and impurity densities. The situation changes when strong forces act on the particles, like the centrifugal force in rotating plasmas. In [34] the density variation along a flux surface at two positions with different major radii, R_1 and R_2, has been described for pure hydrogen plasmas with a trace impurity:

$$\begin{aligned} \frac{n_t(R_2)}{n_t(R_1)} &= exp[\alpha \frac{m_t \omega_{tor}^2 (R_2^2 - R_1^2)}{2T_t}] \\ \alpha &= 1 - \frac{T_e}{T_i + T_e} Z_t \frac{m_i}{m_t} \end{aligned} \tag{3.20}$$

where n, m and T stand for density, atomic mass and temperature, respectively and the subscripts i, e and t for ion, electron and trace impurity. Z_t is the mean ionisation stage of the trace impurity and ω_{tor} the toroidal plasma rotation angular frequency.[4] The largest difference is found between HFS and LFS position of the flux surface. It is significant only when the rotation velocity is in the order of or above the thermal velocity of the tracer, $v_{th,t}$. This can be seen by rewriting (3.20) using $R_0 = (R_2 + R_1)/2$, $r = (R_2 - R_1)/2$ and the toroidal rotation velocity $v_{tor} = \omega_{tor} R_0$:

$$\frac{n_t(R_2)}{n_t(R_1)} = exp[\alpha \frac{v_{tor}^2}{v_{th,t}^2} \cdot \frac{4r}{R_0}] \tag{3.21}$$

Since the thermal velocities for different species at equal temperature behave as the inverse square root of their mass ratio, especially densities of heavy impurities can develop a strong variation along flux surfaces with higher density on the LFS.

When radiation is no longer constant on flux surfaces, 1D-deconvolution cannot describe the equilibrium radiation distribution well. This is the case for SXR radiation for example when heavy ions like tungsten are present in the (relatively fast rotating) plasma, coming either from wall materials or from deliberate impurity seeding. Since future fusion devices will have at least part of the inner wall covered by tungsten, this effect may become relevant. For example for ITER dimensions ($R_0 = 6.2$ m, $a = 2.0$ m) at 10 keV (where the mean charge state of tungsten is around 55 [35]), 10 % higher tungsten density on the LFS than on the HFS can be expected at mid radius for a rotation velocity of $v_{tor} \approx 50$ km/s. At $v_{tor} = 130$ km/s a factor of two can be expected.

[4] For a deuterium plasma with $T_e = T_i$, α varies between 0.5 and 1.

3.8.3 Plain Tomography

An unprejudiced method for reconstructing local quantities is tomography—a method standardly used in medicine. Generally, it is applicable for a set of LOS integrated data in the same plane. In case of SXR diagnostics in tokamaks this is mostly a poloidal plane.

Tomography requires very well relatively calibrated signals with an error in the few percent range and numerous LOS with many intersection points. The radial resolution is mainly defined by the distribution of LOS from one poloidal position (channels in one pinhole camera), while the poloidal resolution depends on the distribution of poloidal observation positions (cameras). With the typical number of SXR LOS in tokamaks (not more than several hundred), the tomography problem is ill-posed: there is an infinite number of solutions which fit the experimental data with same quality. Therefore, additional assumptions are necessary. Different tomography algorithms with different features exist for this purpose, an overview is given e.g. in [36]. From SXR tomography, clear and large-scale structures are distinguishable, while details and absolute quantities like distances or widths have to be treated with care.

Generally, a data preprocessing is possible, e.g. using SVD or Fourier transform, either to reduce noise only, or to select a specific mode to analyse. Also, tomography can be performed on single SVD components like in [11] to visualize their 2D structure. In [36], a combination of SVD and tomography has been applied the other way around: tomography was applied first to smoothen the data followed by SVD.

3.8.4 Rotational Tomography

A variant used for MHD analysis is the rotational tomography, which is restricted to one set of frequency, f, and mode numbers, m and n. The rigid rotation of a mode with f (a movement in time) can be translated into the rotation of the LOS in the poloidal plane. In a circular cylindrical plasma, this results in additional LOS with new directions, which give more equations to determine the 2D emissivity. However, in a tokamak, the modes rotate uniformly only in the flux coordinate system. The additional line integrals are therefore transformed into curved integrals in real space. The transformation relies on the equilibrium reconstruction, especially on the straight field line angle, θ^*. It is restricted to one single helicity, since the transformation depends on the poloidal mode number. Therefore, coupled modes cannot be reconstructed—with the exception of double tearing modes, which have the same helicity.

This method is only applicable when the mode does not change during the rotation. It does not work for a strong variation of the mode amplitude or a strong deviation of the observed quantity from being constant on flux surfaces. The first is the case for strong so-called ballooning character of modes, which means that their amplitude is much higher on the LFS than on the HFS. The latter occurs e.g. for the above mentioned rotation induced radiation asymmetries (Sect. 3.8.2).

3.9 Locked Modes

Locked modes are resistive instabilities that are at rest with respect to the tokamak. Generally, a rotating magnetic island induces mirror currents in the vessel wall, which brake the island. However, driving forces try to maintain a natural rotation velocity of the island, such that in a perfectly symmetric tokamak a residual rotation will remain. Complete locking can result from braking forces of a tokamak error field interacting with the perturbation field of the magnetic island. The error field can either be purely intrinsic or result additionally from external perturbation coils. Large locked modes can brake the plasma such that the plasma rotation is basically stopped.

The most important locked mode is an $n = 1$, typically a (2,1) or (3,1), island, which is often the precursor of a disruption. Its detection in real-time is a very important tool to trigger disruption avoidance or mitigation measures (Chaps. 7 and 8). We can distinguish between detection of locked mode occurance, which is sufficient for some disruption mitigation measures, like massive gas injection, and determination of the exact mode position everywhere in the plasma, which is required for current drive in the island's O-point by ECCD. Detection systems which can determine the mode position are therefore preferable. Furthermore, analysing the locked mode position in various discharge scenarios and with different external magnetic perturbation fields allows to gain information on the tokamak error field [37–39]. This knowledge can be used for a compensation of the error field, thus impeding mode locking.

3.9.1 Detection of Locked Modes

Since locked modes do not rotate in the laboratory frame, the only way to determine them is to detect toroidal or poloidal asymmetries in the total observed quantities (perturbation plus equilibrium). Due to the toroidally mainly symmetric equilibrium, toroidal variations can be detected more robustly. Toroidally distributed arrays of magnetic coils detect locked modes through their magnetic perturbation field, e.g., its radial component

$$\tilde{B}_r^{(m,n)} = \hat{B}_r \cdot \cos(m\theta^* - n\phi + \varphi_{LM}) \tag{3.22}$$

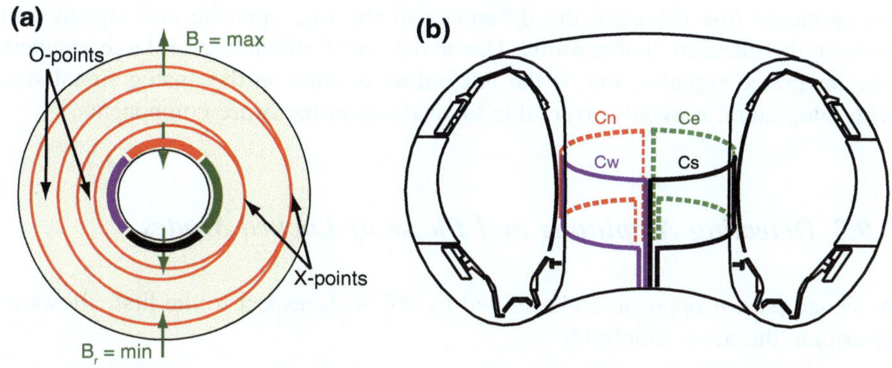

Fig. 3.20 **a** Horizontal cut through the torus at the height of the magnetic axis. The *red lines* indicate the separatrix of a (2,1) mode with the X-points on the *right* and the O-points on the *left side*. The maximal and minimal radial magnetic perturbation field of the mode is half-way between the O and the X-point (*green arrows*). A set of four 90° coils is indicated on the HFS (*fat coloured segments of a circle*). **b** Sketch of four 90° saddle coils to measure $\tilde{B}_r$ on the HFS

Here, φ_{LM} characterizes the locked mode position (φ_{LM}/n is the toroidal angle at $\theta^* = 0$ where $\tilde{B}_r$ is maximum). Detection at different ϕ but same θ eliminates the θ^* dependency.

Instead of small pickup coils, typically large saddle coils, spanning a wide toroidal angle and a significantly smaller poloidal angle, are used in order to increase the signal strength, while keeping the poloidal localization. Figure 3.20 shows a possible installation to measure the radial magnetic field B_r on the high field side for locked mode determination: Four coils (Ce, Cs, Cw and Cn, where e, s, w, n indicate the coil orientation in direction east, south, west and north) that span 90° in ϕ each and are shifted by 90°. Such a configuration is used in ASDEX Upgrade [38].

The signal of each coil integrated in time represents the spatial integral over B_r over the toroidal angle spanned by the coil. Generally, the difference of two measurements that are toroidally 180° apart removes the equilibrium contribution to B_r (not necessarily vanishing in coil measurement, especially in shaped plasmas) and enhances perturbations with odd n number, while even n numbers are completely suppressed. This difference, in the example in Fig. 3.20 $B_r^{ew} = B_r^e - B_r^w$ for the coils Ce and Cw, or $B_r^{ns} = B_r^n - B_r^s$ for Cn and Cs, still depends on the mode orientation, φ_{LM}. However, since modes often lock in similar positions due to the intrinsic error field, one coil pair can be sufficient for locked mode detection, provided the coils are in a good position with respect to the error field.

Since magnetic coils only detect changes in the magnetic field ($\dot{B}_r$), B_r has to be determined by integration in time of the coil signal. This can be technically tricky, because small offsets cause large errors in integrated values. In practice, it is preferable to first calculate the difference of the two opposite coil signals and perform the integration afterwards. This avoids small differences of large numbers (the integrated signals), and, if the integration is done on the analog signal side with integrators, it avoids errors due to variation in hardware components.

3.9.2 Detecting Amplitude and Phase of Locked Modes

A second pair of opposite coils, shifted by 90° with respect to the first, allows to determine the mode amplitude

$$\hat{B}_r = c \cdot \sqrt{\left(B_r^{ew}\right)^2 + \left(B_r^{ns}\right)^2} \tag{3.23}$$

and phase

$$\phi_{LM} = arctan\left(B_r^{ns}/B_r^{ew}\right) + n \cdot \phi_{coil} \tag{3.24}$$

of modes with odd n number, independent of the exact locking position. Here ϕ_{coil} is the centre position of the first coil (Ce in our example). The factor c depends on the toroidal extension of the coils and on their radial distance to the perturbation. Note that the *arctan* function in this calculation has to be redefined such, that its values cover the ϕ-range of $[-\pi, \pi]$ (i.e., $arctan(\alpha)$ is shifted by π $(-\pi)$ if B_r^{ew} is negative and B_r^{ns} is positive(negative)). The detected phase angle, ϕ_{LM}, is the toroidal position at which the perturbation $\tilde{B}_r$ has its maximum at the poloidal position of the coil array, θ_{coil} (in general, if $\theta_{coil} \neq 0$: $\phi_{LM} \neq \varphi_{LM}$). Although this filtering method in principle detects all locked modes with odd n number, virtually the only important locked mode is the $n = 1$ mode, to which we restrict in the following.

In order to determine the mode phase, e.g., the O-point position) for every poloidal position, ϕ_{LM} has to be connected to the mode phase and then mapped along the field lines. Figure 3.21 shows the relation between O-point position and $\tilde{B}_r$ for the common case of increasing plasma current density towards the magnetic axis.[5] At O-point and X-point the radial component, $\tilde{B}_r$, vanishes while the poloidal component, $\tilde{B}_\theta$, has an extremum. A projection of the $\vec{\nabla}\phi$ direction to the drawing plane of Fig. 3.21 reveals if the O-point is between minimal and maximal $\tilde{B}_r$ when going in positive ϕ direction, or vice versa.

[5] The figure shows the standard positive shear case. In reversed shear regions, the island O-point current is in the same direction as the plasma current, thus the perturbation fields are also reversed.

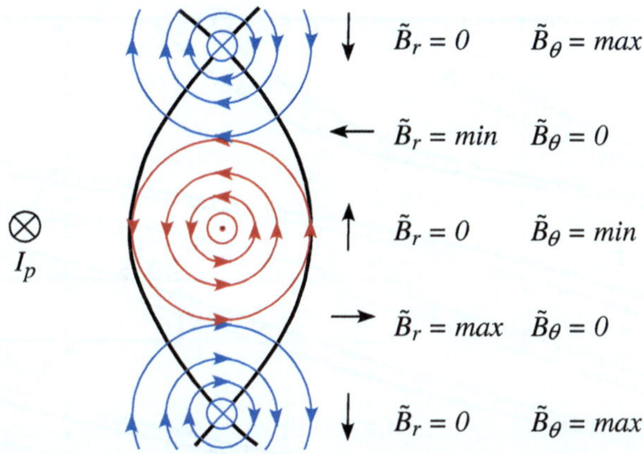

Fig. 3.21 Schematic drawing of a magnetic island in helical coordinates. The *bold line* indicates the island separatrix, the magnetic axis is to the *left*, the measuring coils are to the *right*. The island structure can be described by a perturbation current in the O-point opposite to the plasma current, I_p, and in the X-point in the same direction as I_p. The resulting magnetic perturbation fields are indicated. This figure also shows how to localize the O-point of a rotating mode with Mirnov coils, which is needed for NTM stabilization with modulated ECCD (see Chap. 8): at the O-point, $\tilde{B}_\theta$ has a minimum, therefore $\dot{B}_\theta$ has a zero with positive derivative

The mapping of the mode phases to the (θ, ϕ) plane is shown in Fig. 3.22 for a (2,1) mode. This field line mapping requires an equilibrium reconstruction, and it is restricted to a single helicity. Since $n = 1$ modes of different m can be coupled, the effect of this coupling on amplitude and phase of the locked mode has to be considered. The total locked mode amplitude varies in poloidal direction. The coupling of modes which differ by 1 in m is normally such, that the amplitude is minimal on the HFS and maximal on the LFS [40–42]. A measurement on the LFS should therefore be preferable but is often technically hampered due to heating systems and other installations. Measurement on the HFS should be the worst choice from this aspect. However, to correctly determine the O-point position of a specific mode it is necessary that the measured locked mode phase, ϕ_{LM}, represents this specific mode. Consider for example an $n = 1$ locked mode which is dominantly $m = 2$ with a weaker $m = 3$ component. On the HFS, the combined amplitude is minimal, but the phase is equal to that measured on the LFS, i.e., the mapping along the (2,1) helicity gives the correct (2,1) phase everywhere in the plasma. At all other poloidal positions, the mapping according to the (2,1) helicity of the locked mode phase depends on the relative amplitude of the (2,1) and the (3,1) modes and gives in general incorrect phase information. From this aspect, the HFS position is the second best choice for locked mode detection and other poloidal positions (e.g., top or bottom) are not advisable if phase information is required.

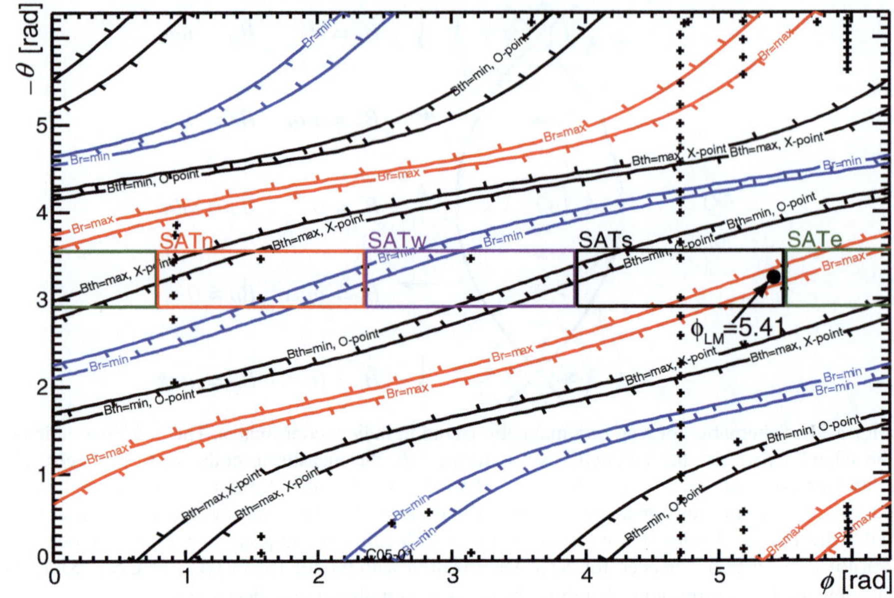

Fig. 3.22 *Field line* mapping of the locked mode phase determined at ASDEX Upgrade on the HFS (*solid black point*) for the $q = -2$ helicity. Shown are the contour lines around the maximum (*red*), minimum (*blue*) and zero (*black*) $\tilde{B}_r$ of the (−2,1) locked mode. The perpendicular ticks at the *contour lines* point towards decreasing $\tilde{B}_r$ (*red* and *blue*) or decreasing $\tilde{B}_\theta$ (*black*). Any position in (θ, ϕ) can thereby be mapped to the locked mode phase. As example, the black crosses indicate the positions of the small pick-up coils for $\dot{B}_\theta$ and $\dot{B}_r$ measurements

3.10 Localization of $q = 1$

Control of sawtooth crashes is a great concern for tokamak physics (see Chap. 4). For this purpose it can be helpful to know the resonant position, $\rho_{q=1}$. Reconstruction of the q profile often is not sufficiently precise in the core plasma to determine the $q = 1$ position. Therefore, additional information is desirable—either from the localization of the crash precursor, a (1,1) mode, or from determining the inversion surface. Although the latter is not directly related to MHD mode observations, we want to describe it briefly.

3.10.1 Localization of the (1,1) Mode

The (1,1) mode in a tokamak is a special case due to its topology, which allows an $m = 1$ perturbation by shifting the flux surfaces rigidly in the (ρ, θ) plane. This results e.g., in the internal (1,1) kink being the only unstable internal kink mode. A

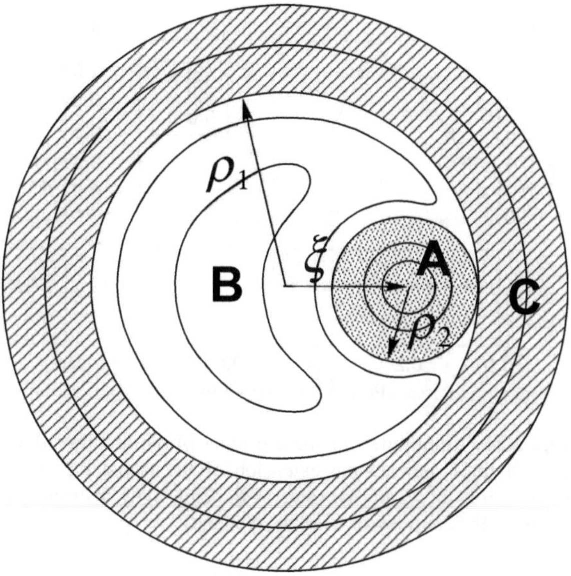

Fig. 3.23 Magnetic structure of a (1,1) island according to [43]. *A* is the shifted core region (radius ρ_2), *B* the island region and *C* the unperturbed plasma outside (radius of the innermost unperturbed flux surface is ρ_1)

(1,1) magnetic island can—in contrast to other tearing modes—develop an island structure that does not lead to a distortion outside the island's X-point. In [43] a model for such a (1,1) island (occurring e.g. during a sawtooth crash) is presented (see Fig. 3.23). With increasing core displacement, ξ, the island grows, the shifted core (ρ_2) shrinks and the X-point moves to larger ρ. Such an island structure is difficult to distinguish from an ideal (1,1) mode in experimental data. For an island of this type, the phase jump in the fundamental Fourier component of the temperature oscillation, which is usually observed for other magnetic islands, only occurs when the O-point is sufficiently hotter than the island's separatrix (see Fig. 3.24). The methods for localization of the rational surface (by the phase jump) and the island width (by the location of signal amplitude maxima) fail for such a (1,1) island structure.

To sum up, in case of the (1,1) mode it is difficult to decide whether it is an ideal or a resistive mode and the resonant surface can basically only be estimated from the extension of the (1,1) amplitude, either from local (e.g., ECE) measurement or from line integrated (e.g., SXR) data.

3.10.2 Sawtooth Inversion

A sawtooth crash is a periodically occurring (with periods in the order of several milliseconds to several seconds) fast process (duration usually below 100 μs) during which the transport is increased such that particle densities and temperatures are basically flattened in the central plasma region. Ideal plasmas do not

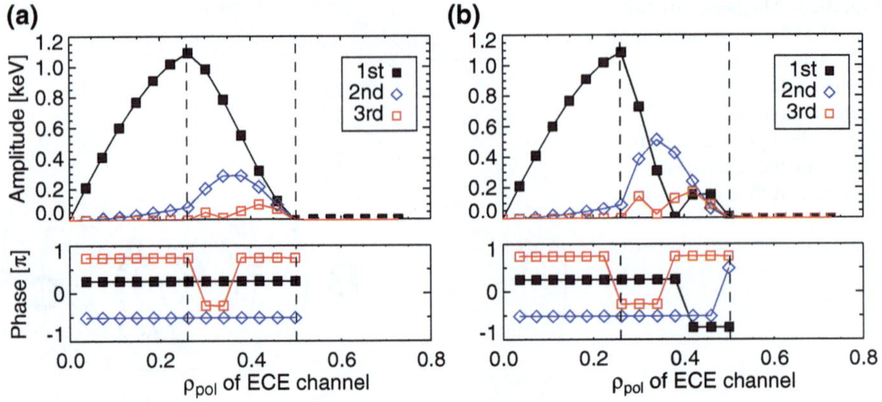

Fig. 3.24 Amplitude and phase profiles of the first three harmonics in the ECE signals. Vertical dashed lines indicate the extension of the island region. **a** flat island **b** hot island. The 2D distributions of T_e are the same as the emissivity distributions in Fig. 3.18 (scaled to reasonable values) (By courtesy of Markus Weiland)

allow such fast transport, so that reconnection is certainly involved. The crash is nearly always preceded by a (1, 1) instability, called the crash precursor. Often also a—rather short-lived—crash postcursor has been observed.

Usually, quantities like SXR radiation, electron density and temperature increase towards the magnetic axis. Thus, a flattening in the central region up to a certain radius, ρ_{mix}, results in decreased values close to the axis and increased values close to but inside ρ_{mix}. In between there is a position where signals remain constant, which is called the inversion surface or inversion radius. Figure 3.25 shows temperature profiles before and after a sawtooth crash and the difference of both (the loss profile), from which the inversion can be easily determined.

The inversion radius is often used as an indication of the $q = 1$ rational surface. It is indeed often close to $\rho_{q=1}$ but not necessarily equal, as has been demonstrated by [44] in cylindrical approximation based on the Kadomtsev model [45]. The authors of [44] have varied the q profile for a fixed temperature, but corresponding calculations can be made for fixed q profile and varying temperature profile. The general result is, that for a measured quantity significantly more(less) peaked than the current profile, the inversion occurs at smaller(larger) radii than $\rho_{q=1}$. Therefore the inversion surface of different quantities (e.g., T_e and SXR radiation) is not necessarily the same, although determined in the same plasma (and therefore same q profile) simultaneously.

For line integrated signals (like SXR intensity) it has to be considered additionally that the inversion for peaked initial profiles is systematically inside the inversion of the local quantity (like SXR emissivity; see Fig. 3.26). The change during a sawtooth crash in a SXR LOS tangent to the inversion surface has zero contribution from the inversion but passes through regions outside the inversion only. The contribution from these is positive, so that this LOS is still outside the

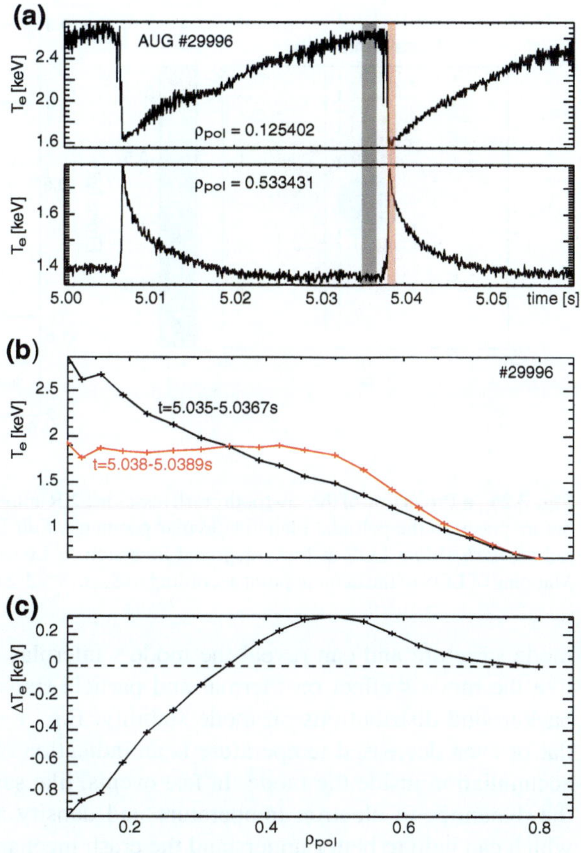

Fig. 3.25 **a** Time traces of some ECE channels showing sawtooth crashes at $t = 5.007\,s$ and $t = 5.038\,s$, **b** Profiles of electron temperature before and after a crash (time ranges indicated in a)) and **c** loss profile (temperatures after minus before the crash). Note that a flat central profile after the crash prevents observation of an instability in this region (a crash postcursor)

inversion as it appears in the line integrated SXR profiles, where each LOS is mapped to its innermost flux surface (see Sect. 3.5.3). This effect is weak for strongly peaked profiles but strong for broader emissivity distributions.

3.11 Further Remarks on Observation Quantities and Diagnostics

In the previous discussion, a mode was considered as a magnetic perturbation that can be observed and identified also from other quantities (temperature, radiation or density) than the magnetic field. The different behaviour of the observation quantities has been regarded as a nuisance that can hamper the mode analysis: a small temperature gradient around a mode reduces the visibility of the mode in the measurement, or a non-flat island temperature leads to an underestimated island width. On the other hand, these differences carry information that goes beyond the

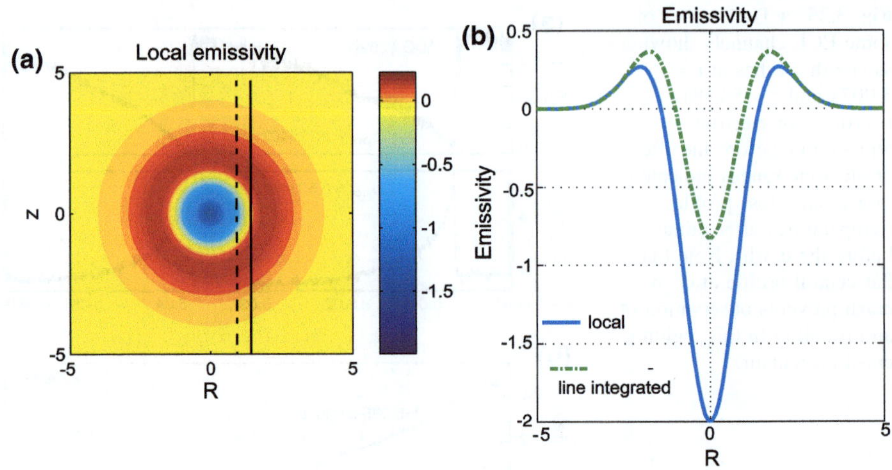

Fig. 3.26 **a** Projection of the sawtooth crash losses in SXR emissivity (local values after crash minus before crash) to the poloidal plane in circular geometry. *Full Line* LOS tangent to local inversion radius, *dashed line* LOS at line integrated inversion. **b** Local versus line integrated emissivity. Mapping of LOS to the tangent point according to Sect. 3.5.3. Schematic values with arbitrary units

mode structure and can reveal the mode's interplay with the background plasma, like the mode's effect on thermal and particle transport and the influence of the background distributions on mode stability. E.g., a strongly radiating island with flat or even decreased temperature is an indication of electron density or impurity accumulation inside the mode. In fast events, like sawtooth crashes, differences in the dynamics of electron temperature and density have been observed [46, 47], which can help to better understand the crash mechanism. Generally, simultaneous observation of the same instability in different quantities is of great interest, also allowing to extract information via correlation and coherence analysis.

Above, we have described mode analysis through radiation, electron temperature and magnetic field perturbations. One of the most important missing quantities in this list is the electron density. A restriction in terms of mode analysis is that electron density—in contrast to electron temperature—can be almost constant in wide regions of the plasma.

Electron density measurements by laser interferometry can easily detect MHD modes, since the measurement is continuous and instantaneous, thus the sampling can be done very fast. But the signals are line integrated and with the very limited number of lines of sight, virtually no spatial information can be extracted.

Microwave reflectometry is a local density measurement and can achieve good radial resolution in regions with sufficient density gradient. There are several examples of mode analysis by reflectometry. In [48] the localization of a tearing mode by the flattened density region is examined, in [49] it is shown that changes in the level of density fluctuations coincide with the flattened density regions in large tearing modes and thus indicate the mode location. The radial structure of

Alfvén Eigenmode instabilities have been examined ([29] and references therein). The fastest current systems take several tens of μs for a radial scan to observe density perturbations over the plasma minor radius, thus they are already in the range for MHD observations. Future developments can still improve this diagnostics.

Thomson scattering measures both electron density and temperature simultaneously and can be a very interesting tool for MHD analysis. The data are taken during a short laser pulse, thus it is a quasi-instantaneous but highly discontinuous method. The individual lasers need tens of ms between pulses and are currently not suitable for standard MHD analysis. However, using a reference signal, like magnetic pick-up coils, for phase-mapping, stationary modes can sometimes be analysed. When multiple lasers are available, as is the case in many tokamaks, they can be fired shortly after each other, followed by a longer break for reloading the lasers. This burst mode has been used for investigation of rotating modes as well as for short events like sawtooth crashes [47, 50]. In [51] a method is presented to use the laser energy more efficiently by a multi-pass system, allowing at present a sample rate of 5 kHz for up to 8 ms.

Each development of new diagnostics and analysis tools and increase in temporal and spatial resolution has led to new insight in MHD instabilities. The diagnostics development will go on in the future. However, in future fusion experiments the specific requirements are different. While for example in ITER many tools will be needed for plasma physics investigations, in DEMO the emphasis will be on basic diagnostics for discharge control purposes. Important features for new diagnostics are insensitivity to neutron and gamma radiation and thermal stresses, and compatibility with restricted space as well as remote handling.

References

1. W.W. Heidbrink, Phys. Plasmas **15**, 055501 (2008)
2. F. Waelbroeck, Nucl. Fusion **49**, 104025 (2009)
3. J. Callen, Nucl. Fusion **51**, 094026 (2011)
4. D. Naujoks et al., Nucl. Fusion **36**, 671 (1996)
5. K.V. Mardia, J.T. Kent, J.M. Bibby, *Multivariate Analysis, Probability and Mathematical Statistics* (Academic Press, Boston, 1979)
6. F. Murtagh, A. Heck, *Multivariate Data Analysis* (Astrophysics and Space Science Library, New York, 1987)
7. H. Akaike, IEEE Trans. Autom. Control **19**, 716 (1974)
8. M. Wax, T. Kailath, IEEE Trans. Acoust. Speech Signal Process. **ASSP-33**, 387 (1985)
9. I.G.J. Classen et al., Rev. Sci. Instrum. **81**, 10D929 (2010)
10. J. Kim, D. Edgell, J. Greene, E. Strait, M. Chance, Plasma Phys. Control. Fusion **41**, 1399 (1999)
11. C. Nardone, Plasma Phys. Controlled Fusion **34**, 1447 (1992)
12. T. Dudok de Wit, Plasma Phys. Controlled Fusion **37**, 117 (1995)
13. M.J. Hole, L.C. Appel, PPCF **49**, 1971 (2007)

14. I.H. Hutchinson, *Principles of Plasma Diagnostics* (University Press, Cambridge, 2002)
15. J. Wesson, *Tokamaks* (Clarendon Press, Oxford, 2004)
16. D. Pacella, R. Bellazzini, A. Brez, M. Finkenthal, RSI **75**, 3926 (2004)
17. D. Pacella et al., RSI **72**, 1372 (2001)
18. B.C. Stratton, M. Bitter, K.W. Hill, D.L. Hillis, J.T. Hogan, Fusion Sci. Technol. **53**, 431 (2008)
19. M.J. Hole, L.C. Appel, R. Martin, Rev. Sci. Instrum. **80**, 123507 (2009)
20. M. Schittenhelm, H. Zohm, ASDEX upgrade team. Nucl. Fusion **37**, 1255 (1997)
21. W.D. D'haeseleer, W.N.G. Hitchon, J.D. Callen, J.L. Shohet, *Flux Coordinates and Magnetic Field Structure* (Springer Series in Computational Physics, Berlin, 1990)
22. V.G. Merezhkin, Sov. J. Plasma Phys. **4**, 152 (1978)
23. D. Testa, A. Fasoli, E. Solano, Rev. Sci. Instrum. **74**, 1694 (2003)
24. V. Igochine, S. Günter, M. Maraschek, ASDEX Upgrade team. Nucl. Fusion **43**, 1801 (2003)
25. J. Kim et al., Plasma Phys. Control. Fusion **43**, 1593 (2001)
26. G. Papp et al., Plasma Phys. Controlled Fusion **53**, 065007 (2011)
27. R. Kaita, S. Sesnic, Rev. Sci. Instrum. **68**, 750 (1997)
28. E. Rebhan, A. Salat, NF **16**, 805 (1976)
29. S. da Graca et al., Plasma Phys. Controlled Fusion **49**, 1849 (2007)
30. P. Piovesan et al., Nucl. Fusion **48**, 065001 (2008)
31. R. Fitzpatrick, Phys. Plasmas **2**, 825 (1995)
32. J.P. Meskat et al., Plasma Phys. Controlled Fusion **43**, 1325 (2001)
33. M. Reich, A. Bock, M. Maraschek, ASDEX Upgrade team. Fusion Sci. Technol. **61**, 309 (2012)
34. J.A. Wesson, NF **37**, 577 (1997)
35. T. Pütterich et al., Plasma Phys. Controlled Fusion **50**, 085016 (2008)
36. M. Anton et al., PPCF **38**, 1849 (1996)
37. A. Zohm, H. Kallenbach, H. Bruhns, G. Fussmann, O. Klüber, Europhys. Lett. **11**, 745 (1990)
38. M. Maraschek et al., Measurement and impact of the $n = 1$ intrinsic error field at asdex upgrade, in *Europhysics Conference Abstracts (CD-ROM, Proceedings of the 40th EPS Conference on Plasma Physics, Espoo, Finland, 2013)*, ed by V. Naulin et al., volume 37D of *ECA*, pp. P4. 127, (Geneva, 2013) European Physical Society
39. J.T. Scoville, R.J. La Haye, Nucl. Fusion **43**, 250 (2003)
40. F. Karger, K. Lackner, G. Fussmann, PPCF **1**, 267 (1977)
41. O. Klüber et al., Nucl. Fusion **31**, 907 (1991)
42. S. Yoshimura, M. Takashi, T. Yasushi, Phys. Plasmas **9**, 3378 (2002)
43. F. Porcelli, E. Rossi, G. Cima, A. Wootton, PRL **82**, 1458 (1999)
44. Z. Pietrzyk et al., Nucl. Fusion **39**, 587 (1999)
45. B. Kadomtsev, Sov. J. Plasma Phys. **1**, 710 (1975)
46. T. Nicolas et al., POP **19**, 112305 (2012)
47. I.T. Chapman et al., PRL **105**, 255002 (2010)
48. M. Manso et al., Local measurement with microwave reflectometry of density profile perturbations due to MHD activity on ASDEX Upgrade, in *Europhysics Conference Abstracts (CD-ROM), Proceedings of the 1998 ICPP & 25th EPS Conference on Controlled Fusion and Plasma Physics, Praha, 1998*, ed by P. Pavlo, vol. 22C (EPS, Petit-Lancy, 1998), pp. 544–547
49. G.D. Conway et al., Locating rational surfaces from reflectometer fluctuations, in *Europhysics Conference Abstracts (CD-ROM, Proceedings of the 29th EPS Conference on Plasma Physics and Controlled Fusion, Montreux, 2002)*, ed by R. Behn, C. Varandas, vol. 26B (EPS,Geneva, 2002), pp. P–4.075
50. K.J. Gibson et al., PPCF **52**, 124041 (2010)
51. M. Yu Kantor, A.J.H. Donné, R. Jaspers, H.J. van der Meiden, T. Team, PPCF **51**, 055002 (2009)

Chapter 4
Sawtooth Instability

Ian T. Chapman

Abstract Sawtooth oscillation results in a periodic relaxation of the tokamak core plasma. These periodic oscillations consist of a quiescent period, during which the density and temperature increase, followed by a rapid collapse in the core pressure, which is often preceded by the growth of a helical magnetic perturbation. The period between these rapid sawtooth collapses is expected to increase in the presence of alpha particles in burning fusion plasmas. However, long sawtooth periods have been observed to increase the likelihood of triggering neoclassical tearing modes (see chapter 8) at lower plasma pressures; these instabilities in turn can then significantly degrade the plasma confinement. Consequently, recent efforts have focussed on developing methods to deliberately trigger short sawtooth periods to avoid seeding NTMs while retaining the benefits of core impurity expulsion. The main sawtooth control tools involve driving localised currents to change the safety factor profile or tailoring the fast ion distribution.

4.1 Introduction

Sawtooth oscillations are periodic relaxations of the core plasma density and temperature [1–3]. These periodic redistributions of the plasma surrounding the magnetic axis were first observed in 1974 [4] and are commonplace in every tokamak. A typical sawtooth cycle is illustrated in Fig. 4.1, showing the three phases: (i) the sawtooth ramp phase during which the plasma density and temperature increase in time; (ii) the precursor phase, during which a helical magnetic perturbation grows until (iii) the fast collapse phase, when the density and temperature in the core drop rapidly. Sometimes magnetic fluctuations are also observed after the sawtooth crash (post-cursors) [5] or even during the ramp phase [6]. When a sawtooth crash occurs, hot electrons transport rapidly to a cooler region of plasma, such that the electron

I.T. Chapman (✉)
CCFE, Culham Science Centre, Abingdon, Oxfordshire OX14 3DB, UK
e-mail: ian.chapman@ccfe.ac.uk

V. Igochine (ed.), *Active Control of Magneto-hydrodynamic Instabilities in Hot Plasmas*, Springer Series on Atomic, Optical, and Plasma Physics 83,
DOI 10.1007/978-3-662-44222-7_4

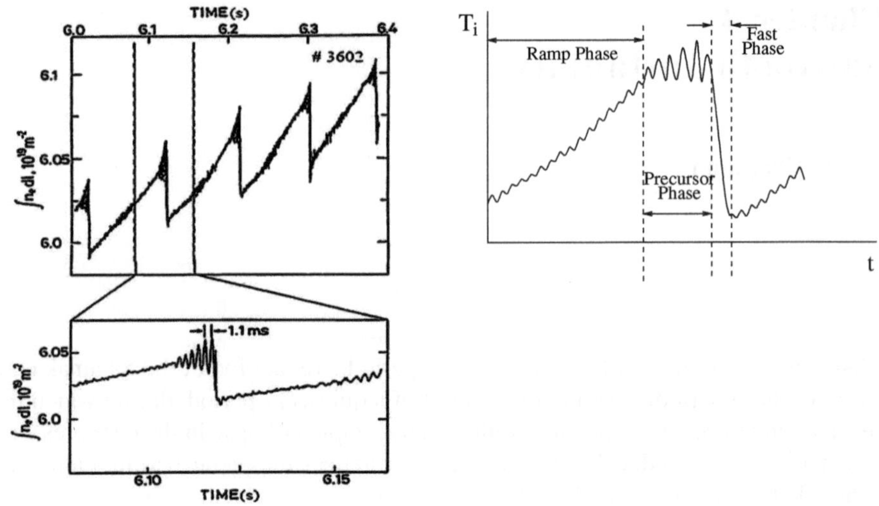

Fig. 4.1 The line-integrated electron density of a JET plasma exhibiting sawtooth oscillations. The sawteeth consist of a ramp phase, then a precursor oscillation followed by the fast collapse

temperature profile is flattened. This rapid drop in the core temperature is accompanied by heating of the edge plasma. The temperature remains constant at the *inversion radius*. Two-dimensional imaging of the collective behaviour during a sawtooth crash [7, 8] shows clearly that the hot plasma core is expelled through a poloidally localised point as magnetic reconnection occurs, as illustrated in Fig. 4.2.

In order to control sawteeth, it is most important to understand the second phase in the cycle—the trigger of the instability growth—which is discussed in detail in Sect. 4.2. Experiments have shown (see [1] and references therein) that the precursor oscillation has the topology of the $n = m = 1$ internal kink mode—a fundamental magnetohydrodynamic (MHD) oscillation of the form $\xi \sim \exp(im\theta - in\phi)$ where m and n are the poloidal and toroidal mode number respectively, ξ is the perturbation to the plasma and θ and ϕ are the poloidal and toroidal angles. The ideal internal kink displacement, which is manifest as the precursor oscillation, takes the form of a tilt and a shift of the core plasma (although it should be noted that tokamak plasmas do sometimes experience precursorless sawteeth [9]). The behaviour of the kink mode can be adequately described by MHD. However, the dynamics of energetic particles (for instance born due to Ion Cyclotron Resonance Heating (ICRH) or Neutral Beam Injection (NBI), or α particles from fusion reactions), which affect the stability of the kink mode, must be treated kinetically and the various stabilising and destabilising terms assessed using the energy principle. In essence this says that if a physically allowable perturbation lowers the potential energy of the plasma, then the wave is unstable [2].

Small sawteeth which have an inversion radius smaller than 40 % of the minor radius and a temperature drop of the order of a fraction of a keV can be tolerated by the plasma [10]. Indeed, such small sawteeth can help to prevent the accumulation of

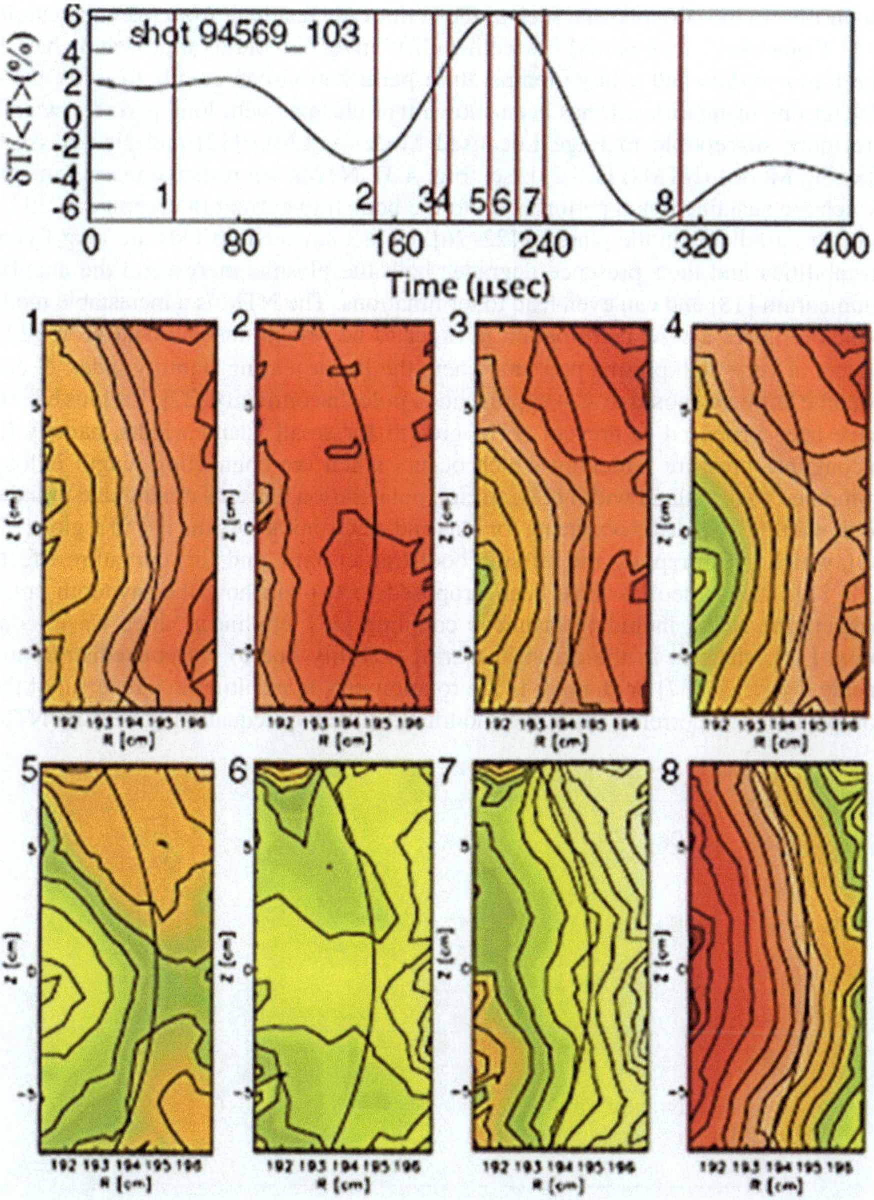

Fig. 4.2 2D images of the sawtooth crash from electron cyclotron emission imaging at the low-field side mid-plane on TEXTOR. As the hot spot swells (shown in *frames 3* and *4*) a sharp temperature point is growing and crosses beyond the inversion radius (marked by the *black arc*). Eventually, the temperature point leads to the reconnection. Initially, it forms an X-point in the poloidal plane (*frame 5*), and heat starts to flow to the outside through a small opening. The initial heat flow is highly collective, and the opening increases up to 15 cm. At the end, the heat accumulates outside the inversion radius, and the poloidal symmetry is recovered. Reprinted figure with permission from Park et al. [7]. Copyright 2006 by the American Physical Society

impurities in the core plasma, such as the helium ash resultant from fusion reactions [11]. Conversely, long period sawteeth with an inversion radius approaching half of the minor radius and a large temperature perturbation can couple to other more deleterious instabilities. It has been shown that plasmas with long period sawteeth are more susceptible to Edge Localised Modes (ELMs) [12] and Neo-classical Tearing Modes (NTMs) [13–21] (see Fig. 4.3). NTMs are resistive tearing modes which are sustained by a perturbation to the bootstrap current (a current caused by pressure gradients in the plasma) [22–26]. Unlike sawteeth, NTMs are long-living instabilities and their presence degrades both the plasma energy and the angular momentum [18] and can even lead to terminations. The NTM is a metastable mode which requires a *seed* perturbation in order to be driven unstable and grow [22], except at very high plasma pressure where the linear tearing stability index Δ' can become large and positive as it approaches a pole discontinuity [27]. Various effects have been proposed to prevent NTM growth for small island widths, namely (i) incomplete pressure flattening which occurs when the connection length is long compared to the island width [28], (ii) ion polarisation currents arising due to finite orbit width $E \times B$ drifts occurring for ions and electrons across the island region [29, 30], which act to replace the missing bootstrap current, and (iii) curvature effects [31, 32]. Many theories have been proposed to explain how the sawtooth crash triggers the NTM, including magnetic coupling [33], nonlinear 'three-wave' coupling [34], changes in the classical tearing stability due to current redistribution inside $q = 1$ [35–37] or changes in the rotation profile resulting in a reversal of the ion polarisation current [19] in the modified Rutherford equation governing NTM

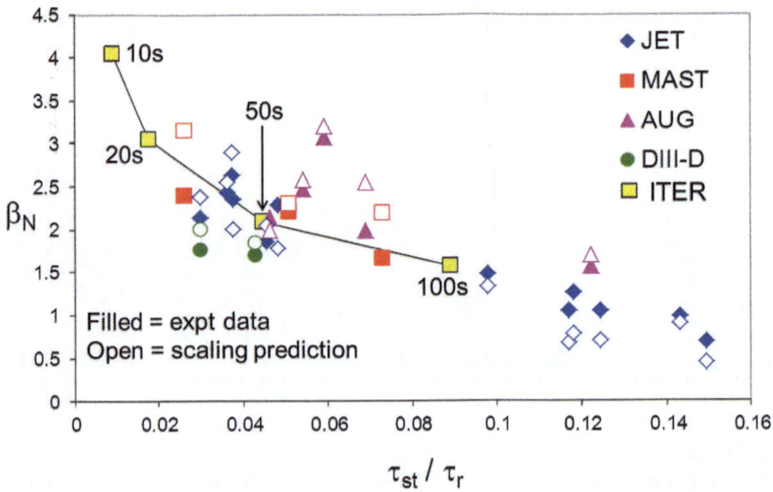

Fig. 4.3 β_N at the NTM onset with respect to the sawtooth period normalised to the resistive diffusion time for ITER-like shape, $q = 1$ radius and injected power normalised in a range just above the P_{LH} threshold. For comparison, ITER baseline scenario is indicated with sawtooth period ranging from 10 to 100 s. Reproduced with permission from "Power requirements for electron cyclotron current drive and ion cyclotron resonance heating for sawtooth control in ITER" [185]

stability [25]. Figure 4.3 shows how the critical β_N at which an NTM is triggered varies with respect to the sawtooth period normalised to the resistive diffusion time. It is evident that as the sawtooth period lengthens, the NTMs are triggered at lower β_N, hence limiting the plasma performance achievable.

It is expected that fusion-born α particles will result in very long sawtooth periods in ITER [38–40]. Indeed, alpha-tail production with ICRF heating of He^4-beam ions in JET confirmed that the energetic α particles result in "monster" sawteeth [41, 42], which are loosely defined as sawteeth with periods longer than the energy confinement time, and hence saturated central plasma temperature. The resultant long period sawteeth are more likely to trigger NTMs, and therefore the control of sawteeth is vital. The two approaches to sawtooth control are to attempt to suppress sawteeth for many energy confinement times (i.e. *stabilise* the kink mode) or to deliberately decrease the sawtooth period (i.e. *destabilise* the kink mode). The baseline scenario in ITER is predicated upon using the sawteeth to reverse the on-axis accumulation of higher-Z impurities that would otherwise cause degradation of energy confinement due to impurity radiation. The approach currently adopted for sawtooth control in baseline scenarios in ITER is to deliberately destabilise the internal kink mode to give frequent, small amplitude sawtooth crashes. A lower limit in the sawtooth period is determined by the slowing down time of the fusion α particles, since the energetic α particles must transfer their energy to the thermal plasma core to ensure continuing fusion reactions.

In Sect. 4.2, recent developments in the theoretical understanding of sawtooth stability are discussed. Models that predict the conditions under which a sawtooth crash will occur are also outlined. Sawtooth control achieved by locally perturbing the current profile is discussed in Sect. 4.3. Recent results exhibiting destabilisation of sawteeth by steerable electron cyclotron resonance heating (ECRH) are presented, including real-time feedback schemes and electron cyclotron current drive (ECCD) control in the presence of energetic ions in the plasma core. These results justified the inclusion of ECCD for sawtooth control in the ITER design [10].

Sawtooth control can also be achieved by changing the gradients of the fast particle distribution. In Sect. 4.4, sawtooth behaviour in plasmas heated by neutral beam injection (NBI) is examined. Understanding how the fast ions affect the sawteeth in these NBI experiments has implications for how sawtooth control can be achieved with ion cyclotron resonance heating (ICRH). ICRH has been used to destabilise sawteeth, as described in Sect. 4.5 through both current drive and kinetic effects. Finally, the application of these sawtooth control techniques in ITER is discussed in Sect. 4.6.

4.2 The Physics of Sawtooth Stability

Since a sawtooth crash is usually accompanied by an $m = n = 1$ kink displacement [43], it is important to understand the factors which influence the stability of the 1/1 internal kink mode. In regions of high current there is a strong poloidal magnetic

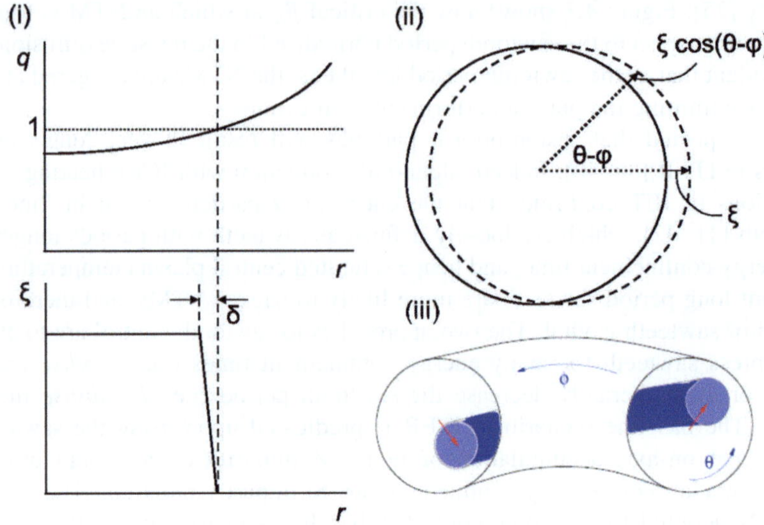

Fig. 4.4 (i) The radial dependence of the $n = m = 1$ internal kink mode, localised at the $q = 1$ surface (ii) the displacement of a circular flux surface for an $m = 1$ perturbation and (iii) the distortion of an $n = m = 1$ kink in a toroidal geometry

field induced by the current. Should the plasma then experience a perturbation, the strong poloidal magnetic field will reinforce the instability and push the plasma further out, extending the "kinking" effect [2]. The $n = m = 1$ internal kink mode takes the radial structure of a top-hat displacement which is rigid within the $q = 1$ surface and zero outside, as depicted in Fig. 4.4 However, sawtooth stability in tokamak plasmas is not determined solely by the fluid drive of the 1/1 internal kink mode; its dynamics are significantly affected by the presence of energetic particles, by sheared flows, by diamagnetic effects, by pressure anisotropy, by complex nonlinear reconnection physics and local effects in the inertial layer around the $q = 1$ surface. All of these effects have implications for the actuators which can be utilised to control sawteeth, as discussed below.

4.2.1 Effect of Energetic Particles

High power ion cyclotron resonance heating (ICRH) experiments in JET showed that the sawtooth instability could be suppressed for many seconds [44] and that long sawtooth quiescent periods were terminated abruptly by a sawtooth collapse that followed the switch off of the ICRH [45]. The inference from this was that the fast ions induced by ICRH provided a stabilising influence upon the 1/1 kink mode, and thus the sawteeth. This was later confirmed using fast ions from NBI [13] and fusion-born α particles [46].

There followed a significant theoretical effort to explain this phenomenon, largely building on the principles developed by Chen et al. [47] to explain the fishbone instability [48]—another $m = n = 1$ internal kink instability driven by gradients in the fast particle pressure. The dispersion relation [49–51] gives

$$i\sqrt{\omega(\omega - \omega_{*i})} \sim \delta W_{MHD} + \delta W_{hf} + \delta W_{hk} \tag{4.1}$$

where δW_{hf} and δW_{hk} are the fluid and kinetic components of the change in the potential energy of the kink mode due to the energetic particles respectively. This equation and $\delta \hat{W}_{MHD}$ is the ideal fluid mode drive [52, 53] was found to have two branches: The first, when $\Re e(\omega) \sim \langle \omega_{dh} \rangle$ and $\langle \omega_{dh} \rangle$ is the bounce-averaged magnetic drift frequency of the hot ions, characterises the fishbone instability [47]. The second, when $\Re e(\omega) < \langle \omega_{dh} \rangle$, relates to the sawteeth. The trapped energetic ions give rise to significant stabilisation providing the lower frequency branch satisfied the condition:

$$\Re e(\omega) \ll \langle \omega_{dh} \rangle \tag{4.2}$$

Kruskal and Oberman [54] also showed that thermal ions can stabilise core MHD instabilities. This was later confirmed using the energy principle including the guiding centre motion of the energetic particles [55–57]. In the next sections we consider the contributions of various energetic particles to the change in the potential energy of the mode (as in 4.1), beginning with bounce-trapped particles.

4.2.1.1 Trapped Energetic Ions

The stabilisation of low-frequency MHD perturbations by trapped fast ions is a consequence of the conservation of the third adiabatic invariant [58, 59]. Porcelli proposed that in a tokamak this third adiabatic invariant of motion [60], Φ_{ad}, is equivalent to the flux of the poloidal magnetic field through the area defined by the toroidal precession of the trapped particle orbit centres. The fast ions can be viewed as a distribution of current loops (in equilibrium with the fast particle pressure) [58]. The current in each of these loops is analogous to the precessional drift frequency of the trapped particle. These loops have a poloidal flux through the area which they enclose. If the plasma experiences an $n/m = 1/1$ displacement, the poloidal flux through this area would change. For instance, the kink depicted in Fig. 4.5 causes the poloidal flux through the area defined by the banana orbit centres to increase. However, the fast ions respond in order to conserve Φ_{ad}, and in so-doing, the loops can contract or expand, or shift and tilt in space. For the case shown in Fig. 4.5, and with the fast ion pressure peaked on-axis, the current loops will contract, which requires an increase in the energy of the fast ions. This energy is taken from the mode, and so the presence of the fast ions results in a stabilisation of the sawteeth.

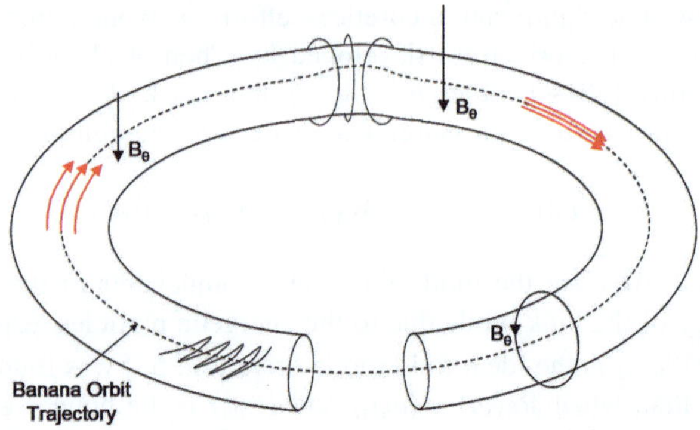

Fig. 4.5 The fast particles can be considered as current loops, whose amplitude is proportional to their precessional drift frequency. The poloidal flux through the area defined by the toroidal revolution of the banana centres is conserved. This figure shows a schematic time evolution depicting a kink displacements, which causes the loops to contract in order to conserve the poloidal flux, taking energy from the mode and so stabilising the sawteeth

By employing the generalised energy principle, the change in the potential energy of the kink mode due to the energetic trapped ions can be calculated, giving [58, 61, 62]

$$\delta W_h = \frac{1}{2}\int \mathrm{d}\Gamma(Mv_\parallel^2 + \mu B)\delta f_h \sum_m \boldsymbol{\kappa}\cdot\boldsymbol{\xi}^{(m)*}(r,t)e^{-i(n\phi-m\theta)} \tag{4.3}$$

where θ is the poloidal angle, $\boldsymbol{\kappa} = \mathbf{b}\cdot\nabla\mathbf{b}$ is the magnetic curvature vector and $\mathbf{b} = \mathbf{B}/B$. This then needs the calculation of the change in the fast particle distribution function, δf_h to be assessed. The fast ion distribution function is typically separated into an equilibrium component, $f_0(\mathcal{E}^0, \mathcal{P}_\phi^0, \mu)$, and two perturbed components, $\delta f_h = \delta f_{hk} + \delta f_{hf}$, a non-adiabatic (kinetic) and an adiabatic (fluid) part respectively. Here, an adiabatic particle has characteristic motion around a closed orbit much faster than the temporal and spatial scales associated with the perturbation (i.e. the perturbation appears as a static modification of the equilibrium). The particle energy ($\mathcal{E}^0 = Mv^2/2$), the canonical momentum ($\mathcal{P}_\phi^0 = MB_\phi v_\parallel/B - e\psi_p$) and the magnetic moment ($\mu = Mv_\perp^2/2B$) are the unperturbed constants of motion, where M is the particle mass, e is the charge and ψ_p is the poloidal flux at the particle position. Analytic theory developed for large aspect ratio circular plasmas [61] can be used to express these contributions to the perturbed distribution function as

$$\begin{aligned}\delta f_{hk} = &\sum_{l=-\infty}^{\infty} \frac{\tilde{\omega} - \Delta\Omega - n\omega_{*h}}{\tilde{\omega} - \Delta\Omega - n\langle\omega_{dh}\rangle + l\omega_b}\frac{\partial f_h}{\partial\mathcal{E}^0} e^{-i(\omega + l\omega_b + n\langle\omega_{dh}\rangle)t} \\ &\times \langle(v_\parallel^2 + \frac{v_\perp^2}{2})\kappa\cdot\xi_\perp e^{i(\omega+l\omega_b+n\langle\omega_{dh}\rangle)}\rangle\end{aligned} \tag{4.4}$$

$$\delta f_{hf} = -\frac{Ze}{M_h}\,\xi \cdot \nabla\psi_p \frac{\partial f_h}{\partial \mathcal{P}^0_\phi} \tag{4.5}$$

respectively, where $\omega_{*h} = (\partial f_h/\partial \mathcal{P}^0_\phi)/(\partial f_h/\partial \mathcal{E}^0)$ is the hot ion diamagnetic frequency, $\omega_b = 2\pi/\tau_b$, τ_b is the poloidal orbit transit time, M_h is the fast ion mass, the dots represent the derivative with respect to time, $\Delta\Omega = \Omega_E(r) - \Omega_E(r_1)$ is the sheared toroidal flow, $\tilde{\omega}$ is the Doppler shifted mode frequency, l is the poloidal quantum number and $\langle\cdots\rangle$ defines an average. The important thing to notice from 4.4 is that a large stabilising contribution can occur when the Doppler-shifted mode frequency is balanced by the characteristic orbit frequencies of the particles, at which point the denominator becomes very small. When this resonance condition is satisifed, the trapped energetic particles have a strong influence on the dynamics of the sawteeth.

Finally, it is worth noting that NBI did not stabilise sawteeth as effectively as ICRH [64]. This is partially because NBI minority ions are far less energetic than ICRH ions, meaning that they were less likely to satisfy the condition for conservation of Φ_{ad}: $\Re e(\omega) \ll \langle\omega_{dh}\rangle$, but also due to greater plasma rotation, flow shear and anisotropy achieved with NBI [63, 65, 66]. NBI also gives rise to a much larger fraction of passing ions, whose effect is discussed below.

4.2.1.2 Passing Energetic Ions

For very energetic ions, the radial drift motion becomes comparable to the radial extent of the kink mode. In this regime, the kinetic contribution to the mode's potential energy associated with the passing fast ions becomes increasingly important when the passing fast ion population is asymmetric in velocity space [67–69]. The contribution from the circulating particles arises primarily from the ions close to the trapped-passing boundary where their orbit widths, Δ_b are large, $\delta W_h^p \sim \Delta_b$. The effect of passing ions is enhanced for large effective orbit widths [70], which is to say, for highly energetic ions (like ICRH or N-NBI in ITER) or for a population with a large fraction of barely passing ions (like NBI in JET). Passing fast ions can destabilise the internal kink mode when they are co-passing and the fast ion distribution has a positive gradient across $q = 1$, or when they are counter-passing, but the deposition is peaked outside the $q = 1$ surface. First let us consider the case of on-axis co-NBI. When a co-passing beam ion is born inside the $q = 1$ surface it experiences an inward ∇B drift, which means that it stays within the $q = 1$ surface and never crosses it. The distribution function has a negative hot particle pressure gradient, $\nabla\langle P_h\rangle < 0$, so this particle will give a stabilising contribution when in the region of good curvature, but will be destabilising when in the region of adverse curvature on the outboard side. These two contributions tend to cancel, and the beam ions inside $q = 1$ (which make up the majority of on-axis NBI ions) do not affect the mode stability. However, the few ions which are born outside $q = 1$ will only pass through the $q = 1$ surface in the

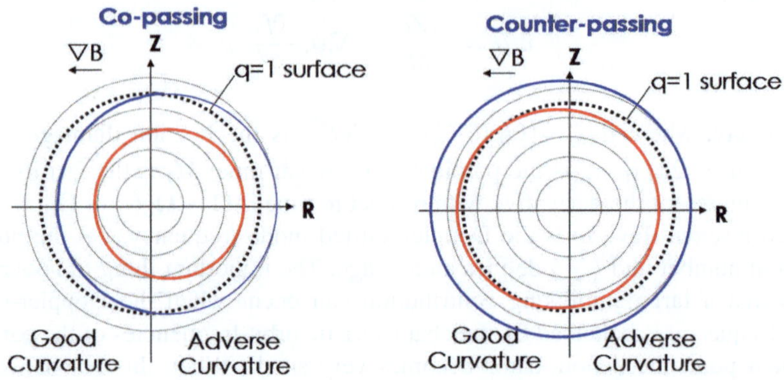

Fig. 4.6 The co-passing ions experience a downwards ∇B-drift, which causes ions born inside the $q = 1$ surface to stay inside $q = 1$, and those born outside it, to only cross the $q = 1$ surface in the region of good curvature. This means that distributions with a fast ion pressure peaked inside $q = 1$ are stabilising, whereas those with a pressure peaked outside $q = 1$ are destabilising. The opposite is true of counter-passing ions

region of good curvature due to the inward ∇B drift. As such, these will give a stabilising contribution to the kink mode (since for $q > 1$, $\xi = 0$). Intuitively the opposite is true for counter-passing ions which have an opposite ∇B drift upward from their flux surface; those ions born inside $q = 1$ will only contribute in the region of adverse curvature. Since they have a negative hot particle pressure gradient, the contribution from counter-passing ions is destabilising. Consequently, the $n/m = 1/1$ internal kink mode is stabilised by co-passing on-axis NBI ions or by counter-passing off-axis NBI ions, but is destabilised by counter-passing on-axis NBI ions or co-passing off-axis NBI ions. This effect is illustrated in Fig. 4.6. This mechanism depends strongly on the local gradient $\partial f_h/\partial P_\zeta$ at the $q = 1$ surface and as such is sensitive to localised heating. This mechanism is described in detail in [68] and [71] with an overview of fast ion effects in [70].

4.2.2 *Effect of Toroidal Rotation*

4.2.2.1 Equilibrium Mass Flow of the Order of the Sound Speed

As well as giving rise to a population of energetic particles, NBI also results in a toroidal rotation of the plasma, sometimes approaching the ion sound speed [72]. At rotation which is a significant fraction of the sound speed, the centrifugal effects can also affect kink mode stability, [73, 74]. Subsequent modelling of the effects of sheared toroidal flows on MHD modes found that rotation approaching the ion sound speed could completely stabilise the ideal $n = 1$ kink mode [75, 76], implying that the longer quiescent periods observed in NBI heated plasmas were not only due to the fast ions, but also the ancillary stabilising effect arising from the beam-driven plasma rotation. It is worth noting that calculations of the stability of the internal kink mode are highly sensitive to including rotation in the formulation of the initial equilibrium [77] as well as the precise density and rotation profiles [78].

4.2.2.2 Flows of the Order of the Diamagnetic Velocity

Even at slow rotation speeds of the order of ω_{*i}, the *shear* of the rotation profile can affect the stabilisation arising from the trapped ions. Conservation of the third adiabatic invariant, Φ_{ad} is only obtained [65] when $\langle\omega_{dh}\rangle + \Delta\Omega - \tilde{\omega} \gg 0$. This can be seen from the denominator in 4.4. Since this condition is more readily satisfied for co-rotation ($\Delta\Omega > 0$), plasmas with co-I_P velocity shear support more effective stabilisation of the kink mode, whereas the stabilising effect is diminished in counter-rotating plasmas ($\Delta\Omega < 0$). The plasma flow will only influence mode stability when $|\Delta\Omega| \sim \omega_{*i}$, which is unlikely to be met in ITER.

4.2.3 Sawtooth Crash Trigger Modelling

The fundamental trigger of the sawtooth crash is the onset of an $m = n = 1$ mode. The dynamics of this instability are constrained by many factors including not only the macroscopic drive from ideal MHD, but collisionless kinetic effects related to high energy particles described in Sect. 4.2.1 and rotation effects described in Sect. 4.2.2, as well as non-ideal effects localised in the narrow layer around $q = 1$. The phenomenology of sawtooth oscillations and their theoretical interpretation is reviewed in [1, 79–81].

Advances in the experimental diagnosis of the sawtooth cycle have led to the proposal of many crash trigger models, including full reconnection [82], resistive two-fluid MHD [83, 84], collisionless kinetic effects [54, 55, 85], accelerated complete reconnection due to nonlinear collisionless effects [86], magnetic stochastisation which led to enhanced perpendicular transport [87], chaos [88], the quasi-interchange model [89], the localised reconnection model [90, 91] and triggering of secondary instabilities [92–96]. In the partial reconnection model, the magnetic surfaces begin to undergo reconnection, just as in the full reconnection Kadomtsev model [82]. However, when the magnetic island reaches a critical width, the core region and critical island region undergo different relaxation processes. The inner core Taylor relaxes, as proposed in [92], whilst the reconstructed surfaces in the island region have the same helical flux, as in full reconnection. This partial reconnection [97] results in the formation of two current sheets, which diffuse rapidly during the ramp phase of the subsequent sawtooth.

A heuristic model developed using linear theory suggests that a sawtooth crash will occur when one of the following criteria is met [38, 96]:

$$-\delta\hat{W}_{core} > c_h\omega_{dh}\tau_A \tag{4.6}$$

$$-\delta\hat{W} > \frac{1}{2}\omega_{*i}\tau_A \tag{4.7}$$

$$-c_\rho\hat{\rho} < -\delta\hat{W} < \frac{1}{2}\omega_{*i}\tau_A \quad \text{and} \quad \gamma_{eff} > \frac{1}{c_*}\sqrt{\omega_{*i}\omega_{*e}} \tag{4.8}$$

where ω_{dh} is the magnetic drift frequency of the hot ions, $\tau_A = \sqrt{3}R/v_A$ is the Alfvén time, c_h, c_ρ and c_* are normalisation coefficients of the order of unity, γ_{eff} is the effective growth rate of the resistive internal kink mode [98] and $\hat{\rho} = \rho_i/r_1$. The change in the kink mode potential energy is defined such that $\delta\hat{W}_{core} = \delta\hat{W}_{MHD} + \delta\hat{W}_{KO}$ and $\delta\hat{W} = \delta\hat{W}_{core} + \delta\hat{W}_h$ where $\delta\hat{W}_{KO}$ is the change in the mode energy due to the collisionless thermal ions [54], $\delta\hat{W}_h$ is the change in energy due to the fast ions. The potential energy is normalised such that $\delta\hat{W} \equiv 4\delta W/(s_1\xi_0^2\epsilon_1^2RB^2)$ The second part of (4.8) can be recast in terms of a critical magnetic shear determined by the pressure gradient, $s_1 > s_{crit}(\omega_{*i})$,

$$s_1 > \max(s_{crit} = \frac{4\delta W}{\xi_0^2\epsilon_1^2RB^2c_\rho\hat{\rho}}, s_{crit}(\omega_*)) \tag{4.9}$$

The linear stability thresholds presented in 4.6–4.9 are useful for understanding when an $n = 1$ internal kink mode will stimulate a sawtooth crash, whilst the nonlinear dynamics of the crash phase and how the profiles are affected are not considered. It is evident from 4.9 that sawteeth can be stimulated by enhancing s_1 (through localised current drive), or reducing δW. Despite its heuristic formulation based solely on linear stability thresholds, the model proposed by Porcelli et al. [38] has had notable success when compared to the observed sawtooth phenomenology in tokamak plasmas [97, 99–101].

4.2.4 Sawtooth Control Actuators

Sawtooth control refers to the ability of an actuator (be it a heating and/or current drive system or plasma shaping control) to alter the likelihood of triggering other MHD instabilities, usually quantified by the sawtooth period. This can be achieved by tailoring the distribution of energetic ions; by changing the radial profiles of the plasma current density and pressure, notably their local gradients near the $q = 1$ surface; by rotating the plasma, or changing the rotation shear local to the $q = 1$ surface; or by heating the electrons inside the $q = 1$ surface. The primary actuators to achieve these perturbations are electron cyclotron current drive (ECCD) (Sect. 4.3), neutral beam injection (NBI) (Sect. 4.4) and ion cyclotron resonance heating (ICRH) (Sect. 4.5).

By driving current just inside the $q = 1$ surface, the magnetic shear at $q = 1$ can be increased, resulting in more frequent sawtooth crashes. This can be understood by considering 4.9 where the linear crash criterion is satisfied when the magnetic shear at $q = 1$ is sufficiently large. Data from TFTR plasmas [102] showed that sawteeth occurred when the magnetic shear at $q = 1$ exceeded a critical value given by collisionless theory [103] strongly supporting the enhancement of the magnetic shear as a sawtooth control actuator. ECCD is the primary sawtooth control actuator in the ITER design [10] due to both the highly localised current density that can be achieved and the ability to provide real time control by changing the launcher angle of the injected EC beam with steerable mirrors. However, some concerns remain regarding the ability of ECCD to destabilise sawteeth in the presence of a significant population of energetic particles inside $q = 1$, as in ITER. This concern is exemplified by 4.8 where the left hand side, ρ/r_1, will be very small since ρ is small and r_1 is predicted to extend towards mid-radius in ITER baseline scenario, whilst the right hand side is likely to have a large δW_h in the numerator due to the presence of the α particles. Fortunately, NBI and ICRH can influence the magnitude of δW directly.

Neutral beam injection affects the change in the potential energy of the internal kink mode in two ways: Firstly, it gives rise to a significant population of energetic particles in the plasma. The predominantly passing fast ions can destabilise the sawteeth when they are injected in the same direction as the plasma current and outside the $q = 1$ surface, or opposite to the plasma current and inside $q = 1$. Secondly, NBI also results in a torque on the plasma, and significant toroidal rotation can stabilise the internal kink mode too. However, due to the rather broad $q = 1$ surface expected in ITER [10], the N-NBI energetic ions are likely to be born inside $q = 1$, even when injected at the most tangential angle of injection [40], meaning that (at least for co-NBI) they will always be strongly stabilising and cannot be used to shorten the sawtooth period.

Conversely, ion cyclotron resonance heating can give rise to a population of energetic particles outside $q = 1$ in ITER. Initial studies of the effects of ICRH on sawtooth behaviour concluded that the (de)stabilisation achieved arose due to the driven current changing the magnetic shear local to $q = 1$. However, recent studies have shown that ICRH sawtooth control persists in plasmas where the driven current is negligible, highlighting that the ICRH destabilisation mechanism also includes strong kinetic effects. For instance, it has been shown [71, 104] that the fast ion population born due to off-axis ^{3}He minority RF-heating scenarios, like the operating scenario proposed for ITER, can give rise to sawtooth destabilisation, even in the presence of core fast ions.

4.3 Current Drive Schemes

When electron cyclotron resonance heating (ECRH) is applied to the plasma, a change in the local current density occurs due to the change in the temperature, and subsequent change in the conductivity. This in turn modified the magnetic shear at $q = 1$, s_1, affecting the likelihood of a sawtooth crash, as seen in 4.9. By adding a toroidal component to the wave vector of the launched EC waves, an ancillary electron cyclotron driven current results either parallel (co-ECCD) or anti-parallel (counter-ECCD) to the Ohmic current, enhancing the potential to alter s_1.

Early experiments to assess the applicability of sawtooth control using current drive schemes focussed on sawtooth stabilisation [9, 105–108]. It was demonstrated that careful placement of the ECCD deposition location could stabilise the sawteeth for the entire gyrotron pulse length on WT-3 (0.03 s) [109], TEXTOR (0.2 s) [110], JT-60U (1.5 s) [111] and ASDEX Upgrade (2.0 s) [112]. Soon after, sawtooth destabilisation was also achieved [113]. ECCD is more effective than ECH (here defined as when the beam injection angle is perpendicular to the magnetic axis) [114, 115], although electron heating does have an indirect effect on the current by changing the local resistivity profile. Furthermore, the effect on the sawteeth can be enhanced by maximising the local ECCD current density rather than the total driven current by narrowing the deposition width [116]. In all cases, significant changes in the sawtooth period occur for very small changes in the deposition location (of the order of the deposition width, typically a few cm) with respect to the location of the $q = 1$ surface [106]. It is this strong localisation of the driven current which makes ECCD a robust sawtooth control actuator, even when the driven current is as little as 1 % of the Ohmic current [117–119]. Sawtooth control using ECCD has been demonstrated in a number of devices including T-10 [120], ASDEX Upgrade [114, 116, 121, 122], TCV [100, 117, 123, 124], JT-60U [111, 125], DIII-D [115], TEXTOR [20, 110], Tore Supra [126–128], FTU [129] and WT-3 [130].

By sweeping the EC deposition location across the $q = 1$ surface (by ramping the magnetic field) it was found that the sawtooth period was highly sensitive to the location of the deposition with respect to the sawtooth inversion radius [107, 108, 131]. Increasing the current inside $q = 1$ increases s_1 and so destabilises the sawteeth, whilst co-ECCD localised just outside $q = 1$ decreases s_1 and so stabilises the sawteeth. Conversely, counter-ECCD just inside $q = 1$ results in stabilisation and just outside $q = 1$ gives rise to destabilisation [100, 114, 118, 119]. This behaviour is demonstrated in Fig. 4.7, showing the sawtooth period with respect to the deposition location of the ECCD in TEXTOR [110], as well as in many other machines [100, 107, 108, 114, 131]. ECCD power scans have also shown that as the driven current is increased, the effect on the sawtooth behaviour is enhanced [114, 132]. Finally, it is worth noting that the control of sawteeth for NTM prevention using ECCD has been demonstrated directly on ASDEX Upgrade. Reference [121] shows that NTMs are avoided at high β_N by complete suppression of the sawteeth using co-ECCD just outside the $q = 1$ surface. Concomitant with the end of the

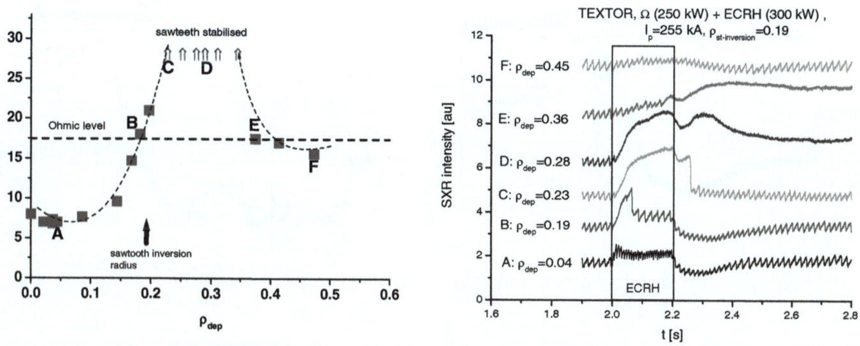

Fig. 4.7 (*left*) The sawtooth period as a function of ECRH deposition location in TEXTOR compared to the inversion radius. The letters refer to discharges shown in the (*right*) figure. Reproduced with permission from "Electron cyclotron resonance heating on TEXTOR" [110]

gyrotron pulse, a sawtooth crash occurred and an NTM was triggered, resulting in a substantial degradation of the plasma performance.

In parallel to the extensive experimental evidence of sawtooth control via current drive, complementary numerical modelling improved the understanding of the physical mechanisms underlying this actuator. Discharges with ECH, and co- and counter-ECCD in TCV have been modelled using the PRETOR-ST code [100, 117]—a transport code including a model for determining when a sawtooth crash will occur based on the linear stability thresholds given in [38]. The modelling shows that whilst co- and counter-current drive have opposite effects, the anti-symmetry is broken by ancillary localised heating. Since the heating acts like co-ECCD, the most efficient destabilisation occurs with co-ECCD and ECH just inside the $q = 1$ surface, whilst the most efficient stabilisation occurs with co-ECCD and ECH outside $q = 1$ [100]. Figure 4.8 shows the sawtooth period in TCV compared to numerical prediction from transport modelling when one ECH beam is oriented just outside $q = 1$ to stabilise the sawteeth and a second beam swept outwards across the inversion radius. The simulations accurately predict the sawtooth period and behaviour during the ECCD sweep, despite using a full reconnection model. Similar ASTRA [133] modelling was used to explain the difference between co- and counter-ECCD on ASDEX Upgrade, once again identifying the change in the magnetic shear as the reason for the change in sawtooth behaviour [114].

However, whilst the experiments using magnetic field ramps to alter the deposition location of the ECCD have significantly improved the understanding of the control mechanisms, a major advantage of current drive schemes is that the angle of inclination of the EC launcher mirrors provides a simple external actuator in a feedback-control loop. Indeed, the uncertainties in the control parameters (such as the launcher aiming and ray-tracing prediction for the resultant driven current) and the plasma equilibrium parameters (such as the $q = 1$ location and plasma position), coupled with the strong sensitivity of the sawtooth period to the deposition location relative to $q = 1$, mean that real-time feedback is a necessity for robust control reliant upon current drive schemes. TCV has demonstrated

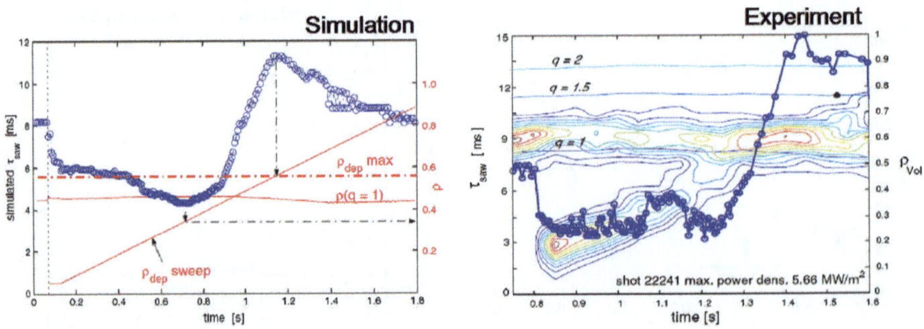

Fig. 4.8 (*left*) Predictions of the sawtooth period from the PRETOR-ST code and (*right*) the measured sawtooth period in TCV with ECH beams directed on-axis to stabilise the sawteeth and an ancillary ECH beam is swept across the inversion radius. The contours in the *right figure* indicate the location of the heat deposition relative to the minor radius, *r*. Reproduced with permission from "Effects of localized electron heating and current drive on the sawtooth period" [100]

sawtooth control in real-time feedback by varying the EC launcher injection angle in order to obtain a pre-determined sawtooth period [134, 135] or to maximise the sawtooth period [136]. Figure 4.9 shows that by changing the launcher angle, and therefore modifying the magnetic shear around $q = 1$, the observed sawtooth period in TCV can be forced to track a requested period. The time lag between the change in the requested period and that achieved is determined by the nonlinear

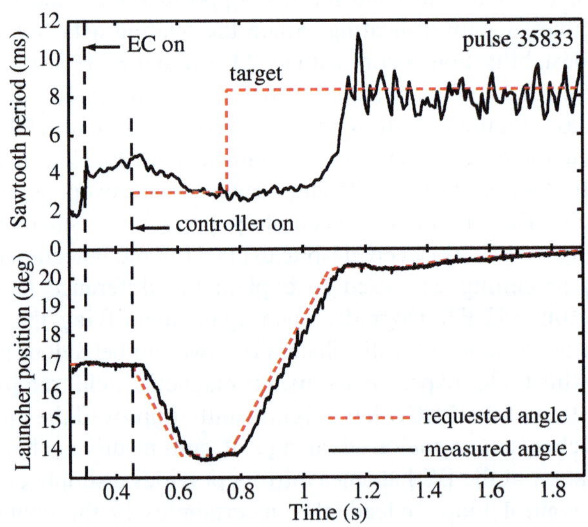

Fig. 4.9 The sawtooth period observed in TCV when real-time ECCD control is applied. The control algorithm successfully achieves and tracks two levels of requested sawtooth period by moving the ECH launcher angle to change the deposition location, and consequently the magnetic shear at $q = 1$ accordingly. Paley et al. [135] published by Institute of Physics Publishing.

plasma response and movement of the $q = 1$ radius. Tore Supra have implemented a 'search and maintain' control algorithm to vary the ECCD absorption location in search of a location at which the sawtooth period is minimised; having achieved this, the controller maintains the distance between the ECCD deposition location and the measured inversion radius despite perturbations to the plasma [128]. Recently, fine control over the sawtooth period has been demonstrated on TCV using either 'sawtooth pacing' via modulated ECCD with real-time crash detection [137], or 'sawtooth locking', where the sawtooth period is controlled even in the absence of crash detection in a reduced region of duty-cycle v pulse-period parameter space [138, 139].

The residual concern for sawtooth control with current drive in ITER is whether changes in s_1 can overcome the stabilisation arising in the presence of energetic particles, as described in Sect. 4.2.1. The change in the magnetic shear may need to be substantial to compete with the kinetic stabilisation of the kink mode, especially if the fast ions arising from concurrent ICRH and NBI heating exacerbate the inevitable effect of the alpha particles. Consequently, recent experiments have attempted to demonstrate destabilisation of sawteeth via driven current despite the presence of energetic particles. Sawtooth destabilisation of long period sawteeth induced by ICRH generated core fast ions with energies ≥ 0.5 MeV has been achieved in Tore Supra, even with modest levels of ECCD power [127]. Figure 4.10

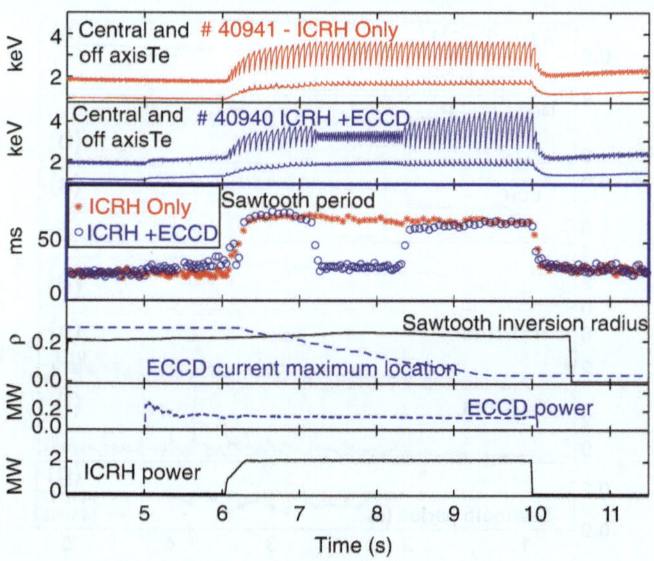

Fig. 4.10 Demonstration of sawtooth destabilisation in the presence of core fast ions in Tore Supra. Two consecutive shots with 2.3 MW of ICRH with and without co-ECCD are shown. The radial ECCD deposition location was scanned from outside the sawtooth inversion radius to the plasma centre. When the deposition was just inside $q = 1$ the sawtooth period dropped to approximately the level of Ohmically heated plasmas, despite the stabilising ICRH ions in the plasma core. Reprinted figure with permission from Lennholm et al. [127]. Copyright 2009 by the American Physical Society

shows the sawtooth period in Tore Supra discharges, with just ICRH heating in the plasma core and with ancillary ECCD swept across the $q = 1$ surface respectively. The ICRH fast ions lead to long sawtooth periods, but the ECCD is able to drop the sawtooth period back to the level of Ohmically heated plasmas when the deposition is optimally located just inside the $q = 1$ surface despite the presence of highly energetic ions. Similar ECCD destabilisation has also been achieved in the presence of ICRH accelerated NBI ions in ASDEX Upgrade [140] as well as with normal NBI fast ions in ASDEX Upgrade [114], JT-60U [125] and HL-2A [141].

More recently sawtooth control using ECCD has even been demonstrated in ITER-like plasmas with a large fast ion fraction, wide $q = 1$ radius and long uncontrolled sawtooth periods in DIII-D [142]. Active sawtooth control using driven current inside $q = 1$ allows the avoidance of sawtooth-triggered NTMs, even at much higher pressure than required in the ITER baseline scenario. Operation at $\beta_N = 3$ without 3/2 or 2/1 neoclassical tearing modes has been achieved in ITER demonstration plasmas when sawtooth control is applied using only modest ECCD power [142]. Such avoidance of NTMs permitting operation at higher pressure than otherwise achievable by application of core ECCD sawtooth control has also been demonstrated in ASDEX Upgrade [143]. Figure 4.11 demonstrates how modest ECCD directed inside the $q = 1$ surface has been utilised in DIII-D to keep the sawtooth period short, thereby avoiding triggering of NTMs,

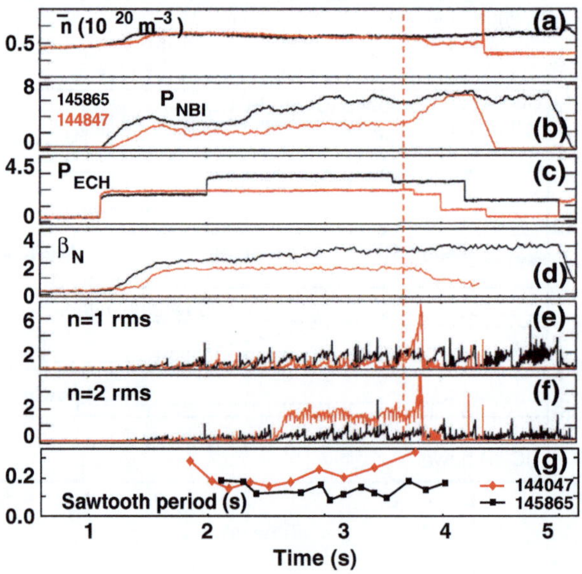

Fig. 4.11 The electron line density, NBI power, ECH power, normalised beta and the rms amplitude of the $n = 1$ and $n = 2$ instabilities in DIII-D discharges 144847 (*red*, no active sawtooth control) and 145865 (*black*, with active sawtooth control). Whilst a 2/1 NTM is triggered at 3.7 s after a sawtooth crash in 144847, when $\beta_N = 1.8$, much higher performance is achieved for much longer with active sawtooth control, allowing $\beta_N \approx 3$. Reproduced with permission from "Sawtooth control using electron cyclotron current drive in ITER demonstration plasmas in DIII-D" [142]

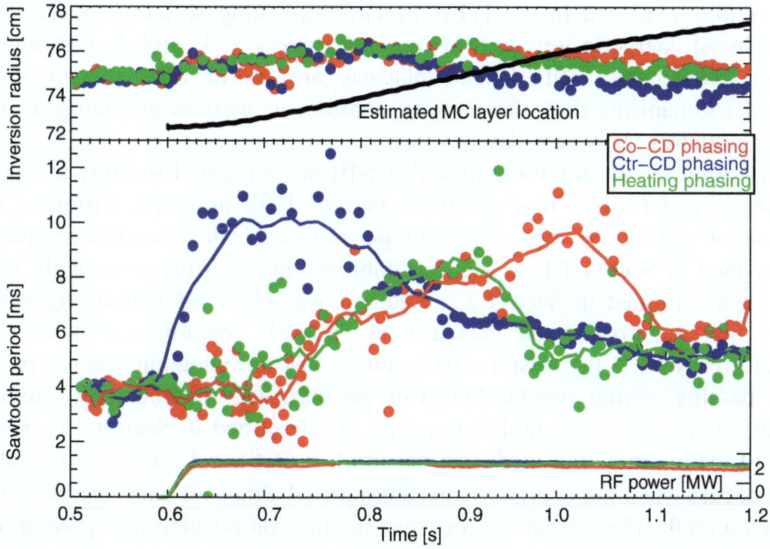

Fig. 4.12 Sawtooth control via changing the magnetic shear by mode conversion current drive demonstrated on Alcator C-Mod. As the mode conversion layer is moved from inside to outside the inversion radius, the sawtooth period is found to increase then decrease for counter-current drive phasing. For both co-propagating ICRH waves and pure heating phasing the sawtooth period decreases then increases. The symmetric phasing results in a co-current due to the strong up-down asymmetry. Reprinted figure with permission from Parisot et al. [148] published by Institute of Physics Publishing. © IOP Publishing. Reproduced with permission. All rights reserved

and consequently achieving higher plasma pressure than would otherwise be possible. This is done in the presence of high core fast ion fractions with strong auxilliary heating, akin to the situation expected in ITER.

Finally, it should be noted that other current drive actuators can also affect sawtooth behaviour. For instance, Lower Hybrid Current Drive (LHCD) has been used to control sawteeth [144–146], as has Mode Conversion Current Drive (MCCD) [147, 148]. Figure 4.12 shows the sawtooth period as a function of the radial location of the mode conversion layer in Alcator C-Mod plasmas as the toroidal field is varied to move the resonance from inside to outside the inversion radius. Just as in the ECCD experiments, the change in the local magnetic shear due to co-(counter-) current MCCD causes the sawtooth period to in(de)crease then de(in)crease respectively as the MCCD deposition is swept from inside to outside $q = 1$.

4.4 Neutral Beam Injection

Neutral beam injection affects sawtooth behaviour through both the introduction of energetic particles as well as the torque applied to the plasma. Having said that, NBI is not considered as a sawtooth control actuator for ITER because the broad

$q = 1$ radius expected in the ITER baseline operating scenario means that the NBI-induced energetic ions will always be inside $q = 1$, and thus strongly stabilising. Nonetheless, NBI-heated plasmas are useful for understanding the physical mechanisms that affect mode stability, as well as providing a tool for sawtooth control in present-day devices.

It has been known for some time that NBI heating could strongly stabilise the kink mode and lead to long sawtooth periods [13], an effect attributed to the stabilisation arising in the presence of a population of core energetic trapped ions, as described in Sect. 4.2.1, as well as stabilisation occurring in strongly rotating plasmas, as outlined in Sect. 4.2.2. Later, it was observed that strong sawtooth stabilisation occurred when heated with 350 keV Negative-ion neutral beam injectors (N-NBI) [149], despite the resulting fast ion population being predominantly passing, giving rise to studies of the influence of circulating ions on the stability of the $n = 1$ internal kink mode. As described in Sect. 4.2.1, destabilisation of the internal kink mode can occur when $\partial f_h(v_\parallel > 0)/\partial r > \partial f_h(v_\parallel < 0)/\partial r$, which occurs when the energetic ions are injected either off-axis ($\partial f_h/\partial r > 0$) and oriented with the plasma current, or when the fast ion population is predominantly on-axis ($\partial f_h/\partial r > 0$) and directed opposite to the plasma current. This effect has been demonstrated by experiments that exhibit an asymmetry in sawtooth behaviour depending upon whether the NBI is injected co-I_p, or counter-current [76, 150, 151]. The asymmetry observed in JET [150] was explained by the competition between the stabilising trapped ions and the destabilising counter-passing ions in the counter-NBI regime, compared with two complementary stabilising effects for co-NBI. Whilst the amplitude of the rotation is strongly sub-Alfvénic, such that gyroscopic effects as outlined in Sect. 4.2.2 play a small role, the flow *shear* at $q = 1$ can be significant, and reduces the stabilising effect of the trapped ions when $\Delta\Omega < 0$, as explained in Sect. 4.2.1. The fact that the finite orbit width of these relatively low energy NBI passing ions could play such an important role in determining sawtooth stability was attributed to the significant fraction of barely passing ions which have a large effective orbit width, and thus strongly influence sawtooth behaviour [70, 71, 152].

Experiments in MAST [76] and TEXTOR [153] also exhibited an asymmetry in the sawtooth period when the injected NBI power was oriented either co- or counter-current, as shown in Fig. 4.13. In these smaller devices, the effect of rotation can become increasingly significant. Whilst the fast ions do have a stabilising influence, the large trapped fraction in spherical tokamaks is stabilising in either co- or counter-NBI regimes. On the other hand, the smaller moment of inertia and high beam power per unit volume in spherical tokamaks result in rotation speeds which approach the ion sound speed [72]. Such strong toroidal rotation results in sawtooth stabilisation, with the minimum in sawtooth period in the counter-NBI regime explained by relative direction of the flow induced by the NBI with respect to the intrinsic rotation of the plasma dominated by the ion diamagnetic drift [76, 154]. In TEXTOR, the sawtooth period reaches a minimum with a low level of co-NBI and a maximum in the counter-NBI regime [153]. This

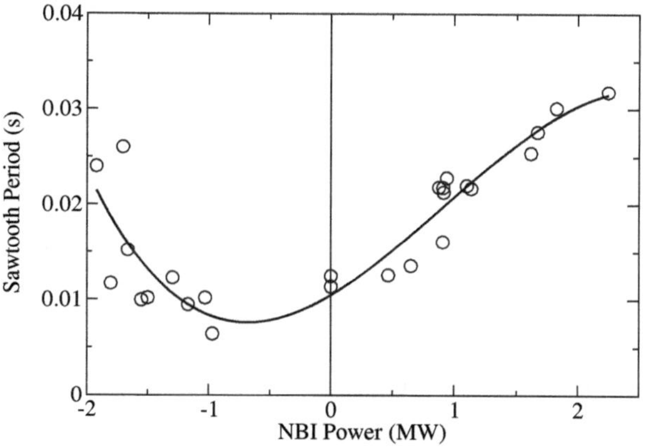

Fig. 4.13 Sawtooth period for MAST discharges as a function of applied NBI power. $I_P \in [680, 740]$ kA, $B_T \in [0.35, 0.45]$ T and $n_e \in [1.6, 2.3] \times 10^{20}\,\mathrm{m}^{-3}$. Negative beam power is taken to represent NBI power in the counter-I_P direction. Reproduced with permission from "The effect of toroidal plasma rotation on sawteeth in MAST" [74]

is due to a competition between the gyroscopic stabilisation of the kink mode and the destabilisation arising in the presence of counter-passing fast ions.

Numerical modelling to assess the effect on the $n = m = 1$ kink mode from both NBI-induced torques and from the resultant fast ion population considered the role played by on-axis co-NBI fast ions in lengthening the sawtooth period in JET. By assessing each of the triggering criteria given in 4.6–4.8 using the PRETOR transport code, Angioni et al. [99] predicted the nonlinear evolution of the sawtooth cycle with good agreement with empirical observations. Furthermore, the assessment of the kinetic effects has improved by applying drift kinetic codes including finite orbit width effects—for instance the Monte-Carlo guiding centre HAGIS code [155] has been applied to JET [40, 151]. Finally, the effect of toroidal rotation on the stability of the internal kink mode has also been assessed numerically [78, 154] and found to be important in determining sawtooth behaviour in fast rotating spherical tokamak plasmas [76].

Following the verification of the importance of passing fast ions in determining sawtooth stability and the observation that different NBI tangency radii led to different sawtooth behaviour [114], experiments were conducted to test whether off-axis co-NBI could be used to deliberately shorten the sawtooth period. The application of off-axis NBI such that the peak of the fast ion population was deposited outside the $q = 1$ surface did indeed result in a destabilisation of sawteeth in JET [156]. Furthermore, sawtooth control using off-axis NBI was also demonstrated in the presence of a concurrent stabilising fast ion population in the plasma core [157]. Figure 4.14 shows that additional off-axis NBI power in JET results in a decrease of the sawtooth period by a factor of three compared to the on-axis NBI only phase, despite an overall increase in β_h leading to stronger stabilisation from the trapped ions. Numerical modelling (also shown in Fig. 4.14)

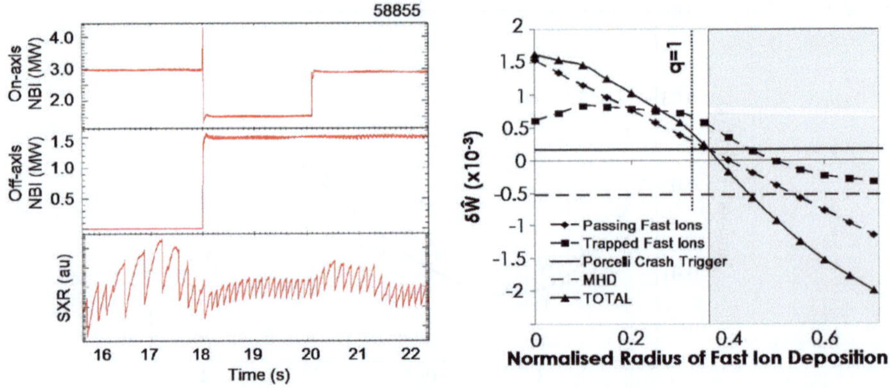

Fig. 4.14 (*left*) The soft X-ray emission and beam heating waveforms for JET shot 58855. The sawtooth period is significantly shorter when the total β_h is kept constant, but some off-axis NBI is used in place of on-axis heating. Further, this discharge also shows that the application of ancillary off-axis NBI can decrease the sawtooth period, despite an overall increase in β_h. Reproduced with permission from "Sawtooth control and the interaction of energetic particles" [157]. (*right*) The potential energy of the internal kink mode as a function of the deposition location of the centre of the fast ion population. When the fast ions are centred just outside the $q = 1$ surface, they destabilise the kink mode and consequently trigger sawteeth more frequently. Equations 4.6–4.9 suggest that a sawtooth crash will occur within the shaded region. Chapman et al. [156] published by Institute of Physics Publishing

confirmed that the passing fast ion effects dominantly determine the sawtooth behaviour.

Confirmation of the dominance of passing fast ion effects came from MAST and ASDEX Upgrade experiments which altered the radial gradient of the fast ion pressure at the $q = 1$ surface, and so changed the destabilising effect. In MAST this was achieved by displacing the plasma vertically such that the beam deposition moves from inside $q = 1$ to well outside [158]. In ASDEX Upgrade, the trajectory of the most off-axis positive ion neutral injector (PINI) was varied in order to move the deposition location of the energetic ions whilst keeping the plasma conditions relatively unchanged. Figure 4.15 shows the beam trajectories for different off-axis PINI inclination and the corresponding sawtooth behaviour. The passing fast ions destabilised the kink mode when the beam deposition location is well outside the $q = 1$ surface. This means that off-axis NBI faces severe limitations as a sawtooth control mechanism since $q = 1$ must be very core localised and even then, the destabilisation from kinetic effects must overcome gyroscopic stabilisation resultant from the beam torque.

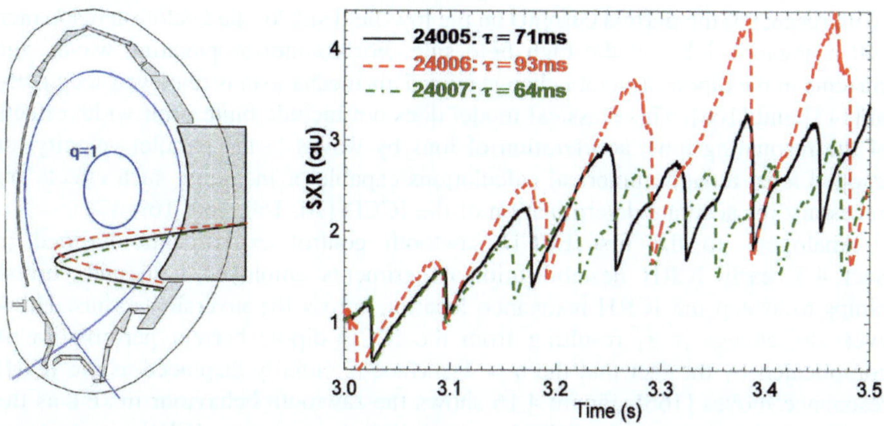

Fig. 4.15 (*left*) The beam trajectories of the off-axis PINI in ASDEX Upgrade as the PINI is tilted on its support. Also shown for comparison is the approximate position of the $q = 1$ surface, and (*right*) the corresponding soft X-ray emission in three ASDEX Upgrade plasmas. The sawtooth period decreases as the beam is injected further off-axis. Discharge 24006 represents the most on-axis NBI heating and shot 24007 is the most off-axis. Reproduced with permission from Chapman et al. [158] Copyright 2009, American Institute of Physics

4.5 Ion Cyclotron Resonance Heating

Early experiments with ICRH focussed on suppressing or delaying the first sawtooth crash, understood by the induced core fast ions stabilising the kink mode. On-axis ICRH was found to result in monster sawteeth, which often triggered NTMs [13, 159]. The long sawtooth periods and giant crashes were shown to be consistent with strong kinetic stabilisation through peaked fast ion pressures [49–51, 160] despite an increase in the destabilising fluid effects associated with the overall increase in pressure [161, 162]. This gave credence to the applicability of kinetic-fluid theory [47, 55] for modelling sawtooth behaviour in plasmas with ICRH. Soon after, experiments with off-axis heating showed that sawtooth destabilisation could also be achieved [113, 163]. Subsequently, control of sawteeth by ICCD has been widely exploited on JET [15, 42, 164–167], using two schemes, viz (i) minority ICCD where a minority ion species resonates with the fundamental cyclotron frequency of the ICRH wave, absorbing the RF power and carrying the fast ion current, and (ii) second harmonic ICCD, where an ion species (not necessarily a minority species) resonates at its second harmonic cyclotron frequency, $\omega = 2\omega_{ci}$ with the RF waves.

Whilst the stabilisation arising from on-axis ICRH was attributed to the trapped fast ion effects [58] and increase in fast ion pressure peaking [49], the destabilisation arising from off-axis ICRH was initially attributed to the non-inductively driven currents. The Fisch model [168] predicts that waves propagating in the co-current direction result in ICCD with a dipole structure with a positive part (with respect to the plasma current) on the low-field side of the cyclotron resonance and a negative lobe on the high field side. For counter-propagating waves, the currents in the dipole structure change sign. This mechanism is reviewed succinctly in [113] and [166]. This classical model does not include finite orbit width effects of the resonating ions, acceleration of ions by waves in the parallel velocity, or trapped ions, though numerical calculations capable of including such effects are necessary for accurate determination of the ICCD [70, 160, 166, 169–171].

Analogous to the first ECCD sawtooth control experiments outlined in Sect. 4.3, early ICRH destabilisation experiments employed field and current ramps to sweep the ICRH resonance location across the inversion radius. However, the change in s_1 resulting from the ICCD dipole current perturbation is complicated by the fact that the $q = 1$ surface is radially displaced as the ICRH resonance moves [166]. Figure 4.16 shows the sawtooth behaviour in JET as the resonance of co-propagating ICRH waves (+90° phasing of the ICRH antenna) and counter-propagating waves (−90°) are swept across the $q = 1$ radius. +90° ICCD results in a strong stabilisation as the resonance is just outside the inversion radius since the shear is reduced, and even more stabilisation as the ICRH moves inside $q = 1$ due to the increase in fast ion pressure. Conversely, the −90° ICCD results in marked destabilisation with the resonance just outside $q = 1$ before an increase in sawtooth period is observed when the ICRH is well inside $q = 1$ and the trapped

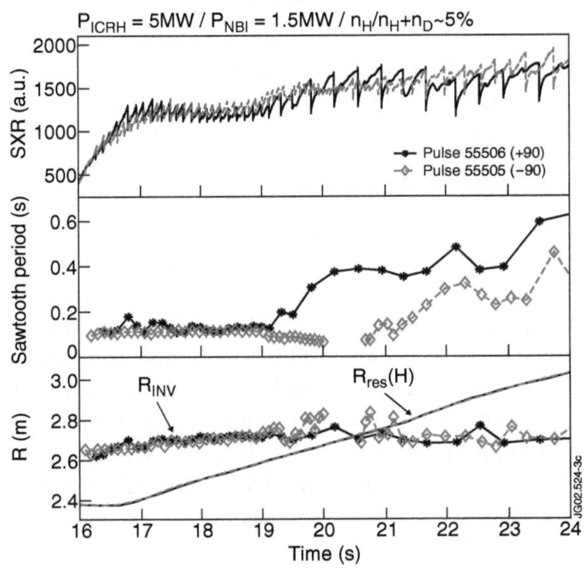

Fig. 4.16 The sawtooth behaviour during JET pules with co-propagating ICRH waves (+90°) and counter-propagating waves (−90°) as the resonance layer is moved across the inversion radius. The sawteeth are destabilised by the −90° ICCD just outside $q = 1$ before the increase in fast ions results in stabilisation when the resonance is well inside $q = 1$. The +90° waves are strongly stabilising when the resonance is both just outside and inside $q = 1$. Reproduced with permission from Mayoral et al. [42] Copyright 2004, American Institute of Physics

fast ion stabilisation begins to dominate. Note that ICRH wave induced pinch in the presence of an asymmetric distribution results in a more peaked fast ion pressure for co-moving waves [160], leading to the observed stronger stabilisation from on-axis heating.

It has also been demonstrated that ICRH is effective in keeping the sawtooth period short in the presence of a substantial core fast-ion population [165]. Figure 4.17 shows the sawtooth behaviour in JET where +90° ICRH is applied in the plasma core, resulting in fast-ion stabilised sawteeth, which are successfully destabilised by concurrent −90° ICCD near the $q = 1$ surface [165, 166]. Furthermore, ICCD control has also been demonstrated in plasmas with even more heating power on-axis from neutral beam injection and much higher β_p, well above the critical threshold for triggering of 3/2 NTMs in the absence of sawtooth control [172].

Detailed wave modelling showed that an ICRH resonance on the high field side gives optimised conditions for the classical Fisch model as the fast ion orbits are closer to the passing-trapped boundary. Conversely, a low field side ICRH resonance results in finite orbit width effects of trapped ions dominating the ICCD [169] decreasing in the magnetic shear near the resonance, independent of antenna phasing, as observed experimentally [167].

Whilst numerical modelling suggested that ICRH sawtooth control occurred primarily due to a change in local shear [113, 164, 166, 167], the sensitivity of sawtooth destabilisation required accuracy of the resonance position with respect to the $q = 1$ surface of less than 1 cm of the $q = 1$ surface in JET [172], which is far more sensitive than expected for control via current drive. Subsequently, Graves et al. showed that the sawtooth control mechanism responsible for localised off-axis toroidally propagating waves is due to the radial drift excursion of the energetic ions distributed asymmetrically in the velocity parallel to the magnetic

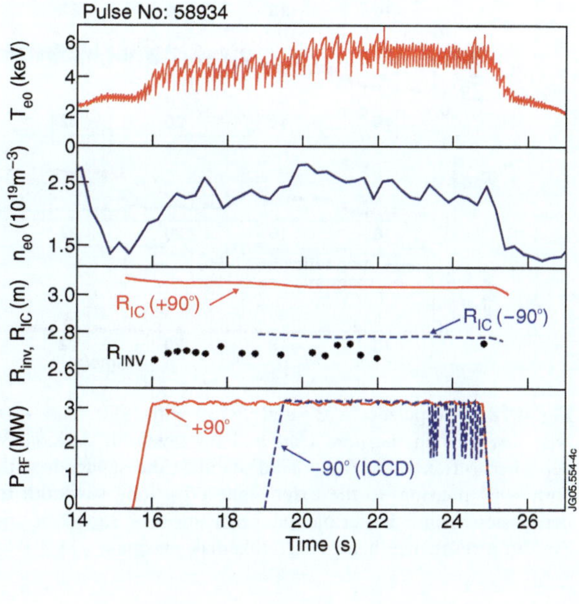

Fig. 4.17 The sawtooth behaviour during JET discharge 58934. When +90° ICRH is applied in the plasma centre the sawtooth period is significantly lengthened. Concurrent −90° ICCD near the $q = 1$ surface results in a destabilisation of the fast ion stabilised sawteeth. Reprinted figure with permission from Eriksson et al. [165]. Copyright 2004 by the American Physical Society

field [70]. Furthermore, this explains why ICRH is more effective than NBI for sawtooth control, since the orbit widths of the fast ions are larger and radial gradients of the parallel asymmetry of the fast ion distribution is stronger. The effect of asymmetry in the distribution was later simulated using SELFO [173] RF wave-field and fast ion distribution function simulations coupled with the drift kinetic HAGIS code, confirming the kinetic mechanism responsible for sawtooth control [152]. Finally, the kinetic mechanism also results in a deep and narrow minimum in the change of the potential energy when the peak of the passing fast ion distribution is just outside the $q = 1$ surface, explaining the extreme sensitivity of the sawtooth behaviour to the deposition location of the ICRH waves.

JET experiments were designed to differentiate between the fast ion and conventional current profile modification effects [70, 71] by using ^{3}He minority heating scheme is employed in a deuterium majority plasmas where the current dragged by the background plasma tends to cancel the ^{3}He current [113, 168, 174] resulting in negligible ICCD. Using such ^{3}He minority heating just outside the $q = 1$ surface led to a strong destabilisation for counter-propagating waves (−90°), overcoming the stabilisation from the core NBI-induced fast ions, and a strong stabilisation for co-propagating waves (+90°), as illustrated in Fig. 4.18.

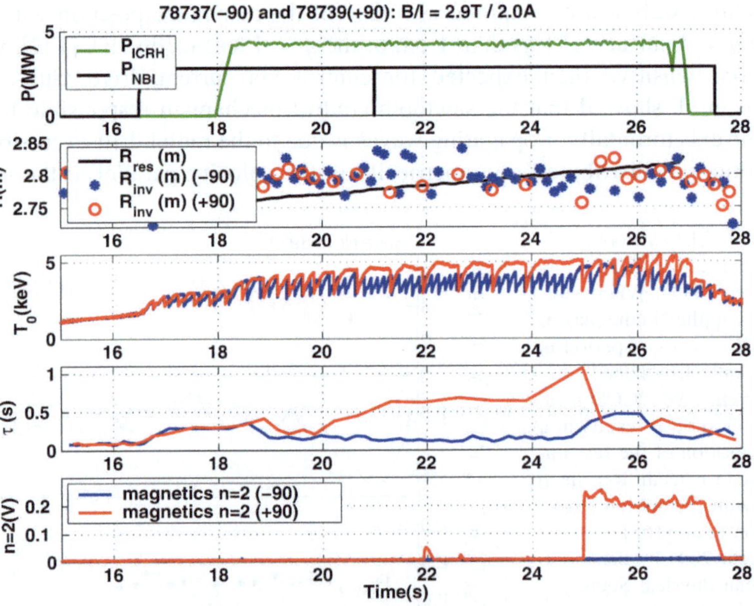

Fig. 4.18 JET pulses 78737 and 78739 with −90° and +90° ICRH off-axis respectively. The NBI fast ions in the core lead to long sawtooth periods. When the He^3 minority heating is deposited off-axis, the fast ions destabilise the sawteeth with −90° phasing, but stabilise them with +90° phasing, to the extent that a 1 s long sawtooth triggers an NTM. Reproduced with permission from "Experimental verification of sawtooth control by energetic particles in ion cyclotron resonance heated JET tokamak plasmas" [71]

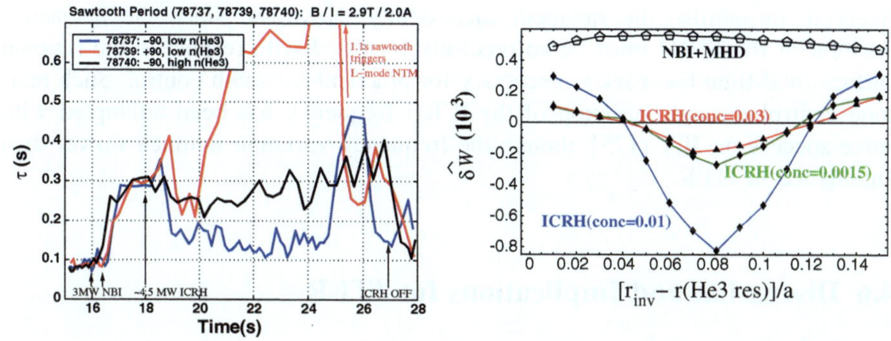

Fig. 4.19 (*left*) JET experiments showing that the fast ions due to He^3 ICRH heating just outside $q = 1$ (de)stabilise the sawteeth for (−) + 90° phasing respectively for low He^3 concentration, but barely affect the sawtooth period at high concentration. This is replicated by drift kinetic modelling (*right*) which shows that the change in potential energy of the kink mode for trace concentration and for 3 % is negligible, whereas for 1 % concentration there is a strong and radially local effect. Reproduced with permission from "Experimental verification of sawtooth control by energetic particles in ion cyclotron resonance heated JET tokamak plasmas" [71]

For the −90° phasing the sawtooth period is reduced to nearly the level of Ohmic sawteeth, whereas +90° increases the sawtooth period significantly, with the longest period of over 1 s triggering an $n = 2$ NTM. Further confirmation of this kinetic mechanism was attained by varying the concentration of the minority ions to change the amplitude of the fast ion mechanism. Figure 4.19 shows the sawtooth behaviour variation with minority ^{3}He concentration in three different JET discharges with −90° phasing ICRH. For $n_{He}/n_e = 0.01$, the destabilising effect of the ICRH energetic passing ions dominates over the stabilising effect of the core NBI ions, with a commensurate reduction in sawtooth period. Conversely, at very low concentration ($n_{He}/n_e = 0.0015$) the minority power absorption is reduced and higher minority ion energies give rise to a broader fast ion distribution and enhanced losses, reducing the impact on sawtooth stability, whilst at high minority density, the reduced effective orbit width means that the effect of the ICRH ions is much smaller than the combined effect of NBI fast ions and fluid drive. This strong sensitivity to minority concentration is also seen in stability calculations shown in Fig. 4.19.

This means that toroidally-propagating ICRH waves could potentially be used for sawtooth control in ITER baseline scenario operation. Studies of the ion cyclotron current drive in ^{3}He minority schemes—as anticipated in ITER [10, 174]—predicted that drag currents will result in negligible driven current [113, 174], meaning that sawtooth control using ICCD was not envisaged for ITER. However, the recent development in understanding the effects of large orbit width passing fast ions near the passing-trapped boundary supported by experimental evidence using ^{3}He minority ICRH in JET, suggests that ICRH can be a useful tool for sawtooth control in ITER. The benefit of this mechanism is that it directly reduces the change in potential energy of the internal kink mode, meaning that a change in the magnetic shear due to concurrent current drive schemes is more likely to destabilise the sawteeth successfully. However, since the resonance position of the ICRH must be so precisely localised with respect to the rational surface, real-time feedback is necessary for practical sawtooth control. Such real-time control through variation of the ICRH frequency has been attempted with some success on JET [175], though the frequency variation is much slower than anticipated in ITER.

4.6 Discussion and Implications for ITER

There has been considerable advancement in both the theoretical understanding and numerical prediction of sawtooth physics and the experimental techniques for sawtooth control, but there remain a number of open questions for sawtooth control in burning plasmas. Amongst these are: (i) What will the natural sawtooth period be in ITER?; and (ii) What is the maximum sawtooth period permissible without triggering an NTM? An early answer to the first question was proposed in [38] where the linear stability thresholds outlined in Sect. 4.2.3 were simulated using a 1–d transport solver, indicating a full reconnection sawtooth period of 100 s with a $q = 1$ radius of 50 % of the plasma minor radius. Subsequently, modelling using TSC [176] with profiles predicted by either the multi-mode model [177] or the Gyro-Landau fluid model GLF23 [178] predicted 50 s sawtooth period for full reconnection and 2–3 times shorter periods for partial reconnection, with a $q = 1$ radius of 42 % of the minor radius. This agrees well with the inversion radius predicted in BALDUR modelling [179]. Finally, time-dependent integrated predictive modelling with the PTRANSP code predicted a sawtooth period much less than 50 s [180]. There is therefore a large range in the predictions from 20 to 100 s, but perhaps more importantly, the issue of whether a sawtooth period in this range will avoid triggering NTMs is currently poorly understood. An empirical scaling of sawtooth-triggered NTMs in current devices provides a rudimentary answer [21], suggesting that at the target plasma pressure for baseline scenario operation, a sawtooth period in the range of 40–70 s will trigger an NTM.

Nonetheless, there have been developments in the capabilities and understanding of the actuators for sawtooth control in ITER even if the operational requirements to avoid NTMs is presently undefined. For instance, the electron cyclotron current drive profiles that can be expected from both the equatorial launcher and the upper launchers have been the subject of much investigation [181–184]. Ray-tracing calculations for the ECCD have allowed ASTRA simulations which include the effect of the fusion-born α particles according to [38, 96] to assess the sawtooth stability in ITER. This led to the prediction that a combination of 13.3 MW of co-ECCD from the equatorial launcher and 6.7 MW from the upper launcher would be able to reduce the sawtooth period by 30 %, or increase it by 50 % with a deposition inside or outside $q = 1$ respectively [185]. The fast ion distribution function arising from both on-axis and off-axis negative-ion neutral beam injection has also been computed [186] using the TRANSP code [187], and the effect of the energetic ions on sawtooth stability has been computed [40, 185]. The N-NBI ions are found to be strongly stabilising to the internal kink, and can only incur destabilisation if the $q = 1$ radius is inside $r = 0.2a$. Finally, whilst numerical modelling of the ICCD expected using ^{3}He minority schemes in ITER predicts the maximum driven current density to be only 0.2–0.5 % of the plasma current density and insufficient for any significant modification to the magnetic shear profile [174], an assessment of the kinetic effects anticipated in ITER suggests that 10 MW of ICRH will be sufficient to negate the stabilising effects arising from the alpha population [185, 188]. Due to the inherent uncertainties in the numerical predictions, it is prudent that a combination of both ICRH and ECCD is planned for controlling sawteeth in ITER. In any case, it will be necessary to have real-time control of these actuators because of both the uncertainties in the control parameters (launcher aiming, ray-tracing, RF frequency) and the equilibrium (plasma position, q profile etc.) and the acute sensitivity of the radial location of the heating or current drive with respect to the rational surface, as highlighted in Sects. 4.3 and 4.5.

An alternative approach to controlling sawteeth is to deliberately maximise the sawtooth period. Indeed, this was originally considered the most desirable route to sawtooth amelioration in the original ITER Physics Basis [10], and was only superseded by destabilisation as anxiety grew about both the ramifications of triggering performance-degrading NTMs and the need for frequent expulsion of the on-axis accumulation of higher-Z impurities that would otherwise radiate energy. Long sawtooth periods are naturally achieved by applying early heating during the current ramp-up phase to increase the conductivity and so slow down the current penetration. Combining this with achieving early ignition will further stabilise the sawteeth due to the α particle stabilisation. ICRH could then be used as an ancillary control tool, with core heating providing a further population of strongly stabilising fast ions. Furthermore, in order to meet the $Q = 10$ goal of ITER baseline scenario, it is desirable to turn off the ECRH power whenever it is not being actively used for mode control. Thus, rather than being constantly required to modify the shear at $q = 1$, an alternative could be envisaged whereby fast ions are used to deliberately stabilise the sawteeth, and before each crash the ECCD is pre-emptively applied near the $q = 2$ surface to stabilise the ensuing NTM [137, 189, 190].

4.7 Summary

In burning plasmas, the significant energetic ion population is likely to result in long sawtooth periods, which in turn are predicted to increase the likelihood of triggering other confinement-degrading instabilities. Consequently, recent experiments have identified various methods to deliberately control sawtooth oscillations in an attempt to avoid seeding NTMs whilst retaining the benefits of small, frequent sawteeth, such as the prevention of core impurity accumulation. The primary methods used to achieve this are (i) tailoring the distribution of energetic ions (ICRH or NBI); (ii) changing the radial profiles of the plasma current density and pressure, notably their local gradients near the $q = 1$ surface (ECCD, LHCD and to a lesser extent ICCD or NBCD); or (iii) heating the electrons inside the $q = 1$ surface (ECRH).

Energetic ions, plasma rotation and the local current density gradients can have a significant effect on the stability of the internal kink mode, thought to underlie the sawtooth phenomenon. Both trapped fast particles and passing fast ions with a large effective orbit width strongly influence sawtooth stability. Circulating ions affect the $n = m = 1$ kink mode through their radial drift excursion when distributed asymmetrically in the velocity parallel to the magnetic field. When these effects are combined, numerical modelling has been able to explicate the sawtooth behaviour observed with different heating and current drive actuators in a number of tokamaks. In order to destabilise the internal kink mode by tailoring the fast ion phase space, and so stimulate a sawtooth crash, the radial gradient of the passing fast ion distribution must be such that $\partial f_h(v_\parallel > 0)/\partial r > \partial f_h(v_\parallel < 0)/\partial r$, which occurs when the energetic ions are injected either off-axis ($\partial f_h/\partial r > 0$) and oriented with the plasma current, or when the fast ion population is predominantly on-axis ($\partial f_h/\partial r > 0$) and directed opposite to the plasma current.

When changing the plasma current density to control sawteeth, increasing the current just inside $q = 1$ increases s_1 and so destabilises the mode, whilst co-CD localised just outside $q = 1$ decreases s_1 and so stabilises the sawteeth. Conversely, driving counter-current just inside $q = 1$ results in stabilisation and just outside $q = 1$ gives rise to destabilisation. Destabilisation of sawteeth has been achieved using steerable electron cyclotron resonance heating which include real-time feedback schemes and robust sawtooth control despite the presence of energetic ions in the plasma core. Dramatic changes in sawtooth stability can also be achieved by the application of off-axis ICRH, both through changes to the magnetic shear, and perhaps dominantly, through establishing a strong radial gradient in the passing fast ion population just outside the $q = 1$ surface.

Whilst the present explanation of the physics of sawtooth oscillations remains incomplete, various robust control schemes have been established and are now well understood. Consequently, there is reasonable confidence that a strategy for sawtooth control in burning plasmas will be refined.

Glossary

I. Greek Symbols

β	Plasma beta, ratio of pressure to magnetic pressure, $\beta = 2\mu_0 p/B^2$
ϵ	Inverse aspect ratio, $\epsilon = r/R_0$
ϵ_1	Inverse aspect ratio of $q = 1$ surface, $\epsilon_1 = r_1/R_0$

η	Resistivity
γ	Linear growth rate
γ_{ad}	Adiabatic index, ratio of specific heats
Γ	Phase space
λ	Pitch angle, $\lambda = v_\parallel/v$
μ	Magnetic moment, $\mu = Mv_\perp^2/2B$
μ_0	Permeability
Φ	Scalar potential
Φ_{ad}	Third adiabatic invariant of motion
ϕ	Toroidal angle
$\dot{\phi}$	Toroidal precession frequency
$\rho_{i,e}$	Ion/Electron Larmor radius, $\rho_{i,e} = Mv_\perp/eB$
$\hat{\rho}$	Average Larmor radius, $\hat{\rho} = \sqrt{\rho_i^2 + \rho_e^2}$
ψ_t	Toroidal flux
ψ_p	Poloidal flux
ψ	Radial coordinate in toroidal geometry
ψ_*	Helical flux
ψ_1	Flux at the $q = 1$ surface
θ	Poloidal angle
ω	Mode frequency
$\tilde{\omega}$	Doppler shifted mode frequency, $\tilde{\omega} = \omega - \Omega_\phi(r_1)$
ω_b	Bounce frequency, $\omega_b = 2\pi/\tau_b$
ω_c	Cyclotron frequency, $\omega_c = eB/M$
ω_A	Toroidal Alfve′n frequency, $\omega_A = v_A/R_0$
ω_{*i}	Diamagnetic frequency of thermal plasma ions, $\omega_{*i} = (T_i dp_i/dr)/(eBp_i r_1)$
ω_{*h}	Diamagnetic frequency of hot ions, $\omega_{*h} = (\partial f_h/\partial P_\phi^0)/(\partial f_h/\partial \mathcal{E}^0)$
ω_{dh}	Drift frequency of hot ions, $\omega_{dh} \approx cE_h/4eBR_o r_1$
$\langle\omega_{dh}\rangle$	Bounce averaged magnetic drift frequency of hot ions
Ω_ϕ	Toroidal plasma rotation

Ω_E	Toroidal plasma rotation caused by electric potential, $\Omega_E = q\Phi'/rB_0$
$\Delta\Omega$	Flow shear, $\Delta\Omega = \Omega_E(r) - \Omega_E(r_1)$
ξ	Fluid displacement
ξ_0	Fluid displacement at the magnetic axis
ξ_1	Fluid displacement at the $q = 1$ surface
ξ_a	Fluid displacement at the plasma edge
τ_A	Alfve'n time, $\tau_A = \sqrt{3R}/v_A$
τ_s	Sawtooth period
τ_E	Energy confinement time
τ_η	Resistive diffusion time
χ	Pitch angle, $\chi = v_\perp^2 B_0/v^2 B$
ζ	Toroidal coordinate, $\zeta = q\theta - \phi$
$\dot{\zeta}$	Toroidal precession frequency

II. Roman Symbols

a	Minor radius of the plasma edge
$\mathbf{A}$	Vector potential
B_ϕ	Toroidal magnetic field strength
B_θ	Poloidal magnetic field strength
$\mathbf{B}$	Magnetic field
e	Charge of particle
$\mathbf{E}$	Electric field
$\mathcal{E}_i$	Energy of *i*th particle
f_h	Hot minority ion distribution function
f_i	Thermal ion distribution function
f_0	Initial distribution function
$\delta f_{h,i}$	Perturbed hot or thermal ion distribution function
δf_{hk}	Perturbed fast ion distribution function due to kinetic effects
δf_{hf}	Perturbed fast ion distribution function due to fluid effects
$\mathbf{j}$	Current density

$\mathbf{k}$	Wave vector
m	Poloidal mode number
M	Particle mass
n_e	Electron number density
n_i	Ion number density
n	Toroidal mode number
p	Plasma pressure
$P_{\zeta,\phi}$	Canonical angular momentum
q	Safety factor, $q = 1/2\pi \int B_\phi/RB_\theta \mathrm{d}s$
R	Major radius
R_0	Major radius at magnetic axis
r	Minor radius
r_1	Minor radius at $q = 1$ surface
s	Magnetic shear, $S = r/qdq/dr$
s_1	Magnetic shear at $q = 1$ surface
T_i	Temperature of ions
T_e	Temperature of electrons
v_A	Alfve′n speed, $v_A = B_0/\sqrt{\mu_0\rho_0}$
$\mathbf{v}_h$	Fast particle velocity
v_ϕ	Toroidal speed of plasma
$\mathbf{v}$	Particle velocity
$\mathbf{v}_{\|\|}$	Particle velocity parallel to the magnetic field
$\mathbf{v}_\perp$	Particle velocity perpendicular to the magnetic field

III. Potential Energy Terms

δW	Perturbed potential energy
δW_{MHD}	Perturbed potential energy due to MHD terms only
δW_{KO}	Perturbed potential energy due to collisionless thermal ions
δW_{core}	$\delta W_{core} = \delta W_{MHD} + \delta W_{KO}$
δW_h	Perturbed potential energy due to collisionless energetic ions

δW_{hf}	Perturbed potential energy due to hot ion adiabatic terms only
δW_{hk}	Perturbed potential energy due to hot ion non-adiabatic terms only
δW^{t}	Perturbed potential energy due to trapped hot ions only
δW^{p}	Perturbed potential energy due to passing hot ions only
$\hat{\delta} W$	Potential energy normalised as per [61], $\hat{\delta} W = \mu_0 \delta W / 6\pi^2 \xi_0^2 \varepsilon_1^4 R_0 B^2$

References

1. R.J. Hastie, Astrophys. Space Sci. **256**, 177 (1998)
2. J. Wesson, in *Tokamaks* (Oxford Science, Oxford, 1997)
3. J.P. Friedberg, in *Ideal Magnetohydrodynamics* (Plenum Press, New York, 1987)
4. S. Von Goeler, W. Stodiek, N. Sauthoff, Phys. Rev. Lett. **33**, 1201 (1974)
5. D.J. Campbell et al., Nucl. Fusion **26**, 1085 (1986)
6. Y. Nagayama et al., Nucl. Fusion **36**, 521 (1996)
7. H.K. Park et al., Phys. Rev. Lett. **96**, 195003 (2006)
8. H.K. Park et al., Phys. Rev. Lett. **96**, 195004 (2006)
9. I. Furno et al., Nucl. Fusion **41**, 403 (2001)
10. ITER Physics Basis, Nucl. Fusion **47**, S1 (2007)
11. M.F.F. Nave et al., Nucl. Fusion **43**, 1204 (2003)
12. M.F.F. Nave et al., Nucl. Fusion **35**, 409 (1995)
13. D.J. Campbell et al., Phys. Rev. Lett. **60**, 2148 (1988)
14. R.J. La Haye et al, in *Proceedings of 24th Europe Conference on Berchtesgarden*, vol. 21A Part III (European Physical Society, Geneva, 1997), pp. 1121–1244
15. O. Sauter et al., Phys. Rev. Lett. **88**, 105001 (2002)
16. A. Gude, S. Günter, M. Maraschek, H. Zohm, Nucl. Fusion **42**, 833 (2002)
17. R.J. Buttery et al., Nucl. Fusion **44**, 678 (2004)
18. R.J. Buttery et al, in *20th IAEA Fusion Energy Conference, Villamoura* EX/7-1 (2004)
19. R.J. Buttery et al., Nucl. Fusion **43**, 69 (2003)
20. E. Westerhof et al, in *Proceedings of 14th Joint Workshop on Electron Cyclotron Emission and Electron Cyclotron Heating* (Heliotopos Conferences Ltd, Santorini, 2006) p. 38
21. I.T. Chapman et al., Nucl. Fusion **50**, 102001 (2010)
22. R. Carrera, R.D. Hazeltine, M. Kotschenreuther, Phys. Fluids **29**, 899 (1986)
23. Z. Chang et al., Phys. Rev. Lett. **74**, 4663 (1995)
24. R.J. La Haye, Phys. Plasmas **13**, 055501 (2006)
25. O. Sauter et al., Phys. Plasmas **4**, 1654 (1997)
26. H. Zohm, Phys. Plasmas **8**, 2009 (2001)
27. D.P. Brennan et al., Phys. Plasmas **10**, 1643 (2003)
28. R. Fitzpatrick et al., Phys. Plasmas **2**, 825 (1995)
29. H.R. Wilson et al., Phys. Plasmas **3**, 248 (1996)
30. A.I. Smolyakov, A. Hirose, E. Lazzaro, G.B. Re, J.D. Callen, Phys. Plasmas **2**, 1581 (1995)
31. M. Kotschenreuther, R.D. Hazeltine, P.J. Morrison, Phys. Fluids **28**, 294 (1985)
32. H. Lütjens, J.F. Luciani, X. Garbet, Phys. Plasmas **8**, 4267 (2001)
33. C.C. Hegna, J.D. Callen, R.J. La Haye, Phys. Plasmas **6**, 130 (1999)
34. M.F.F. Nave et al., Nucl. Fusion **43**, 179 (2003)
35. H. Reimerdes et al., Phys. Rev. Lett. **88**, 105005 (2002)
36. P. Maget et al., Plasma Phys Control Fusion **47**, 357 (2005)
37. H.R. Koslowski et al., Nucl. Fusion **40**, 821 (2000)

38. F. Porcelli, D. Boucher, M. Rosenbluth, Plasma Phys. Control. Fusion **38**, 2163 (1996)
39. B. Hu, R. Betti, J. Manickam, Phys. Plasmas **13**, 112505 (2006)
40. I.T. Chapman et al., Plasma Phys. Control. Fusion **49**, B385 (2007)
41. M.J. Mantsinen et al., Phys. Rev. Lett. **88**, 105002 (2002)
42. M.L. Mayoral et al., Phys. Plasmas **11**, 2607 (2004)
43. R.J. Hastie et al., Phys. Fluids **30**, 1756 (1987)
44. J. Jacquinot et al., Plasma Phys. Control. Nucl. Fus. Res. **1**, 449 (1986)
45. D. Start et al, in *Proc 29th Annual Meeting of APS Division of Plasma Physics*, vol. 159 (San Diego, 1987) p. 286
46. M.F.F. Nave et al., Nucl. Fusion **42**, 281 (2002)
47. L. Chen, R.B. White, M.N. Rosenbluth, Phys. Rev. Lett. **52**, 1122 (1984)
48. K. McGuire et al., Phys. Rev. Lett. **50**, 891 (1983)
49. R.B. White, P. Rutherford, P. Colestock, M. Bussac, Phys. Rev. Lett. **60**, 2038 (1988)
50. R.B. White, M. Bussac, F. Romanelli, Phys. Rev. Lett. **62**, 539 (1989)
51. B. Coppi et al., Phys. Rev. Lett. **63**, 2733 (1989)
52. M.N. Bussac et al., Phys. Rev. Lett. **35**, 1638 (1975)
53. A. Martynov, J.P. Graves, O. Sauter, Plasma Phys. Control. Fusion **47**, 1743 (2005)
54. M. Kruskal, C. Oberman, Phys. Fluids **1**, 275 (1958)
55. T. Antonsen, B. Lane, J. Ramos, Phys. Fluids **24**, 1465 (1981)
56. G. Fogaccia, F. Romanelli, Phys. Plasmas **2**, 227 (1995)
57. R.J. Hastie, T.C. Hender, Nucl. Fusion **28**, 585 (1988)
58. F. Porcelli, Plasma Phys. Control. Fusion **33**, 1601 (1991)
59. J. Van Dam, M. Rosenbluth, Y. Lee, Phys. Fluids **25**, 1349 (1982)
60. T. Northrop, E. Teller, Phys. Rev. **117**, 215 (1960)
61. B. Briezman, J. Candy, F. Porcelli, H. Berk, Phys. Plasmas **5**, 2326 (1998)
62. F. Porcelli, R. Stankiewicz, W. Kerner, H. Berk, Phys. Plasmas **1**, 470 (1994)
63. J.P. Graves, O. Sauter, N. Gorelenkov, Phys. Plasmas **10**, 1034 (2003)
64. F. Porcelli et al, in *15th EPS Conference on Control. Fusion and Plasma Heating, Dubrovnik* **12B/1** 377 (1990)
65. J.P. Graves, R.J. Hastie, K.I. Hopcraft, Plasma Phys. Control. Fusion **42**, 1049 (2000)
66. J.P. Graves, Phys. Plasmas **12**, 090908 (2005)
67. S. Wang, T. Ozeki, K. Tobita, Phys. Rev. Lett. **88**, 105004 (2002)
68. J.P. Graves, Phys. Rev. Lett. **92**, 185003 (2004)
69. K. Ya, V.S. Marchenko, R.B. White, Phys. Plasmas **12**, 022501 (2005)
70. J.P. Graves, I.T. Chapman, S. Coda, T. Johnson, M. Lennholm, Phys. Plasmas **17**, 056118 (2010)
71. J.P. Graves et al., Nucl. Fusion **50**, 052002 (2010)
72. R.J. Akers et al., Nucl. Fusion **42**, 122 (2002)
73. F.L. Waelbroeck, Phys. Plasmas **3**, 1047 (1996)
74. C. Wahlberg, A. Bondeson, Phys. Plasmas **7**, 923 (2000)
75. R. Kleva, P. Guzdar, Phys. Plasmas **9**, 3013 (2002)
76. I.T. Chapman, T.C. Hender, S. Saarelma, S.E. Sharapov, R.J. Akers, N.J. Conway, Nucl. Fusion **46**, 1009 (2006)
77. C. Wahlberg, I.T. Chapman, J.P. Graves, Phys. Plasmas **16**, 112512 (2009)
78. I.T. Chapman, J.P. Graves, C. Wahlberg, Nucl. Fusion **50**, 025018 (2010)
79. S. Migliuolo, Nucl. Fusion **33**, 1721 (1993)
80. I.T. Chapman, Plasma Phys. Control Fusion 53, 013001 (2001)
81. ITER Physics Basis, Nucl. Fusion **39**, 2137 (1999)
82. B.B. Kadomtsev, Sov. J. Plasma Phys. **1**, 389 (1976)
83. M.N. Bussac, D. Edery, R. Pellat, J.L. Soule, *Plasma Phys. Controlled Nucl. Fusion Res.* (IAEA Vienna) **1**, 607 (1976)
84. G. Ara et al., Ann. Phys. **112**, 443 (1978)
85. S. Migliuolo, F. Pegoraro, F. Porcelli, Phys. Fluids B **3**, 1338 (1991)
86. M. Ottaviani, F. Porcelli, Phys. Rev. Lett. **71**, 3802 (1993)

87. A.J. Lichtenberg, K. Itoh, S.I. Itoh, A. Fukayama, Nucl. Fusion **32**, 495 (1992)
88. V. Igochine, O. Dumbrajs, H. Zohm, A. Flaws, Nucl. Fusion **47**, 23 (2007)
89. J. Wesson et al., Plasma Phys. Control. Fusion **28**, 243 (1986)
90. Y. Nagayama et al., Phys. Plasmas **3**, 1647 (1996)
91. T. Munsat et al., Nucl. Fusion **47**, L31 (2007)
92. C.G. Gimblett, R.J. Hastie, Plasma Phys. Control. Fusion **36**, 1439 (1994)
93. M.N. Bussac, R. Pellat, J.L. Soule, M. Tagger, Phys. Lett. **105A**, 51 (1984)
94. M.N. Bussac, R. Pellat, Phys. Rev. Lett. **59**, 2650 (1987)
95. Y. Nishimura, J.D. Callen, C.C. Hegna, Phys. Plasmas **6**, 4685 (1999)
96. T.K. Chu, Nucl. Fusion **28**, 1109 (1988)
97. G. Bateman, C.N. Nguyen, A.H. Kritz, F. Porcelli, Phys. Plasmas **13**, 072505 (2006)
98. O. Sauter et al, in *Theory of Fusion Plasmas, Proceedings of Joint Varenna-Lausanne International Workshop, Varenna* (AIP) p. 403 (1998)
99. C. Angioni et al., Plasma Phys. Control. Fusion **44**, 205 (2002)
100. C. Angioni, T. Goodman, M. Henderson, O. Sauter, Nucl. Fusion **43**, 455 (2003)
101. M. Choi et al., Phys. Plasmas **14**, 112517 (2007)
102. F.M. Levinton et al., Phys. Rev. Lett. **72**, 2895 (1994)
103. F. Porcelli, Phys. Rev. Lett. **66**, 425 (1991)
104. J.P. Graves et al., Nat. Commun. **3**, 624 (2012)
105. G. Bobrovskii et al., Sov. J. Plasma Phys. **13**, 665 (1987)
106. B.N. Kuvshinov, P.V. Savrukhin, Sov. J. Plasma Phys. **16**, 353 (1990)
107. K. Hanada et al., Phys. Rev. Lett. **66**, 1974 (1991)
108. R.T. Snider et al., Phys. Fluids B **1**, 404 (1989)
109. K. Hanada et al., Phys. Fluids B **4**, 2675 (1992)
110. E. Westerhof et al., Nucl. Fusion **43**, 1371 (2003)
111. Y. Ikeda et al., Nucl. Fusion **42**, 375 (2002)
112. A. Muck et al, in *29th EPS Conference on Controlled Fusion and Plasma Physics, Montreux,* vol. 26B, P1.037 (2002)
113. V.P. Bhatnagar et al., Nucl. Fusion **34**, 1579 (1994)
114. A. Mück, T.P. Goodman, M. Maraschek, G. Pereverez, F. Ryter, H. Zohm, Plasma Phys. Control Fusion **47**, 1633 (2005)
115. R.I. Pinsker et al., Bull. Am. Phys. Soc. **48**, 128 (2003)
116. A. Manini et al, in *32nd EPS Conference on Plasma Physics*, Tarragona, vol. **29C**, P4.073 (2005)
117. O. Sauter et al., Phys. Plasmas **8**, 2199 (2001)
118. M.A. Henderson et al., Fus. Eng. Des. **53**, 241 (2001)
119. H. Zohm et al., Nucl. Fusion **43**, 1570 (2003)
120. V.V. Ahkaev et al., Nucl. Fusion **35**, 369 (1995)
121. M. Maraschek et al., Nucl. Fusion **45**, 1369 (2005)
122. A. Manini et al, in *14th Joint Workshop on ECE and ECRH*, Santorini (2006)
123. T.P. Goodman et al., Nucl. Fusion **43**, 1619 (2003)
124. J.P. Graves et al., Plasma Phys. Control. Fusion **47**, B121 (2005)
125. A. Isayama et al., J. Plasma Fus. Res. Series **5**, 324 (2002)
126. M. Lennholm et al, in *17th Topical Conference on Radio Frequency Power in Plasmas*, vol. 933, p. 401 (2007)
127. M. Lennholm et al., Phys. Rev. Lett. **102**, 115004 (2009)
128. M. Lennholm et al., Fus. Sci. Tech. **55**, 45 (2009)
129. S. Cirant et al, in *17th IAEA Fusion Energy Conference,* Yokohama CDP/07 (1998)
130. M. Asakawa et al, in *17th IAEA Fusion Energy Conference,* Yokohama CDP/06 (1998)
131. Z.A. Pietrzyk et al., Nucl. Fusion **39**, 587 (1999)
132. H. Reimerdes et al., Plasma Phys. Control. Fusion **42**, 629 (2000)
133. A. Polevoi, S. Medvedev, V. Mukhovatov, A. Kukushkin, Y. Murakami, M. Shimada, A. Ivanov, ITER confinement and stability modelling. J. Plasma Fusion Res. **5**, 82–87 (2002)
134. J.I. Paley et al., Nucl. Fusion **49**, 085017 (2009)

135. J.I. Paley et al., Plasma Phys. Control. Fusion **51**, 055010 (2009)
136. J.I. Paley, F. Felici, S. Coda, T.P. Goodman, Plasma Phys. Control. Fusion **51**, 124041 (2009)
137. T.P. Goodman et al., Phys. Rev. Lett. **106**, 245002 (2011)
138. M. Lauret et al., Nucl. Fusion **52**, 062002 (2012)
139. G. Witvoet et al., Nucl. Fusion **51**, 103043 (2011)
140. V.G. Igochine et al, In prep for Nucl. Fusion (2010)
141. Y. Liu et al, *Private Communication* (2010)
142. I.T. Chapman et al., Nucl. Fusion **52**, 063006 (2012)
143. I.T. Chapman et al., Plasma Phys. Control. Fusion **55**, 065009 (2013)
144. A. Ekedahl et al., Nucl. Fusion **38**, 1397 (1998)
145. F.X. Söldner et al., Nucl. Fusion **34**, 985 (1994)
146. T.K. Chu et al., Nucl. Fusion **26**, 666 (1986)
147. S. Wukitch et al., Phys. Plasmas **12**, 056104 (2005)
148. A. Parisot et al., Plasma Phys. Control. Fusion **49**, 219 (2007)
149. G. Kramer et al., Nucl. Fusion **40**, 1383 (2000)
150. M.F.F. Nave et al., Phys. Plasmas **13**, 014503 (2006)
151. I.T. Chapman, S.D. Pinches, J.P. Graves, L.C. Appel, R.J. Hastie, T.C. Hender, S. Saarelma, S.E. Sharapov, I. Voitsekhovitch, Phys. Plasmas **14**, 070703 (2007)
152. J.P. Graves, I.T. Chapman, S. Coda, L.G. Eriksson, T. Johnson, Phys. Rev. Lett. **102**, 065005 (2009)
153. I.T. Chapman, S.D. Pinches, H.R. Koslowski, Y. Liang, A. Kramer-Flecken, M. de Bock, Nucl. Fusion **48**, 035004 (2008)
154. I.T. Chapman, G.T.A. Huysmans, A.B. Mikhailovskii, S.E. Sharapov, Phys. Plasmas **13**, 062511 (2006)
155. S.D. Pinches, L.C. Appel, J. Candy, S.E. Sharapov, H.L. Berk, D. Borba, B.N. Breizman, T.C. Hender, K.I. Hopcraft, G.T.A. Huysmans, W. Kerner, Comput. Phys. Commun. **111**, 133 (1998)
156. I.T. Chapman et al., Plasma Phys. Control. Fusion **50**, 045006 (2008)
157. I.T. Chapman et al., Nucl. Fusion **49**, 035006 (2009)
158. I.T. Chapman et al., Phys. Plasmas **16**, 072506 (2009)
159. C.K. Phillips et al., Phys. Fluids B **4**, 2155 (1992)
160. L.-G. Eriksson et al., Phys. Rev. Lett. **81**, 1231 (1998)
161. J.P. Graves et al., Phys. Rev. Lett. **84**, 1204 (2000)
162. K.G. McClements et al., Phys. Plasmas **3**, 2994 (1996)
163. D.F.H. Start et al, in *Proceedings of International Conference on Plasma Physics (Innsbruck)*, vol. 16C Part II (European Physical Society, Geneva, 1992) p. 1521
164. E. Westerhof et al., Nucl. Fusion **42**, 1324 (2002)
165. L.-G. Eriksson et al., Phys. Rev. Lett. **92**, 235004 (2004)
166. L.-G. Eriksson et al., Nucl. Fusion **46**, S951 (2006)
167. M. Mantsinen et al., Plasma Phys. Control. Fusion **44**, 1521 (2002)
168. N.J. Fisch, Rev. Mod. Phys. **59**, 175 (1987)
169. T. Hellsten, J. Carlsson, L.-G. Eriksson, Phys. Rev. Lett. **74**, 3612 (1995)
170. J. Carlsson, T. Hellsten, J. Hedin, Phys. Plasmas **5**, 2885 (1998)
171. M. Mantsinen et al., Phys. Rev. Lett. **89**, 115004 (2002)
172. S. Coda et al, in *Proceedings of 34th EPS Conference on Plasma Physics* (Warsaw, 2007) P5.130
173. J. Hedin et al., Nucl. Fusion **42**, 527 (2002)
174. M. Laxå back, T. Hellsten, Nucl. Fusion **45**, 1510 (2005)
175. M. Lennholm et al., Nucl. Fusion **51**, 073032 (2011)
176. S.C. Jardin, M.G. Bell, N. Pomphrey, Nucl. Fusion **33**, 371 (1993)
177. G. Bateman et al., Phys. Plasmas **5**, 2355 (1998)
178. R. Waltz et al., Phys. Plasmas **4**, 2482 (1997)
179. T. Onjun, Y. Pianroj, Nucl. Fusion **49**, 075003 (2009)

180. R.V. Budny et al., Nucl. Fusion **48**, 075005 (2008)
181. C. Zucca et al, in *Theory of Fusion Plasmas, Joint Varenna-Lausanne International Workshop, Varenna (AIP)*, vol. 1069, p. 361 (2008)
182. M. Henderson et al, in *3rd IAEA TM on ECRH Physics and Technology in ITER* (Como, 2005) p. 143
183. W.A. Houlberg et al., Nucl. Fusion **45**, 1309 (2005)
184. N.B. Maruschenko, H. Maassberg, Y. Turkin, Nucl. Fusion **48**, 054002 (2008)
185. I.T. Chapman et al., Nucl. Fusion **53**, 066001 (2013)
186. R.V. Budny, Nucl. Fusion **42**, 1383 (2002)
187. R.V. Budny et al., Nucl. Fusion **32**, 429 (1992)
188. I.T. Chapman et al., Plasma Phys. Controlled Fusion **53**, 124003 (2011)
189. O. Sauter, M.A. Henderson, G. Ramponi, H. Zohm, C. Zucca, Nucl. Fusion **52**, 025002 (2010)
190. F. Felici et al., Nucl. Fusion **52**, 074001 (2012)

Chapter 5
Edge Localized Mode (ELM)

Yunfeng Liang

Abstract The next generation of fusion machines like ITER and DEMO will need a reliable method for controlling the periodic transient expulsion of a considerable amount of energy onto the plasma-facing components caused by instabilities at the plasma edge. The good plasma confinement in these tokamak devices will result in a steepened pressure profile at the plasma edge. When the pressure gradient exceeds a critical value, so-called edge- localized modes (ELMs) are destabilized. These modes feature a periodic fast collapse of the edge pressure, a sudden loss of the confinement, and a subsequent release of heat and particles onto plasma-facing components. The associated transient heat loads might cause excess erosion and lead to a strong reduction of the plasma-facing component lifetime. In this chapter, an overview of recent development of several ELM control methods for next-generation tokamaks, e.g., ITER is given. Some key physics issues related to the mechanism of ELM control are discussed.

5.1 Introduction

The discovery of the high confinement mode (H-mode), which is characterized by the formation of a transport barrier at the edge of the plasma, was made at the ASDEX Tokamak [1]. The transport barrier creates a strong pressure gradient at the plasma edge called the edge pedestal. The H-mode increases the plasma energy confinement time by around a factor of two compared to the Low confinement mode (L-mode). This discovery constitutes a great step towards achieving the higher temperatures and pressures needed to create ignition conditions. The standard tokamak H-mode is foreseen as the baseline operating scenario of a future

Y. Liang (✉)
Forschungszentrum Jülich GmbH, Association EURATOM-FZ Jülich,
Institut Für Energieforschung—Plasmaphysik, Trilateral Euregio Cluster,
52425 Jülich, Germany
e-mail: y.liang@fz-juelich.de

V. Igochine (ed.), *Active Control of Magneto-hydrodynamic Instabilities in Hot Plasmas*, Springer Series on Atomic, Optical, and Plasma Physics 83,
DOI 10.1007/978-3-662-44222-7_5

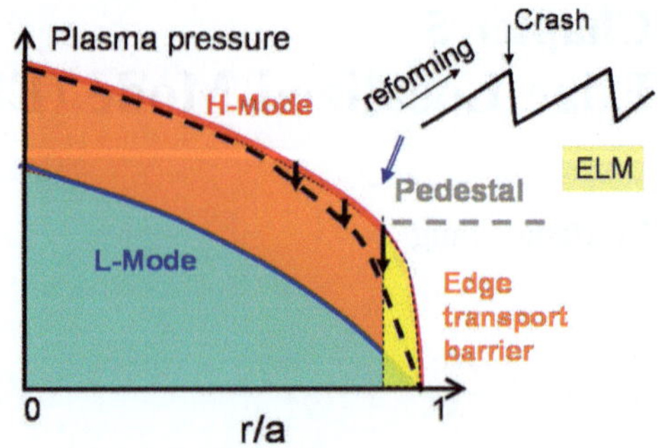

Fig. 5.1 Pressure profiles of L and H- Mode plasmas

fusion machine, e.g. ITER [2]. However, as another consequence of this discovery, a steep plasma pressure gradient and associated increased current density at the edge pedestal could exceed a threshold value to drive magnetohydrodynamic (MHD) instabilities referred to as Edge Localized Modes (ELMs) [1, 3].

In a standard H-mode plasma, ELMs occur repetitively and the edge pedestal collapses towards a shallower pressure gradient within a few hundred microseconds. An expulsion of large amounts of heat and particles onto the plasma facing components from the confined plasma occurs and later the edge pedestal recovers again to a steep gradient, as shown in Fig. 5.1. Each ELM is characterized by an increase in the radiation shown in the D_α line emissions, a burst of magnetic activity. The increase in D_α indicates an increase in edge recycling and can be used to give a measurement of the inward particle flux. ELM events can lead to large transient heat and particle loads on the plasma facing components as well as reducing the pedestal energy confinement by $\sim$10–20 %.

Using results from various current devices, an extrapolation of the heat and particles deposited on the wall components has been carried out for ITER. Since the exact physics and scaling is unknown, the predicted ELM energy loss ranges from $\sim$5 to $\sim$22 MJ. It is expected that approximately half of this energy will reach the wall and be deposited over a region of $\sim$1 m^2, known as the wetted area. Thus, the surface energy density is suggested to be 2.5–11 MJm^{-2} which is $\sim$5–20 times higher than acceptable for the planned first wall components, primarily made of tungsten or carbon fibre composites, which can receive a maximum of 0.5 MJm^{-2}. Therefore, it is important to find mitigation/suppression solutions for ELMs.

The research of ELMs is also of high interest generally, as it involves both linear and non-linear relaxations, requires knowledge of microscopic and macroscopic processes in a volatile plasma with a large magnetic field, and includes higher dimensional effects such as turbulence and 3-dimensional distortions. This understanding enhances similar research into the mechanisms occurring at the edge of stars, for example solar flares.

In this Chapter, a brief introduction on the present common understanding of ELM physics is given in Sect. 5.2. An overview of recent developments of several existing ELM control methods for next-generation tokamaks, e.g., ITER is given in Sects. 5.3–5.5. Some key physics issues related to the mechanism of ELM control are described.

5.2 Physics of Edge Localized Mode in Tokamaks

5.2.1 ELM Types

Edge localized modes have been studied on a wide range of tokamaks including: Alcator C-MOD [4], ASDEX-U [5], COMPASS-D [6], DIII-D [7], EAST [8], JET [9], JFT- 2 M [10], JT-60U [11, 12], MAST [13], NSTX [14], TCV [15], and TEXTOR [16]. The magnetic fluctuations, the short time scales of the ELM growth, and the proximity of the plasma to an MHD stability limit when an ELM occurs, all point towards MHD being able to explain the ELM onset. The phenomenology varies depending on the size and shape of the plasma making it necessary to distinguish between different types of ELM. Connor [3], Suttrop [17] and Zohm [18] have summarized these observations and attempted to create a classification of which the main points will be outlined here.

Three main criteria are used to classify ELMs: the dependence of ELM repetition frequency on the heating power (the energy flux through the separatrix), the occurrence of magnetic precursors, and MHD stability analysis, although reference [18] argues that this third criterion is insufficient across different machines.

- **Type-I ELMs**: The D_α radiation shows large isolated bursts and, therefore, Type-I ELMs are also called 'large' or even 'giant' ELMs. These events occur in regimes which have good confinement but expel a large amount of energy. The repetition frequency of type-I ELMs is $\sim$10–100 Hz. As the heating power is increased, the ELM repetition frequency also increases. The energy loss of a Type-I ELM, ΔW_{ELM}, is also much larger than that of other ELM types, being up to $\sim$20 % of the pedestal energy reported in the international database [19]. Unacceptably high transient heat loads onto the plasma facing components (PFCs) are expected in a burning fusion plasma with a type-I ELMy H-mode [20].
- **Type-II ELMs**: To date, Type-II ELMs [21–24] are observed only in strongly-shaped plasmas, i.e. with high elongation and triangularity of the plasma cross-section. Further, the plasma density needs to be rather high. The magnitude of the ELM bursts is lower and the frequency is higher than that of type-I ELMs, while the confinement stays almost as good. Sometimes, type-II ELMs are called 'grassy' ELMs. Compared with a Type-I ELMy plasma, enhanced magnetic turbulence has been seen in the inter-ELM phase of Type-II ELMy H-mode plasmas. Although Type-II ELMs show potential for steady-state operation of a tokamak with good confinement, and efficient impurity exhaust while not damaging the divertor plates, they appear only in a narrow operational window, and it is still unclear whether Type-II ELMs will be possible to achieve in a burning fusion plasma. On JET, it is hard to establish a stationary pure Type-II ELMy H-mode plasma with a single null configuration, but mixed Type-I/II ELMs are often seen in high triangularity and high density H-mode plasmas.

- **Type-III ELMs**: The bursts are small and frequent. Therefore, another name for type-III ELMs is 'small' ELMs. Type-III ELMs appear when the plasma resistivity is rather high (i.e. the edge temperature is rather low). The ELM repetition frequency is found to decrease with increasing heating power. Although the energy lost in a single type-III ELM is significantly lower than in a type-I ELM, the rather high overall energy transport leads to a stronger degradation of the energy confinement of the plasma compared to other ELM types.

In addition to the three conventional ELM types, there are still other different ELM types, such as compound ELMs, Type-V ELMs, observed in different devices. Several small/no ELM regimes such as EDA (enhanced D_α H-mode), grassy ELM, HRS (high recycling steady), QH-mode (quiescent H-mode), type-II and V ELMs with good confinement properties have been obtained in Alcator C-Mod, AUG, DIII-D, JET, JFT-2 M, JT-60U and NSTX.

Type-II ELMs with small ELM amplitude have been found in DIII-D with a large plasma elongation ($\kappa > 1.8$), a high edge safety factor ($q_{95} \sim 7$) and a high triangularity ($\delta \sim 0.5$) [21]. On AUG and JET, Type-II ELMs have been observed in highly shaped plasmas at high density [22, 25, 26]. The grassy ELM regime has been found in JT-60U at lower collisionality in high poloidal beta (β_p) plasmas with a high δ [27, 28]. The grassy-like ELMs have been also observed in H-mode plasmas with $\beta_p > 1.7$, $q_{95} \sim 7$, and $\delta > 0.4$ on JET and AUG [29, 30]. On NSTX, Type-V ELMs are observed in high density, high performance discharges, and they are characterized by a short-lived $n = 1$ pre-cursor mode rotating counter to the plasma current [31]. All these small ELM regimes show that ELM energy losses are evaluated as less than 5 % of the pedestal stored energy.

The QH-mode regime was originally observed in DIII-D [32] and then also produced in AUG [33], JT-60U [34] and JET [35]. QH-mode plasmas, which do not have ELMs, can be sustained with good confinement levels comparable to those observed in the standard ELMy H-mode. The formation of a strong rotation shear in the edge pedestal is thought to be the key to obtaining QH-mode [34, 12, 36]. EDA H-mode was found in Alcator C-Mod after boronization of the first wall [37]. HRS H-mode, which is similar to the EDA regime, has also been observed in JFT-2 M after boronization [38]. Both EDA and HRS H-mode are characterized by an enhanced D_α signal as a consequence of particle losses due to edge MHD and density fluctuations, such that a steady H-mode can be sustained without ELMs.

The characteristics of edge fluctuations and activities of ideal MHD stability leading to small/no ELMs are summarized in [39]. Some of the small ELM regimes identified at high collisionality, such as EDA, Type V, Type-III and Type-II, have been extended into lower collisionality regimes, of increasing relevance to ITER, in recent years. However, the extrapolation and uncertainty are still large, so it remains unclear whether they can be accessed on ITER.

5.2.2 Understanding Edge Localized Modes

Over the last two decades considerable theoretical work has been performed, both analytically and through modelling calculations, to improve the theoretical background. Ideal MHD modes driven by the steep current and pressure gradients at the edge transport barrier are regarded as the most likely candidates to explain the origin of ELMs. From stability calculations performed on the basis of experimental data three types of ideal MHD instabilities can be expected at the edge transport barrier:

- **Localized peeling modes**, which are driven by the edge plasma current density, do not rely on toroidicity as a drive.
- **Edge ballooning modes**, which are driven by the edge plasma pressure gradient, with largest amplitude on the outboard "bad curvature" side, and very small amplitude on the inboard side.
- **Coupled peeling-ballooning modes**, which are driven by the steep edge pressure gradient and consequently large edge bootstrap current.

5.2.2.1 Localized Peeling Modes

The term "peeling" in literature is first mentioned by Frieman [40] as a test function for the radial displacement which describes the occurrence of an instability when a resonance condition is met, $m - nq_a = 0$ for $m > 1$. The peeling mode is destabilized by the finite edge current density and is dependent on the location of the closest rational surface to the plasma edge in the vacuum. The pressure gradient gives both a stabilising effect through the magnetic shear and a destabilising effect through the bootstrap current. This is essentially the same as the well-known external kink mode. The difference is that the kink mode is driven by the derivative of the parallel current density, whilst the peeling mode is driven by the torque created by a finite value of the current density at the plasma edge and no current in the vacuum region. In addition, the peeling mode has a higher localisation than the external kink due to its sensitivity on the outer rational surface. The effect of the distance from the plasma edge to the rational surface leads to strong dependencies on the q profile and the tokamak geometry. A stability criterion for the peeling mode in a toroidal MHD equilibrium surrounded by a vacuum with a continuous pressure profile is formed in reference [41]. This showed that the peeling modes would theoretically be unstable when the ELMs occur. Manickam [42] also argues that the external kink and the more localized peeling mode are strong candidates for driving ELMs. In terms of experimental evidence, the PBX-M machine observed an MHD precursor to an ELM, identified as an ideal external kink mode [43].

The effect of a divertor on the peeling stability needs to be considered. Theoretically at the separatrix, created by the divertor magnetic topology, $q \to \infty$.

This will have a large effect on the peeling stability as multiple resonant q values will be covered by the peeling mode. The behaviour of the unstable peeling mode growth rate at the last closed flux surface of the plasma has been examined analytically [44, 45]. Modelling focusing on the X-point region is also being conducted [46, 47]. It was found that although the peeling drive is always present, the growth rate tends to zero and the mode becomes marginal in the presence of the separatrix. However, It is still unclear how the q profile acts in a real world situation and so it is general procedure to use the value of the effective edge safety factor at 95 % of the normalized poloidal flux, q_{95}, for tokamaks with a divertor configuration.

5.2.2.2 Edge Ballooning Modes

The ballooning mode arises from the curvature of the tokamak geometry. Modes with high toroidal mode number *n* modes, localized around their resonant surfaces, feel different curvature effects as they follow the helical field lines. On the high field side the curvature effect is stabilising, whereas for the low field side the curvature is seen to have a destabilising effect. The average of these effects is found to be stabilising for a plasma with a low pressure gradient. However, if the pressure gradient becomes too high, then the average of the curvature becomes a destabilising drive leading to ballooning modes. The stability of ballooning modes can be approximated from a balance of the driving term from the pressure gradient and the stabilising effect of the energy required for field line bending.

The value of the magnetic shear is also important for ballooning mode stability. At values of high magnetic shear the mode is stabilized. On the other hand, at very low values of shear "second stability access" is granted. This additional region of stability is not an obvious result, as a higher pressure would increase the drive of the ballooning modes. However, at high pressure a strong distortion of the equilibrium magnetic flux surfaces occurs, which increases the local pitch and decreases the shear at the LFS. The increase in the local pitch causes the plasma to spend more time in the good curvature region, which as stated is stabilising. The decrease of the shear creates a large region of negative shear, which apart from being stabilising itself, also pushes the zero shear point away from the LFS to a more stable part of the plasma. The combination of these effects is responsible for the second stability access. This is advantageous not only due to the extra stability granted but also because it produces significant reductions in the required toroidal field and plasma current.

At the edge of the plasma an extra boundary condition must be in place as the ballooning mode can not extend into the vacuum region. This breaks the symmetry of the envelope of influence of the mode on the surrounding surfaces and creates a more localized perturbation. Non-linear ballooning theory [48] suggests that the mode will be broad along the field line but narrow perpendicular to it. As a result the ejected particles will have a filament structure which will narrow and twist to squeeze between adjacent magnetic field lines. This filament is then predicted to have a radially explosive behaviour but could remain partially connected to the plasma core acting as a conduit for further particles to flow along.

On MAST, during an ELM crash, multiple peaks have been observed in the ion saturation current, measured using the reciprocating Langmuir probe [49], suggesting more than one structure rotating around the plasma. The energy of these structures, calculated from the change of the density profiles, would be small amount ($\sim$ <10 J) and would only account for a fraction of the total ELM energy loss. Thus, these structures must not be isolated blobs but remain partially connected to the plasma and act as a particle conduit, i.e. a filament from core to SOL. These predictions have been verified experimentally on the MAST tokamak [50, 51]. This was found when comparing high resolution Thomson scattering density profiles over an ELM event, which shows that ELMs have little effect on the inboard side but cause a large reduction in the density gradient on the outboard side. Immediately following this density gradient drop, poloidally localized density structures, moving radially away from the plasma, were seen, indicating propagating structures rather than a diffusion of particles. A high speed visible camera backs up these observations showing clear elongated structures along the field lines. Filament structures associated with ELMs have also been observed on ASDEX Upgrade [52].

5.2.2.3 Coupled Peeling-Ballooning Modes

It has been put forward that a spectrum of peeling modes are unstable in the L-mode creating a large amount of anomalous transport [53]. As the plasma enters H-mode the collisionality increases stabilising the majority of the peeling modes, thus reducing anomalous transport; although a few peeling modes may remain marginally unstable. The transport barrier is now formed allowing the pedestal pressure to increase which in turn increases the ballooning stability parameter. When this reaches the ballooning stability limit it can either cause the onset of a ballooning mode, or the plasma can stay on the ballooning stability threshold whilst on a slower diffusive time scale the bootstrap current increases, due to the increasing pressure and decreasing collisionality which would allow a greater trapped particle fraction. When the bootstrap current has risen sufficiently to reach the peeling mode stability threshold, there is a possibility that the harmonics of the unstable peeling and ballooning modes couple, creating a large crash in the pedestal as seen for the large ELMs. Stability analysis of the coupled peeling-ballooning modes has indeed shown the plasma to reach the ballooning boundary and hold there until the peeling instability condition is met before the ELM crash occurs with intermediate toroidal mode numbers [54]. These coupled peeling-ballooning modes are complex and interesting phenomena as the bootstrap current, shear and pressure all play dual roles of stabilising and destabilising. Thus a useful representation of these modes is their potential energy δW for a radial displacement ξ [55]

$$
\begin{aligned}
\delta W = \pi \int_0^{\psi_a} d\psi \oint d\theta \Bigg\{ & \frac{JB^2}{R^2 B_p^2} \left| k_{\|} \xi \right|^2 + \frac{R^2 B_p^2}{JB^2} \left| \frac{1}{n} \frac{\partial}{\partial \psi} (JBk_{\|} \xi) \right|^2 \\
& - \frac{2J}{B^2} \frac{dp}{d\psi} \left[|\xi|^2 \frac{\partial}{\partial \psi} \left(p + \frac{B^2}{2} \right) - \frac{i}{2} \frac{I}{JB^2} \frac{\partial B^2}{\partial \theta} \frac{\xi^*}{n} \frac{\partial \xi}{\partial \psi} \right] \\
& - \frac{\xi^*}{n} JBk_{\|} \left(\frac{\partial \sigma}{\partial \psi} \xi \right) + \frac{\partial}{\partial \psi} \left[\frac{\sigma}{n} \xi JBk_{\|}^* \xi^* \right] \Bigg\},
\end{aligned} \tag{5.1}
$$

where B_p is the poloidal component of the magnetic field B, p is the plasma pressure, $k_{\|}$ is the parallel length to the magnetic field over which the displacement varies, R is the radius of curvature $\sim R_0$, the major radius of the torus, and θ, ϕ, and ψ are the toroidal coordinate system.

Here the first two terms are associated with field line bending, the third term calculates the pressure gradient drive including curvature effects, and the last two terms give the current density gradient drive where I is the plasma current, J is the current density, and $\sigma = \frac{I}{B^2}\frac{\partial p}{\partial \psi} + \frac{\partial I}{\partial \psi}$.

Considering each of the two modes individually, (5.1) states clearly the different terms involved. The peeling mode is highly localized at the plasma edge and has little coupling between its Fourier modes, so the field line bending effects disappear. The pressure gradient acts as a stabilising effect when the plasma experiences good curvature, whereas the current drive acts to destabilize the mode. For a ballooning mode quite the opposite is true and it is found that the field line bending term is enhanced by multiple Fourier harmonics coupling over the rational surfaces and thus acts as a stabilising force. In this case the curvature is described as bad. The current gradient drive becomes less important at the large n associated with ballooning modes.

It should be noted that Peeling-ballooning modes are decoupled when the triangularity is increased, since the magnetic shear depends highly on the shape of the plasma cross section, allowing access to second stability. Thus a "Bean" shape poloidal cross section would allow access to the second stability region [56].

5.2.3 ELM Stability Diagram

Although some of the features are common to all ELMs, there are also distinctive differences. Regarding the theoretical peeling-ballooning stability limit, the ELM stability diagram can be summarized as shown in Fig. 5.2. Here, the peeling-ballooning stability limit depends on the maximum value of the normalized pedestal pressure gradient [57],

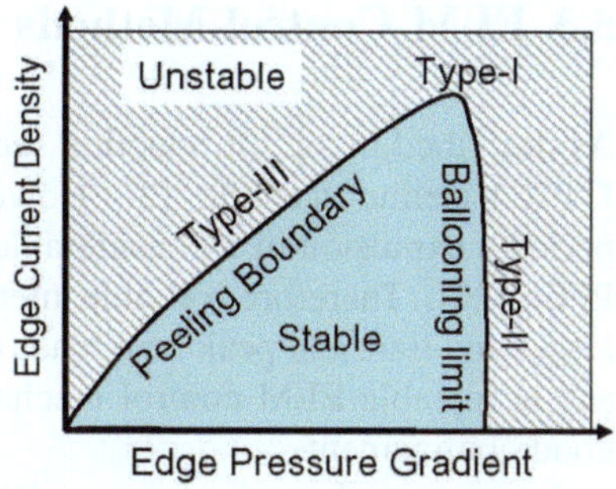

Fig. 5.2 ELM stability diagram

$$\alpha = -\frac{\mu_0}{2\pi^2}\frac{\partial p}{\partial \psi}\frac{\partial V}{\partial \psi}\left(\frac{V}{2\pi^2 R_0}\right)^{1/2}, \tag{5.2}$$

(where V is the plasma volume, p is the pressure, ψ is the poloidal magnetic flux and R_0 is the major radius of the plasma), and a normalized pedestal current density j_N^{ped}, which is taken to be the peak value of the parallel current density in the pedestal region normalized by the average parallel current density in the pedestal. In an H-mode plasma, the edge plasma current density has a substantial bootstrap current component which is parallel to the magnetic field. In a toroidal magnetic confinement system, the momentum transfer between trapped and passing electrons is balanced with that between passing electrons and ions. The bootstrap current, j_b, is generated from the difference in velocity between the passing ions and the passing electrons. This bootstrap current can be expressed as [58]

$$j_b = -\frac{\varepsilon^{1/2} n}{B_\theta}[2.44(T_e + T_i)\frac{1}{n}\frac{\partial n}{\partial r} + 0.69\frac{\partial T_e}{\partial r} - 0.42\frac{\partial T_i}{\partial r}], \tag{3}$$

Here, ε is the inverse aspect ratio r/R of the plasma minor radius (r) to the major radius (R), B_θ is the poloidal magnetic field, and n, T_e, T_i are the plasma density, electron and ion temperatures respectively. A more precise estimation of the bootstrap current can be made by simulating particle orbits [59].

The development of efficient MHD stability codes such as ELITE [60, 61] and MISHKA [62] has allowed detailed quantification of peeling-ballooning stability boundaries [63] and extensive and largely successful comparisons with observation [64–67]. To date, the peeling-ballooning ELM stability diagram has been commonly used to explain the trigger mechanism for those three conventional ELM types. For Type-I ELMs, the edge instability is both pressure and edge current density driven and close to the corner of the peeling-ballooning stability diagram or even beyond it. However, Type-II ELM instability is pressure driven and close to the "ballooning" limit, while Type-III ELM instability is current density driven and close to the peeling limit.

5.3 ELM Control Methods

As described in the introduction section, Type-I ELMy H-mode is foreseen as the ITER baseline scenario [2]. However, Type-I ELMs lead to an unacceptable periodic expulsion of a considerable fraction of the stored energy content onto the PFCs [20]. Therefore, reliable methods for the control of Type-I ELMy power losses and transient peak heat loads on the PFCs are required for ITER. In addition, any acceptable ELM control mechanism should ensure that the plasma is kept in good confinement.

Experimental results from smaller machines have shown that by tailoring the plasma shape and the gas injection, H-mode regimes [22] with small ELMs can be accessed. However, on larger machines such as JET, this is only possible over a limited range of plasma parameters [29]. Active methods of ELM control with the goal of reducing the power loading are therefore required.

To date, investigation of Type-I ELM control is mainly directed into three different strategies:

(i) **Radiating dispersion**: Dispersing the ELM energy loss by radiation before it is deposited at the PFCs.
(ii) **ELM suppression**: stabilizing the ELM instability by means of controlling either the pedestal pressure gradient or the edge current density below the peeling-ballooning ELM stability limit.
(iii) **ELM mitigation**: destabilizing the ELM instability, thus increasing ELM frequency and reducing the ELM energy losses, by applying either steady-state or transient perturbations at the plasma edge.

In the last ten years, several active methods, including (i) radiating divertors (impurity gas puffing) [68–70], (ii) magnetic triggering (vertical kicks) [71], (iii) pellet pace-making of ELMs [72], and (iv) resonant magnetic perturbation (RMP) fields [73, 74], have been developed for Type-I ELM suppression/mitigation.

5.3.1 Radiating Dispersion

Impurity gas seeding is considered as the primary technique for decreasing the inter-ELM heat loads onto the divertor, and large radiation fractions in the SOL and divertor (P_{rad}/P_{out} higher than 0.5 for ITER [75] and 0.9 for DEMO [76], where P_{out} is power exhausted to the SOL) are required. On the another hand, both increases in the ELM frequency and reduction in the ELM peak heat loads onto the divertor, so-called ELM mitigation, have been observed with impurity radiation in a Type-I ELMy H-mode plasma.

5.3.1.1 Type-I ELM Mitigation with Radiative Divertor

On JET, argon and nitrogen have been injected into Type-I ELMy H-modes up to radiative power fractions of 65 % to avoid a transition into Type-III ELMy H-mode or accumulation of argon in the plasma core [77]. The reduction in ELM energy due to radiative dissipation is about 20 % on the outer divertor target and about 25 % on the inner divertor target. Typically the confinement enhancement factor ($H98(y, 2)$) [2] is decreased from 1.0 to ~0.87–0.97, depending on the radiative power fraction and the plasma density [77]. As a consequence the energy deposited onto the divertor target is then decreased. Further increase in the radiative power fractions above 65 % causes a transition from Type-I ELMs to Type-III ELMs [78–80], the so-called radiating Type-III ELMy H-mode, and the plasma confinement becomes even worse ($H98(y, 2) < 0.85$).

In JT-60U, power handling by neon and argon seeding has been investigated in Type-I ELMy H-mode and reversed shear plasmas [81–85]. Good confinement ($H98(y,2) \geq 0.85$) was maintained up to a high density fraction of $\overline{n_e}/n_{GW} \sim 0.8 - 0.9$ (n_{GW} is the Greenwald density) and a high radiation fraction of $P_{rad}/P_{out} = 0.7$–0.9, but it was sustained for only ~2 s in a standard Type-I H-mode plasma, and then radiative collapse occurred. Sustainment of a high radiation for a longer period with impurity seeding has been achieved in ELMy H-mode plasma with an internal transport barrier (ITB). The best performance of the energy confinement of $H98(y,2) = 0.88 - 0.95$ with a large radiation fraction of 0.75–0.95 was obtained for the combination of the Ar and Ne seeding case. Similar to the JET observations, a large radiation loss in the main plasma caused a change in the ELM characteristics from Type-I to Type-III.

5.3.1.2 Open Questions

On JET, to date, a strong radiative dissipation effect with a reduction in ELM energy of more than 50 % is only observed in plasmas with small ELMs ($\Delta W_{ELM} < 25$ kJ) [77, 86]. This experimental observation is consistent with the prediction of edge transport modelling. The calculation results from the edge transport codes EDGE2D/EIRENE [87, 88] show that the nitrogen in the divertor region is ionized to higher ionization stages, which radiate less effectively. For ITER, the modelling by B2-EIRENE [89] predicts that significant radiative dissipation of the large Type-I ELM energy is unlikely. Only for very small ELMs ($\Delta W_{ELM} < 0.6$ MJ), the ELM energy loss is possibly dissipated by impurity seeding on ITER. Therefore, a combination of a radiating divertor with other active ELM control methods is essential for applicability to ITER.

5.3.2 Vertical Kicks

Actively controlled vertical kicks (fast vertical movements of the plasma column with pre-programmed frequency and amplitude) have demonstrated that the ELM frequency can be locked to the frequency of the externally imposed $n = 0$ magnetic perturbation, enabling ELM control (frequency and size). Here, n is the toroidal mode number. The fast modulating $n = 0$ magnetic perturbation is induced by a set of vertical stabilization coils with controllable frequency and amplitude. This method was first developed in TCV [71], and has also been successfully applied in AUG [90], JET [91] and NSTX [92].

5.3.2.1 ELM Control Using Vertical Kicks

In TCV, the fast vertical movement is provoked by the positional control coils inside the vacuum vessel. Experiments on the TCV tokamak showed synchronous modulation of the ELM activity, with ELMs occurring in bursts only during a rapid upward motion of the plasma approaching ~1 cm, in a single null Ohmically heated type-III ELMy H-mode plasma [71]. The modulation frequency was around 50 Hz. Phase synchronization was found between the ELM cycle and the external perturbation. The ELM frequency was found to track scans in the external driver frequency about its unperturbed value over a frequency range that increased with the amplitude of the perturbation. It was also found that scaling of the ELM amplitude with the inter-ELM period was preserved in the presence of the perturbation, so that the ELM amplitude could be controlled by altering the ELM period.

On AUG, active ELM control with vertical kicks has now also been demonstrated in the ITER-relevant type-I ELMy regime [90]. The experiments showed the ELM frequency becoming identical to the driving frequency in steady state for an applied motion of only about twice the value caused by an intrinsic ELM event. Figure 5.3 shows a typical example from this experiment. The ELM frequency is raised and locks to the frequency of imposed vertical (position z,$v \equiv dz/dt$) movement of the plasma column. The modest impact of magnetic triggering on particle and energy confinement has been seen when the ELM frequency increases from 43 to 56 Hz. A reduction in ELM-imposed energy losses is not pronounced, which might be due to the small increase in the ELM frequency by a factor of 1.3 only.

On NSTX, scenarios for ELM pacing with rapid vertical kicks induced by external coils have been developed for the first time in an ST plasma [92]. Similar to TCV results, ELMs are most likely to be triggered as the plasma moves up by ~2 cm [92]. When the frequency of pre-programmed kicks is set to 30 Hz, synchronization of the ELM cycle with the kicks has been observed. However, when the frequency of the kicks increases to 45 Hz, the ELMs were observed to become quite rapid (~60 Hz), though the synchronization with the plasma motion was unclear. These rapid ELMs resulted in a substantial reduction in the per-ELM

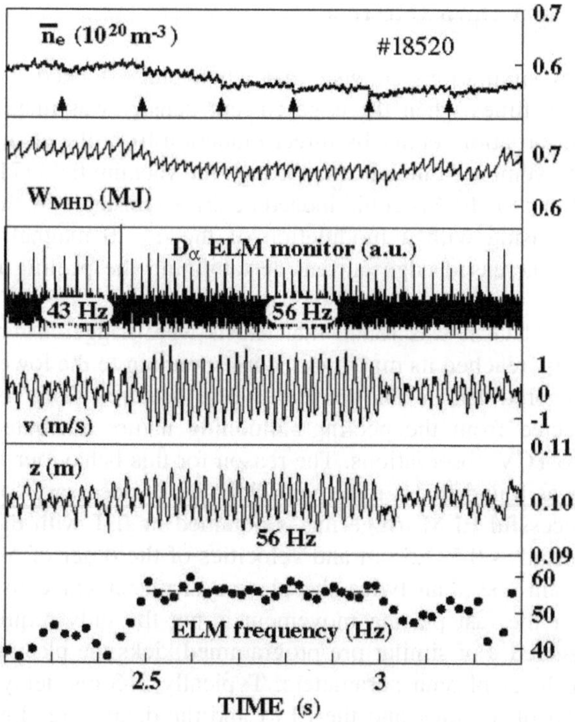

Fig. 5.3 Demonstration of magnetic ELM triggering in a type-I ELMy H-mode. The time traces plotted from top to bottom are line averaged density, plasma stored energy, D_α signal measured from the divertor region, plasma vertical motion velocity and position, and ELM frequency, respectively. The ELM frequency is raised and locks to the frequency of the imposed vertical movement of the plasma column. During the driving phase (2.5–3.0 s) only a modest effect on particle and energy confinement is found. The arrows indicate sawtooth crashes. From [90]

energy loss. Similar to the observation from AUG, the triggered ELMs with a kick frequency of 30 Hz on NSTX show type-I features.

Recently, ELM control with vertical kicks has been successfully applied in JET [91, 93]. The fast vertical movement of the plasma is controlled by the vertical stabilization controller, which has been modified to allow the application of a user defined voltage pulse (so called kick) at an adjustable frequency. Presently, the maximum value of the kick frequency is ~60 Hz due to technical constraints. Similar to the results observed from AUG, JET experimental results show that plasma kicks moving the plasma down towards the X-point can generate high frequency, synchronous ELMs in standard Type-I ELMy H-modes. With an application of the vertical kicks the pedestal density reduces by ~20 % when the ELM frequency is increased from 15 to 40 Hz, while a modest impact on the pedestal temperature is seen. This causes the edge pressure gradient to be reduced. The reduction in ELM size (up to a factor of 3) is accompanied by a minor (<10 %) reduction in the stored energy.

5.3.2.2 Physics and Open Questions

On TCV, DINA simulations showed that when phase locked, the ELMs were found to occur at times when the edge current density was increased under the action of the perturbation, either by direct induction from the changing current in the coil, or by movement of the plasma through the vacuum field [71]. Therefore, it was considered that the ELMs could indeed be driven unstable by the rapid upward motion of the plasma with a modulation of the $n = 0$ magnetic perturbation, attributed to the increased edge current destabilizing the peeling mode (current-driven modes).

However, on AUG, it was found that ELMs were triggered when the plasma downshift velocity reached its maximum, corresponding to the lowest edge current value. The triggered ELMs show clear type-I features. This is the opposite behaviour expected from the peeling-ballooning nature attributed to the ELM boundary and to TCV observations. The reason for this behaviour is not yet clear.

On JET, it was found that a minimum kick size is necessary for the trigger to occur [94]. Successful ELM triggering is obtained in JET with displacements of the current centroid $\sim$0.5–1.5 cm and velocities of the order of 5–10 m/s. Those values still remain less than twice the plasma displacement caused by intrinsic ELMs. However the fast plasma movement is not the only requirement for the ELM to be triggered. For similar pre-programmed kicks the plasma response also depends on the local plasma parameters. Typically 2–3 ms delays are observed between the start of the kick and the ELM and the delays are slightly higher for plasmas with higher pedestal temperature. An increase in the edge temperature will increase the current penetration time. This observation indicates that the modification of edge currents by the induced field and/or change in the plasma equilibrium might be the possible role for the ELM trigger. The sign of the induced edge current perturbation during vertically downward kicks is still under investigation on JET. To date, the precise physics mechanism is still unknown.

5.3.3 Pellet Pace-Making

Shallow injection of pellets has been thought to be a tool for increasing the Type-I ELMs frequency to the pellet injection rate, so called pellet pace-making of ELMs, and consequentially reduce the power load per ELM on the divertor target. Following pioneering experiments on AUG [95], this method has been also studied on JET [96] and DIII-D [97], and it is considered as one potential tool for ITER ELM control [98]. Recently, the first results of ELM pacing using small spherical lithium granules injected mechanically into H-mode discharges were reported on EAST [99].

5.3.3.1 ELM Control Using Frozen Pellet Injection

On AUG, the feasibility of ELM control by continuous injection of small frozen deuterium pellets from the high field side into H-mode discharges has been demonstrated [95]. Two examples with (#15420) and without (#15520) pellet injection are shown in Fig. 5.4. In this experiment, a pellet velocity of 560 m/s and a size of about 6×10^{19} D-atoms were selected. To avoid over-fuelling of the core plasma, a moderate repetition rate close to 20 Hz was chosen. It was found that the ELM frequency was increased and controlled to the imposed pellets injection frequency. Moreover, the impact of the high frequency triggered ELMs on plasma density and stored energy is obviously smaller.

On JET, ELM control experiments with pellet injection were performed. The pellet size was ~4 mm (about 4×10^{21} D atoms), and the injection velocities were 150–300 m/s. Although this pellet injector was not originally designed for ELM pacing, the preliminary experimental results demonstrated that prompt triggering of ELMs by pellet injection takes place on a relatively large device [96]. It was found that despite the fuelling effects of the large pellets, ELMs are triggered at any time in the ELM cycle. The pellet triggered ELM appears not to be different from an intrinsically occurring one in terms of energy loss. Experimental results also show that only a small fraction (less than 1 %) of the fuelling pellet mass is ablated when

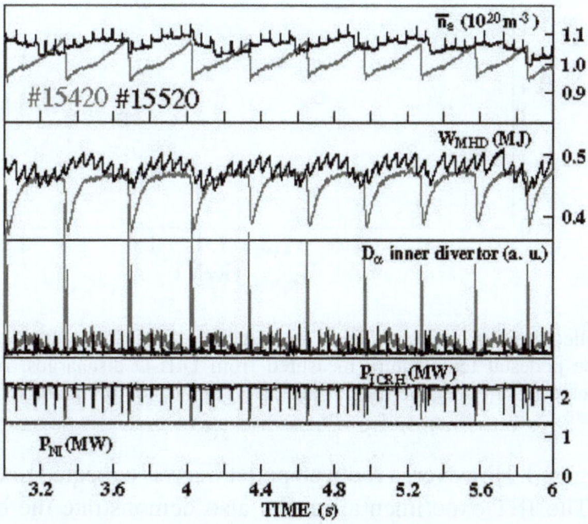

Fig. 5.4 Global temporal evolution of identical discharges without (#15420, *grey*) and with (#15520, *black*) pellet forced ELM control on AUG. The time traces plotted from top to bottom are line averaged density, plasma stored energy, D_α signal measured from the inner divertor region, and heating power from ICRH and NBI. A low frequency (~2 Hz) modulation of the plasma stored energy appears in the discharge with pellets injection. This corresponds to slow periodic core profile variations (this is not directly due to sawteeth, the frequency of sawteeth is slightly above 19 Hz in this discharge). The driving mechanism of this phenomenon appearing in the pellet ELM-pacing phase is not yet understood. From [95]

Fig. 5.5 The temperature on a flux surface just inside the separatrix. The density contour of twice the central density is shown in *yellow*. The initial density perturbation was injected into the pedestal on the *left-hand side* in the figure. From [101]

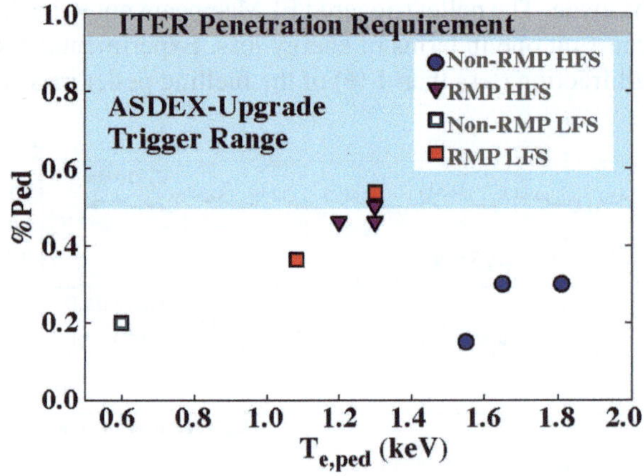

Fig. 5.6 The pellet position where an ELM is triggered, plotted as a fraction of the pedestal height, versus the pedestal temperature measured from DIII-D discharges. All the pellets are 1.8 mm and injected from either the inner wall (HFS) or outside midplane (LFS). From [97]

an ELM is triggered. However, a reduced pellet mass is expected to cause a reduced ablation rate. The JET experimental results also demonstrate the benefits of high field side pellet injection: the penetration to the pedestal top is achieved with a lower pellet mass [96].

Recently, pellets ELM pacing has been applied for an ITER-like shape H-mode plasma on DIII-D. In this experiment, deuterium pellets are injected with a frequency of 14 Hz by alternating pellets between two barrels (one from vertical port at the LFS and one from outside midplane). It was found that an ELM can be triggered as a local event and occurs just as the pellet reaches the separatrix

(± 1 cm) [100]. With pellet injection, the ELM frequency increases by a factor of 5 from 5 to 25 Hz. ELM energy losses were reduced from ~10 to ~3 % of total stored energy with this increase in ELM frequency. The resulting ELM frequency was larger than the pellet frequency. This indicates both a direct ELM trigger by each pellet and an indirect effect on the overall pedestal stability to ELMs from the multiple pellets. In comparison with the non pellet case, the energy confinement time was reduced by ~10 % in the case with pellet paced ELMs. No plasma density increase from the pellets was observed.

5.3.3.2 Physics of Pellet Pace-Making

Progress in understanding the physics of pellet-triggered ELMs is emerging from simulations using the nonlinear MHD code, JOREK [101]. The simulation of pellets injected into the H-mode pedestal shows that high pressure develops in the high density plasmoid, in this case the maximum pressure is ~5 times the pressure on axis. There is a strong initial growth of the low-n modes followed by a growth phase of the higher-n ballooning-like modes. The coupled toroidal harmonics lead to one single helical perturbation centred on the field line of the original pellet position as seen in Fig. 5.7, and there is some experimental evidence for this from JET [102] and AUG [103].

On AUG, it has been observed that ELMs were triggered in less than 200 μs after pellet arrival at the plasma edge. Only a fraction of the pellet had been ablated, forming a rather localized, three-dimensional plasmoid, at the time an ELM was triggered. The localized plasmoid drove the edge unstable well before the deposited mass was spread toroidally. The most probable location of the seed perturbation was found to be at the middle of the pedestal, where the plasma pressure gradient is large. The experimental results also show that the MHD signature of the ELMs appeared about 50 μs after the pellet reached the seed position [103].

On DIII-D, ELM pacing with fuelling-size pellets (containing 2×10^{20} atoms) has been achieved [97]. In this experiment, the fuelling pellets can be injected from either the high field side (HFS) or the low field side (LFS) with a slow speed of 100–150 m/s [104]. The pellet position, where an ELM was triggered, is plotted as a function of the pedestal temperature in Fig. 5.6 for some different operational conditions. Here, the pellet location is defined as a fraction of the pedestal height. It has been found that the ELMs can be triggered well before the fuelling pellet reaches half way up the pedestal. This result is somewhat in contrast to the observations from AUG where the most probable location of the HFS injected pellets to trigger an ELM was found to be at the middle of the pedestal [103].

5.3.3.3 Open Questions

Further experiments on AUG also demonstrated that over-fuelling of the core plasma with either a high pellet injection frequency or a large pellet size could

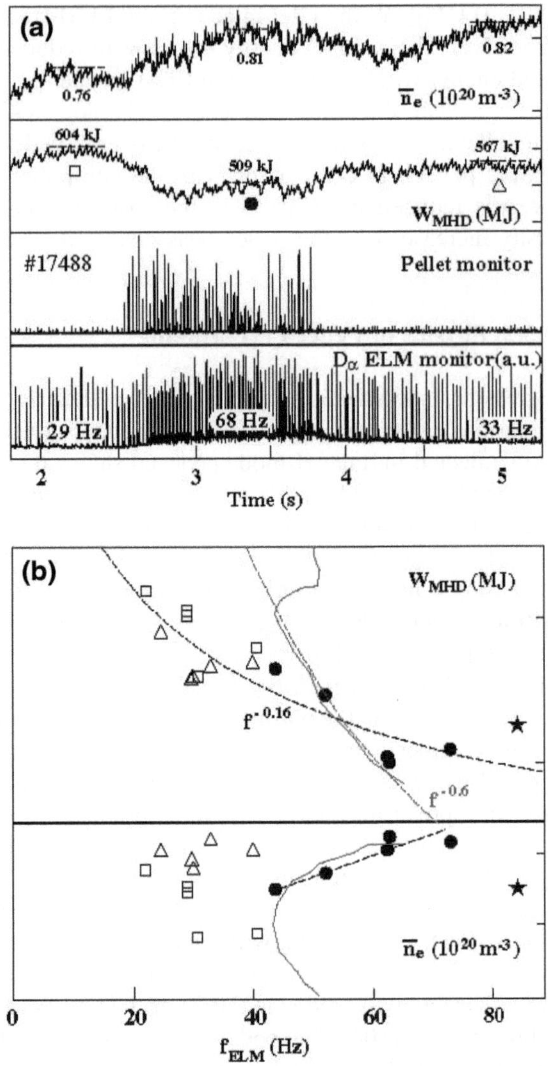

Fig. 5.7 **a** Temporal evolution of plasma density, stored energy, pellet and ELM monitor signals in a type-I ELMy discharge containing a pellet pace-making sequence (averaged ~68 Hz) and fuelling rate of ~17 × 10^{20} D/s. **b** Data compiled from a pellet injection frequency scan showing a mild degradation of confinement with increasing ELM frequency for pellet pace-making. From [72]

cause increasing the convective energy losses and then impact on the energy confinement [72]. One example from this experiment is shown in Fig. 5.5. A clear increase in the density was observed in the pellet injection phase. Reduction of the stored energy and increase in ELM frequency with respect to the pre-pellet phases is visible. Hence, optimization of the pellet size to avoid over-fuelling is important for ELM control with pellet pace-making. These results also indicate that shallow injection is needed for ELM pacing. On the other hand, it is still to be proven whether the concept will work for bigger machine sizes.

On DIII-D, a pellet dropper has been developed for testing the ability of small slow pellets to achieve ELM pacing [105]. The pellet dropper has obtained a

50 Hz injection rate and a 1 mm pellet size. Since pellets are dropped into a funnel that guides them into a tube directed to the top of the plasma using gravity as the accelerator, it results in slow injection speed, <10 m/s, to minimize penetration into the core plasma. Initial results with the pellet dropper indicate that fast ions deflect the pellets toroidally before penetrating deep enough to trigger ELMs [97]. However, a pellet dropper type device might be difficult to apply in ITER due to lack of vertical injection/dropping ports. Future experiments from the dropper on DIII-D will help to investigate the optimized pellet size and speed needed to trigger ELMs reliably in ITER.

In addition, a question, whether alternative materials for pellets could be used for ITER ELM control, is under consideration. On EAST, ELMs have been triggered successfully and in a controlled manner using low velocity Li granules during H-mode discharges [99]. Triggering of ELMs was accomplished using a simple rotating impeller to inject sub-millimeter size granules at speeds of a few tens of meters per second into the outer midplane of EAST. During the injection phase, ELMs were triggered with near 100 % efficiency. A wide range of granule penetration depths was observed, and a substantial fraction of the injected granules appeared to penetrate up to 50 % deeper than the nominal EAST H-mode pedestal width. This result confirmed that ELMs can be triggered using the injection of something other than frozen hydrogenic/deuterium pellets and allows for the contemplation of lithium or beryllium-based ELM pace-making on future fusion devices.

With respect to ITER, the pellet pace-making of ELMs is still to be investigated in the relevant parameter region of, e.g. lower collisionality or much higher pedestal temperature. In addition, an enhancement factor of the intrinsic ELM frequency by a factor of almost 20 is necessary. On the other hand, synergism of pellet fuelling and external ELM triggering seems quite possible. Using low-Z impurity pellets would allow for the decoupling of the ELM-triggering process from the plasma-fuelling process. Indeed, pellet pace-making might assist other ELM control techniques, helping to compensate the density pump-out caused by RMP ELM control/suppression.

5.3.4 Resonant Magnetic Perturbation Fields

Active control of ELMs by RMP fields offers an attractive method for next-generation tokamaks, e.g. ITER. The results obtained from DIII-D, JET, MAST, KSTAR, AUG and NSTX tokamaks have shown that magnetic field perturbations can either completely suppress ELMs [73], trigger small ELMs during ELM free periods, or affect the frequency and size of the type-I ELMs in a controllable way, preserving good global energy confinement [74, 106, 107].

5.3.4.1 Type-I ELM Suppression with RMPs

The first successful demonstration of the ELM suppression technique was reported from DIII-D, where the in-vessel coils (I-coils) have been employed. The I-coils consist of 12 single-turn loops, six above and six below the midplane (up-down symmetric) mounted on the low field side of the vessel. For the ELM suppression experiments, the upper and lower loops are operated with either the same current polarities (even parity) or opposite current polarities (odd parity), and induce a static perturbation field with a toroidal mode number n = 3.

On DIII-D, the experimental results show that the effectiveness of ELM suppression with n = 3 fields depends on q_{95} as well as plasma edge collisionality. In high collisionality ($\nu_e^* \approx 1$) H-mode plasmas [108], Type-I ELMs are suppressed and replaced by small intermittent events with a coherent amplitude modulation of 130 Hz during the application of odd parity n = 3 fields. In low collisionality ($\nu_e^* \leq 0.2$) H-mode plasmas, ELM suppression without small intermittent events is obtained with either even or odd parity n = 3 fields. To date, ELM suppression in those experiments on DIII-D is only observed in a narrow q_{95} window ranging from 3.5 to 3.9 with an even parity n = 3 field and ~7.2 with an odd parity n = 3 field. Outside this q_{95} range Type-I ELMs are mitigated (ELM frequency increased and ELM size decreased) by the applied n = 3 fields. These results indicate a resonant condition on the amplitude of RMPs for ELM suppression.

Furthermore, ELM suppression with n = 3 fields has been also observed in an ITER similar shaped (ISS) high triangularity ($\overline{\delta} = 0.53$) plasma with ITER relevant collisionalities $\nu_e^* \leq 0.2$ [109]. An example discharge with a q_{95} scan from this experiment is shown in Fig. 5.8. ELM suppression with an n = 3 field was observed within a resonant q_{95} window from 3.52 to 3.62. In comparison with a low trangularity ($\overline{\delta} = 0.26$) plasma, the width of the resonant q_{95} window, Δq_{95}, for ELM suppression is reduced by a factor of 3 from 0.3 to 0.1 in ISS plasmas. No clear shift of the resonant q_{95} window was observed between low and high triangularity plasmas. The ELM suppression threshold, which is defined as the minimal coil current required to suppress ELMs, increases by ~25 % in ISS plasmas with respect to that in low δ plasmas. With a vacuum assumption (no plasma response to the perturbation fields taken into account), the width of the island overlap region ($\Delta\Psi$) at the plasma edge has been calculated for an ISS case and a low δ case. It was found that the width of the island overlap region required for ELM suppression is the same. However, 25 % more I-coil current is needed to achieve this width in the ISS plasma. This is mainly due to the edge magnetic shear being higher in the ISS plasma (Figs. 5.9, 5.10 and 5.11).

Suppression of Type-I ELMs has also been established at high collisionality in many other devices, including AUG using an internal off mid-plane coil set (called B-coils) [110], KSTAR using a resonant n = 1 perturbation [111], and JET with the ITER-like wall (ILW) and an n = 2 perturbation [112]. The remaining small ELMs observed in those experiments have negligible ELM energy loss and ELM-peak heat flux onto the divertor plates.

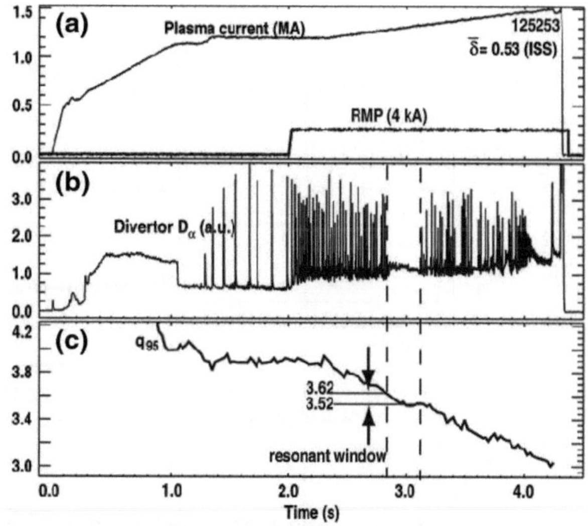

Fig. 5.8 **a** Plasma and I-coil current, **b** lower divertor D_α signal showing ELM suppression window and **c** q_{95} resonant window during a 4 kA RMP pulse in an ITER similar shape plasma with an average triangularity of 0.53 from a DIII-D ELM control discharge (#125253). From [109]

It should be noted that although the suppressed ELM state in high density/collisionality on AUG has many similarities to that observed in DIII-D, the access conditions are significantly different. First of all, type-I ELM suppression can be obtained with n = 1, 2, and 4 magnetic perturbations that are either resonant or non-resonant with a wide range of q_{95} on AUG [113]. The lack of ELM mitigation at n = 0 indicates that a helical perturbation is required, however, there is no apparent correlation with the amplitude of the resonant field component. Secondly, there is a window in edge density in which both resonant and non-resonant magnetic perturbations lead to type-I ELM suppression on AUG [110].

On JET, with a carbon first wall, no clear effect of either n = 1 or n = 2 fields on the ELMs was observed at high collisionality [114]. However, with the ILW suppression of type-I ELMs with an n = 2 perturbation has been achieved [112].

5.3.4.2 Type-I ELM Mitigation with RMPs

Experiments on JET have shown that type-I ELMy plasmas can be actively controlled by the application of static low n = 1 or 2 external magnetic perturbation fields produced by four external error field correction coils (EFCCs) [115] mounted far away from the plasma between the transformer limbs [74, 114, 116, 117]. An overview of a JET ELM control pulse is shown in Fig. 5.12. In these experiments, a type-I ELMy H-mode plasma with a high triangularity shape ($\delta_U = 0.45$ and $\delta_L = 0.4$) was sustained by neutral beam injection. The electron collisionality at the pedestal is ~0.2. The Chirikov parameter calculated using the experimental parameters and the vacuum approximation of the perturbation field is about 0.85 at $\Psi^{1/2} = 0.95$. During the EFCC phase the D_α signal measuring the

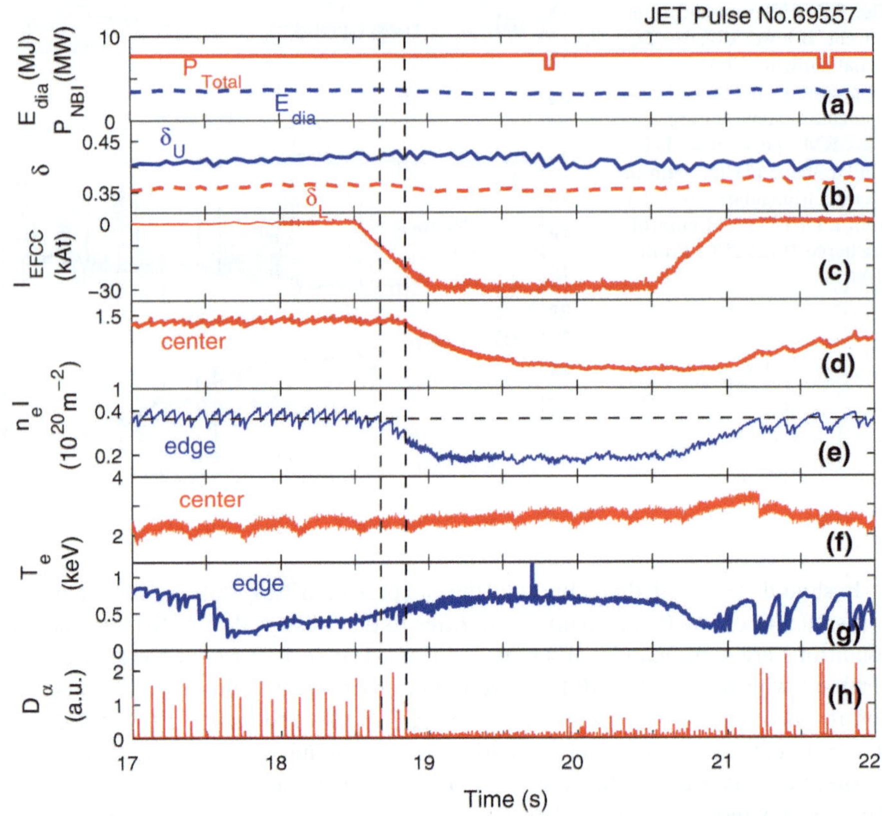

Fig. 5.9 Overview of a typical ELM control experiment on JET. The traces are **a** the total input power, P_{Total}, and the total stored energy, E_{dia}, **b** upper and lower plasma triangularity, δ_U, δ_L, **c** EFCC coil current, the line-integrated electron densities $n_e l$, measured with an interferometer along two lines of sight, **d** one close to the magnetic axis and **e** the other near the pedestal top, electron temperature **f** in the core and **g** near the pedestal top, **h** the D_α signal measured at the inner divertor. From [114]

ELMs showed a strong reduction in amplitude. The ELM frequency increased from ~30 to ~120 Hz, while the periodic energy loss due to the ELM crashes normalized to the total stored energy, $\Delta W/W$, measured by a fast diamagnetic loop, indicates a strong reduction from ~8 % to values below the noise level (<2 %) of the diagnostic. A modest drop (a few per cent) in the total stored energy has been observed during the ELM control phase with the EFCCs. However, when normalized to the $IPB98(y,2)$ scaling the confinement time shows almost no reduction.

Reduction of both the peak particle and heat fluxes during the ELM crash has been observed in ELM control experiments with the application of an n = 1 field on JET [114, 117, 118]. The heat fluxes were measured by Langmuir probes embedded in the divertor tiles and a fast IR camera viewing the divertor target

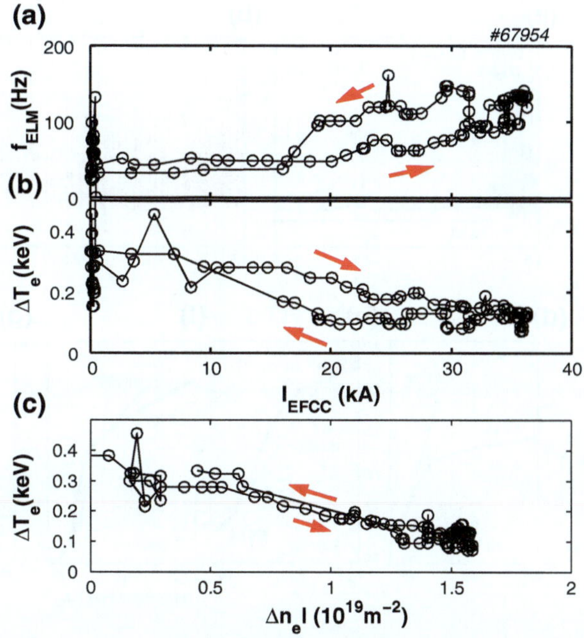

Fig. 5.10 **a** Frequency of the ELMs, f_{ELM}, and **b** the amplitude of ΔT_e as a function of I_{EFCC}. **c** Dependence of ΔT_e on the density drop due to the pump-out effect. From [74]

plates. The reduction in heat flux is mainly due to the drop in particle flux rather than the change in the electron temperature. A reduction in the particle flux has also been observed on the outboard limiter. The fraction of ELMs with a larger particle flux reduces significantly, which benefits the life time of the JET limiter. In addition, a clear reduction of the amplitude of the ELM spikes in the total radiation signal measured by bolometry during ELM mitigation with an n = 1 field has also been observed. These results can be explained in the case of less erosion of carbon from the target when the ELM size is reduced by means of application of the n = 1 field.

JET experimental results also demonstrate that ELM frequency and size can be actively controlled by adjusting the amplitude of the perturbation field/coil current [74]. The dependence of both, f_{ELM} and the amplitude of ΔT_e, on I_{EFCC} is shown in Fig. 5.13. This dependence is found to be different between ramp-up and ramp-down of I_{EFCC} which could be due to a hysteresis effect or the non-stationary nature of the experiment.

Active control of type-I ELMs with n = 1 fields has been developed toward more ITER-relevant configurations and parameters in a wide operational space of plasma triangularity (δ up to 0.45), q_{95} (3–4.8) and normalized beta (β_N up to 3.0) [116, 119] on JET. A similar wide operational window of q_{95} has also been obtained for ELM control with n = 2 fields [116] on JET.

ELM control is also seen with n = 2 fields from external coils in MAST and with n = 3, 4 and 6 fields from in-vessel coils in MAST [106] and with n = 3 fields in DIII-D [120]. On MAST, an increase of the ELM frequency from 500 to

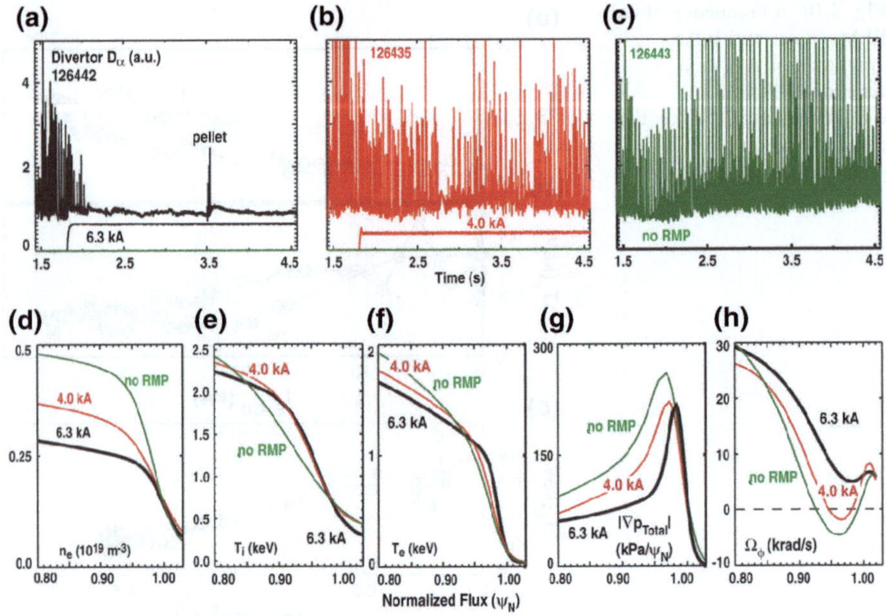

Fig. 5.11 Lower divertor D_α signals showing the ELM characteristics in similar ISS plasmas with n = 3 I-coil currents of **a** 6.3 kA, **b** 4.0 kA and **c** 0 kA. Pedestal profiles showing the **d** density, **e** ion temperature, **f** electron temperature, **g** absolute value of the total pressure gradient and **h** C6 + toroidal rotation for the three I-coil currents (6.3, 4.0 and 0 kA) shown in **a**, **b** and **c**. From [109]

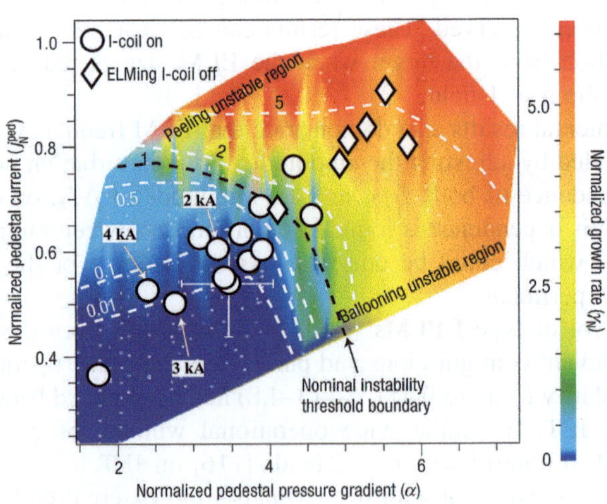

Fig. 5.12 The normalized Peeling-Ballooning mode growth rates of an ELM unstable H-mode plasma (*diamonds*) and RMP-induced ELM-free (*circles*) plasma. From [73]

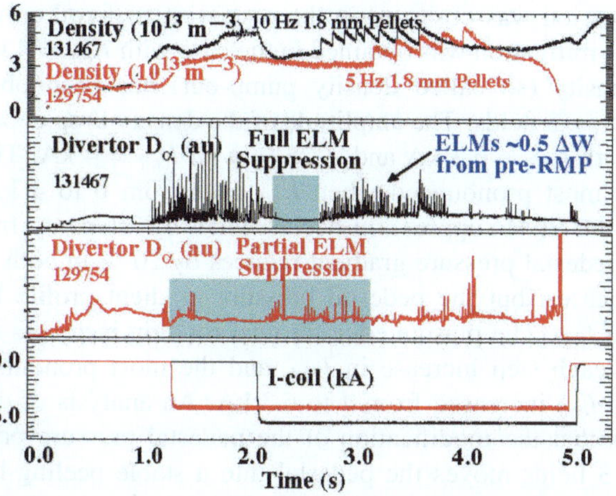

Fig. 5.13 Two cases of 1.8 mm fuelling pellets injected from the inner wall into H-mode plasmas with RMP applied. The time traces plotted are plasma density (*top row*), D_α signals (the *second* and the *third rows*), and currents applied in I-coils (*bottom*). In the fully ELM suppressed case (no. 131467) with 10 Hz pellets, frequent ELMs are observed after the pellets start. The partial ELM suppression case ((no. 129754) with a lower coil current) shows only a few ELMs with the 5 Hz pellets. From [97]

700 Hz has been observed during the application of n = 2 fields induced by the external coils. In this experiment, the Chirikov parameter is greater than 1 where $\sqrt{\Psi_{pol}} > 0.91$, i.e. the region of flux space from the edge of the plasma that is thought to be stochastic satisfies $\Delta\Psi_{pol} > 0.17$ based on vacuum field calculations.

On DIII-D, outside the q_{95} resonance window or below the RMP amplitude threshold for ELM suppression, the DIII-D results show that Type-I ELMs are mitigated. The q_{95} operating window for ELM control with n = 3 fields on DIII-D appears to be much wider than for ELM suppression.

An additional advantage of RMP coils is that they could also provide a technique for ELM pacing. Experiments on NSTX show reliable ELM triggering each time the n = 3 RMP coil current is pulsed, with no apparent loss of stored energy [121]. Similar results are also observed on DIII-D, which demonstrates that a factor of two increase in the ELM frequency can be achieved. However, at higher ELM frequencies, it appears that the triggered ELM energy loss is not inversely proportional to the ELM frequency.

5.3.4.3 RMP Effects on the Pedestal Profiles and Stability

RMP effects on the pedestal profiles have been studied in order to compare ELM suppression with stability theory on DIII-D [109]. Figure 5.9 shows a comparison of the ELM behaviour and pedestal profiles with various I-coil currents in ISS plasmas.

ELM suppression was observed in the case with an I-coil current (I_{coil}) of 6.3 kA, while ELM mitigation was obtained in the case with $I_{coil} = 4$ kA. Reduction of the plasma density (so called density pump-out) has been observed with the application of n = 3 fields. The amplitude of the density drop depends on I_{coil}, and it is ~40 % with $I_{coil} = 6.3$ kA and ~25 % with $I_{coil} = 4$ kA. The change in ion temperature is most pronounced when I_{coil} goes from 0 to 4 kA, however, the bifurcation of the ELM suppression occurs when I_{coil} steps up from 4 to 6.3 kA. The maximal pedestal pressure gradient reduces by 20 % in both $I_{coil} = 4$ kA and $I_{coil} = 6.3$ kA cases, but the pedestal pressure gradient profile becomes narrow with increasing I_{coil}. The plasma edge toroidal rotation increases in the co-current direction with each step increase in I_{coil}, and the most pronounced change was observed when I_{coil} increased from 4 to 6.3 kA. An analysis of the edge pedestal stability shows that the modification of the pedestal pressure profile with application of n = 3 fields moves the pedestal into a stable peeling-ballooning mode operating region as seen in Fig. 5.10 [73]. This observation indicates that ELM suppression with RMP fields occurs due to a direct effect of the RMP on the plasma edge transport rather than an interaction with the ELM eigenmode.

On JET, the electron density at the pedestal top decreased by ~20 % due to the so-called density pump-out [122] during the application of the n = 1 field, while the pedestal electron temperature increased keeping the pedestal pressure almost constant. However, the pedestal pressure gradient obtained from the derivative of the fitted curve shows that the maximum pressure gradient in the profile is decreased by 20 % during the application of the n = 1 field, and the edge pressure barrier is 20 % wider [123]. This is an effect mostly ascribable to the strong decrease in the n_e pedestal height with an almost unvaried width. The minimal amplitude of perturbation field required for an increase in ELM frequency, the so-called ELM mitigation threshold, has a higher value than the density pump-out threshold. In addition, previous JET experimental results also show that the ELM mitigation threshold increased with decreasing q_{95} [116]. Stability analysis of controlled ELMs suggests that the operational point with an n = 1 perturbation field moves from the intermediate-n peeling-ballooning boundary to the low-n peeling boundary, and the radial width of the most unstable mode is reduced from ~3 % down to ~1 % of the normalized minor radius [124].

Compensation of the density pump-out with additional fuelling has been performed during DIII-D ELM suppression experiments. Partial compensation of density pump-out during RMP ELM suppression can be achieved by means of gas puffing. However, strong fuelling with gas puffing or with pellet injection can bring small ELMs back. One example is shown in Fig. 5.11. Thus further optimization of the RMP technique is needed to be able to suppress ELMs fully during pellet fuelled H-mode scenarios for application to ITER [97].

On JET, compensation of the density pump-out has been also investigated using either gas fuelling or pellet injection in low triangularity H-mode plasmas [114, 117, 125]. Although the ELM frequency stays high with n = 1 fields, no recovery of stored energy is observed. An optimized fuelling rate to compensate or the density pump-out effect has been identified, and it depends on the plasma configuration.

5.3.4.4 Non-resonant Magnetic Braking

Plasma rotation braking effects with applied perturbation fields have been observed on DIII-D, NSTX, and JET. It is well known that the plasma rotation is a significant concern for the control of MHD instabilities in a tokamak, because of its stabilizing effects on resistive wall modes [126] and neoclassical tearing modes [127]. Plasma rotation is also important for increasing the field penetration threshold [128], and hence improving the error field tolerance in tokamaks. Therefore, understanding the plasma braking mechanism with an RMP field becomes an important issue for optimising the application of magnetic perturbations.

Neoclassical toroidal viscosity (NTV) theory has been developed to describe the effects of non-axisymmetric magnetic perturbations on the plasma rotation in the collisionless [129] and plateau [130] regimes. Here, the names of the regimes indicate the dependence of the transport on the collisionality (ν). The collisionless regime ($\nu/\varepsilon < \sqrt{\varepsilon}/\omega_t$, where ε is the amplitude of the $\cos\theta$ component of the magnetic field, θ is the poloidal angle, $\omega_t = v_t/R_0 q$ is the transit frequency, v_t is the thermal velocity, R_0 is major radius of magnetic axis, and q is the safety factor) can be further divided into two main regimes: the $1/\nu$ regime and the ν regime [129]. Determination of $1/\nu$ ($|q\omega_E| < \nu/\varepsilon < \sqrt{\varepsilon}/\omega_t$) and ν ($|q\omega_E| > \nu/\varepsilon$) regimes is according to the relationship between the values of the collisionality and $\vec{E} \times \vec{B}$ drift frequency $\omega_E = E_\rho/\rho B_0$ (where E_ρ is the radial electron field, B_0 is the magnetic field strength on the magnetic axis and ρ is the normalized flux-surface label). The breaking of the toroidal magnetic symmetry due to the application of the non-axisymmetric magnetic perturbations causes a nonambipolar radial particle flux and hence the NTV. A generalized analytic solution of NTV and the influence of the NTV torque on the field penetration process has been investigated [131]. A neoclassical 'offset' rotation has been predicted using the NTV theory [129, 132].

Recently, good agreement between the calculated torque from NTV theory in the $1/\nu$ regime and the observed torque induced by the n = 3 fields has been reported on NSTX [133]. The observed rotation damping time on DIII-D was close to the NTV damping time in the 1/ν regime, whereas the NTV damping time in the ν regime is at least two orders of magnitude larger [134]. However, it was shown that the plasma in DIII-D is mainly in the ν regime. An increase in the NTV torque in the ν regime has been predicted by including the collisional boundary layer effect (the boundary between the trapped and untrapped particles) [135]. The NTV torque from the boundary layer contribution scales like $\sqrt{\nu}$. The variation of the magnetic field strength should be calculated in the distorted magnetic flux surface [129]. Hence, one should take a Lagrangian variation of the magnetic field strength in NTV torque calculations. It was found that the Lagrangian variation of the magnetic field strength is about one order higher than the Eulerian part [131]. The damping rate of the plasma rotation calculated based on NTV theory using a Lagrangian variation of the magnetic field strength agrees well with that observed on NSTX and DIII-D [136]. The existence of the neoclassical offset rotation with an n = 3 perturbation field was observed on DIII-D [137]. On JET, similar plasma braking effects have been observed with n = 1 and n = 2 external fields when the same EFCC coil current was applied [125]. The torque profile induced by the n = 1 field has been measured, and it has a global profile. The maximal value of this torque is in the plasma core region and it is about half of the neutral beam injection torque. The calculation shows that the plasma is mainly in the v regime in the plasma core, but is close to the transition between the $1/v$ and v regimes. A discrepancy between the observed torque and the torque predicted by NTV theory has been found. Although, both the boundary layer effect and using a Lagrangian variation of the magnetic field strength can reduce this discrepancy. However, the magnitude of the NTV torque calculated in the v regime is still at least one order smaller than the observed one [129, 138].

5.3.4.5 Strike Point Splitting

On MAST, the 3D "manifold" structures, which are induced by an interaction between the applied 3D RMP fields and the 2D plasma equilibrium fields, have been observed using filtered visible imaging [139]. These manifold structures are particularly complex near to the X-point. The manifolds form lobes that are stretched radially both outwards and inwards. Some of these lobes can intersect the divertor target and result in the strike point splitting often observed during RMP experiments [140]. A clear correlation is observed between the size of the lobe length and the change in ELM frequency, which may suggest that the lobes themselves are having a direct impact on the stability of the edge plasma to peeling ballooning modes [141].

Recently, mitigation of type-I ELMs was observed with the application of an n = 2 field in H-mode plasmas on JET with the ILW. In moderate collisionality ($v^*_{e,ped} \sim 0.8$) H-mode plasmas, similar to previous results with the C-wall, both an increase in the ELM frequency and density pump-out were observed during the application of the n = 2 field. There are two new observations compared with the C-wall results. Firstly, the effect of ELM mitigation with the n = 2 field was seen to saturate so that the ELM frequency did not further increase above a certain level of n = 2 magnetic perturbations. Secondly splitting of the outer strike point during the ELM crash was seen, resulting in mitigation of the maximal ELM peak heat fluxes on the divertor region.

Splitting of the outer strike point (SP), appearing as multiple peaks in the ELM heat flux profiles measured by a fast IR camera along the outer divertor plate as shown in Fig. 5.14, has been observed during the application of the n = 2 fields on JET with the ILW. Similar findings have been reported on DIII-D in the presence of n = 3 RMP fields [142]. These multiple peaks in the heat flux profile are observed only during a mitigated ELM crash when a certain I_{EFCC} threshold is reached. The preliminary results indicate that this I_{EFCC} threshold for the appearance of splitting of the outer SP during the ELM crash is a similar level as that occurring for the saturation effect of the plasma response.

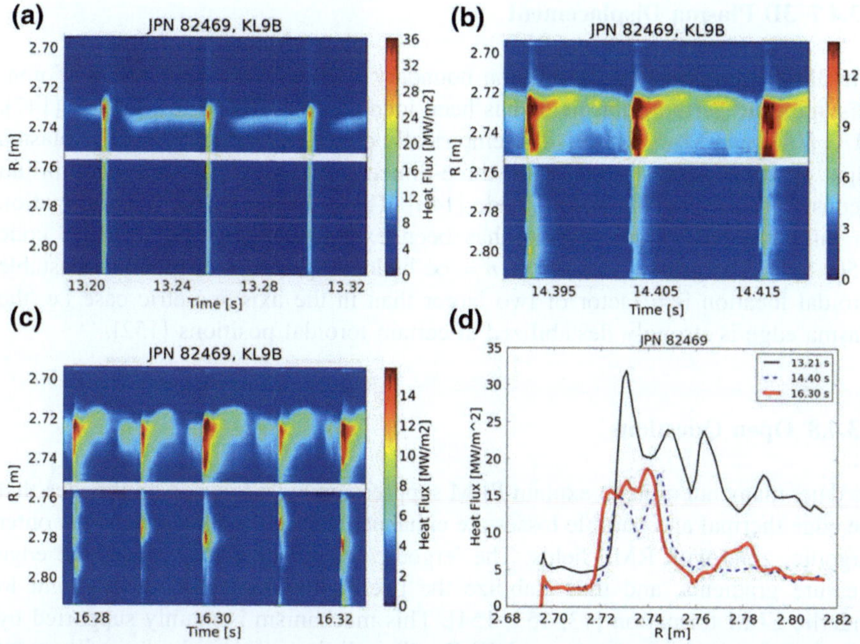

Fig. 5.14 Extended time traces of the heat flux distribution on the outer divertor plate in the phases (*upper left*) without n = 2 field, (*upper right*) with $I_{EFCC} = 44$ kA and (*lower left*) with $I_{EFCC} = 88$ kA. (*lower right*) ELM peak heat flux profiles along the outer divertor. From [112]

5.3.4.6 Multiple Resonances Effects

Multiple resonances in f_{ELM} as a function of the edge safety factor q_{95} have been observed for the first time with an applied low n (= 1, 2) field on the JET tokamak [143, 144]. Without an n = 1 field applied, f_{ELM} increases slightly from 20 to 30 Hz by varying the q_{95} from 4 to 5 in a type-I ELMy H-mode plasma. However, with an n = 1 field applied, a strong increase in f_{ELM} by a factor of 4–5 has been observed with resonant q_{95} values, while the f_{ELM} increased only by a factor of 2 for non-resonant values. The Chirikov parameter calculated using the experimental parameters and the vacuum approximation of the perturbation field indicates that the ergodisation zone may only appear at the far plasma boundary ($\sqrt{\Psi} > 0.95$). The mechanism of edge ergodisation, which is used to explain the results of the ELM suppression with n = 3 fields on DIII-D, may explain the global effect of the n = 1 field on f_{ELM} on JET, but it cannot explain the multi-resonance effect observed with the low n fields. A model, which assumes that the ELM width is determined by a localized relaxation triggered by an unstable ideal external peeling mode, can qualitatively predict the observed resonances when low n fields are applied [145].

5.3.4.7 3D Plasma Displacement

The 3D displacement of the plasma boundary caused by an application of non-axisymmetric perturbation fields has been identified on JET [146], DIII-D [147], AUG [148] and MAST [149]. The empirically observed corrugation of the plasma edge position agrees well with three-dimensional ideal plasma equilibrium reconstruction using the VMEC code [146]. The influence of the 3D corrugation on infinite-n ballooning stability has been examined using the COBRA code [150, 151]. The growth rate of the $n = \infty$ ballooning modes at the most unstable toroidal location is a factor of two larger than in the axisymmetric case i.e. the plasma edge is strongly destabilized at certain toroidal positions [152].

5.3.4.8 Open Questions

To date, many attempts to explain ELM suppression have focused on the idea that the edge thermal and particle losses are enhanced due to the formation of an outer 'ergodic' zone with RMP fields. The 'ergodic' boundary would reduce the edge pressure gradients, and thus stabilize the peeling-ballooning modes thought to underlie ELM formation [73, 153, 154]. This mechanism is mainly supported by two experimental results from DIII-D: (i) splitting of the inner strike-point observed during the RMP ELM suppression phase [155, 156]; and (ii) spin-up of the edge plasma rotation in the co-current direction and changing of the plasma edge electric field to a more positive value due to larger losses of electrons than ions with an ergodic boundary [157]. However, either bulk plasma or diamagnetic rotation [158, 159] can screen the RMP fields from the resonant magnetic flux surfaces. Many calculations of the Chirikov parameter [160] or overlapping of resonant magnetic islands employ a vacuum assumption, which neglects the plasma response (rotational screening effect and equilibrium effect).

Recent experimental results from ELM suppression in Hybrid H-mode plasmas on DIII-D show that small ELMs can appear when the edge safety factor is outside the resonance window or when the H-mode pedestal is perturbed, unrelated to P–B stability [161]. This result indicates that a 3D equilibrium modification may be important for understanding the effect of RMPs on the edge pedestal stability [162]. On the other hand, changes in the edge-plasma profiles during the RMP ELM suppression phase are indicative of a significant alteration in the particle balance with a relatively small change in the energy transport. However, this result can not be explained by stochastic transport theory.

On DIII-D, large Type-I ELMs were also suppressed by $n = 3$ fields induced by a single row of off-midplane in-vessel coils in plasmas with an ISS at the ITER pedestal collisionality ($\nu_e^* \sim 0.1$), and low edge safety factor ($q_{95} = 3.6$) [163]. The perturbation spectrum induced by a single row of coils is different to that with both upper and lower coils. Based on the results from DIII-D, the correlation of island overlap width and ELM suppression is observed as a useful figure of merit

(Chirikov parameter larger than 1 at $\sqrt{\Psi} \geq 0.925$) to guide the design of the ITER RMP coils [153, 164–168]. The width of the region in the plasma edge with good overlap of the RMP magnetic islands (from vacuum field calculations) is an ordering parameter for the maximum ELM size during the RMP for either RMPs from one row or two poloidally separated rows of internal n = 3 RMP coils, although outliers in the ordering point to important contributions from additional physics mechanisms.

In addition, recent experimental results from both MAST (n = 3, in-vessel off-midplane coils) [106] and JET (n = 1 and 2, external midplane coils) [74] show that ELM suppression is not achieved even with a Chrikov parameter larger than 1 at $\sqrt{\Psi_{pol}} > 0.925$. Those results suggest that ELM suppression using RMPs may also depend on the perturbation spectrum (not only the mode number, but also the ratio of the resonant to the non-resonant components). Although complete ELM suppression at low collisionality with RMPs is only obtained on a single device (DIII-D), the application of RMPs on other machines has either triggered ELMs in otherwise ELM free H-mode periods or increased the ELM frequency in regularly ELM-ing discharges.

Although the mechanism of ELM control with RMPs is not fully understood yet, it has been examined in a wide operational window in many different devices. Further optimisation of the magnetic perturbation with less reduction of the plasma performance and understanding the underlying physics are essential for future investigations.

5.3.4.9 Future RMP ELM Control/Suppression Experiments

The coil systems in different devices employ differing designs, e.g. internal off-midplane coils (DIII-D, MAST) and external midplane coils (JET, MAST, NSTX), and provide different poloidal, m, and toroidal, n, mode number spectra as well as differing radial profiles. Nevertheless, there are common observations like plasma density pump-out and magnetic rotation braking. To date, complete ELM suppression at low collisionality has only been achieved within a narrow operational window of q_{95} by using either n = 2 or 3 fields induced by in-vessel off-midplane coils in DIII-D. Guided by the DIII-D results, a set of ELM control coils has been designed for ITER [153].

To provide a more reliable method for ELM control/suppression, substantial ongoing efforts are presently contributing to two major topics: **the application** and **the physics mechanism**.

- **Application**: This includes (i) the influence of the RMPs on the ELM frequency and size, ELM peak heat loads, plasma confinement; (ii) the operational window, which is defined by the locked mode threshold and the minimal perturbation necessary for an increase in ELM frequency; (iii) rotation braking with magnetic perturbations; (iv) density dump-out compensation; (v) application for ITER-like scenarios.

- **Physics mechanism**: This includes (i) the field penetration process including edge ergodisation, field screening by the plasma rotation and the 3D equilibrium effect; (ii) ELM stability; (iii) Dynamics of edge pedestal profiles; and (iv) multi-resonance effects.

To investigate the physics mechanism of RMP ELM control/suppression, several new RMP coil systems on the world's tokamaks are planned. DIII-D proposes to have 36 new coils (3 rows, and each row has 12 coils) on the inboard wall [120]. This system is very flexible for adjusting the perturbation spectrum. The maximal toroidal mode number of the perturbation fields is 6. When used in combination with the present two rows of coils above and below the outer midplane, the new inboard coils would ultimately allow variation of the RMP radial and poloidal localization plus the capability to separately rotate either n = 3 or n = 4 RMPs toroidally for tests of field penetration and heat flux spreading models [120]. On EAST, a set of 2 × 8 ELM control coils including the ability to rotate the perturbation up to n = 3 has been installed at the end of 2013. ELM control experiments with n = 2, 3 and 4 fields are planned in the coming experimental campaign on EAST. Two rows of internal coils above the mid-plane are being proposed for JET (one with 8 coils, the other with 24) [169]. This system is flexible enough to be able to adjust both the Chirikov parameter and the magnetic perturbation spectrum independently in ITER-relevant scenarios. NSTX is also proposing [170] two rows of internal coils above and below the mid-plane, each one having 12 coils. Feasibility studies of installing in-vessel coils on NSTX and JET are currently ongoing. Future experiments from those devices offer the possibility of examining the underlying mechanism of complete ELM suppression and ELM control.

5.4 New Control Schemes

5.4.1 SMBI ELM Mitigation

ELM mitigation by Supersonic Molecular Beam Injection (SMBI) has been observed on HL-2A [171], KSTAR [172] and EAST [173]. When SMBI cold particles are deposited within the pedestal, consequently, the ELM frequency increases and the ELM amplitude decreases for a finite duration after SMBI. Increases in frequency of $f_{ELM}^{SMBI}/f_{ELM}^{0} \sim 2 - 3.5$ are achieved on HL-2A. In principle, the penetration depth of SMBI is shallower than that of pellet injection, therefore SMBI need not immediately provoke an ELM burst. The results from KSTAR suggest that shallow particle deposition by SMBI changes the local characteristics of the pedestal transport dynamics and the local density profile structure, and consequently alters the ELM dynamics. In addition, pellet injection is highly coherent with individual ELM triggering but the SMBI pulses are not directly correlated with individual ELMs.

5.4.2 ELM Mitigation with Lower Hybrid Waves

Recent results from the Experimental Advanced Superconducting Tokamak (EAST) [174] show that lower hybrid waves (LHWs) provide an effective means of mitigating or suppressing ELMs (seen in Fig. 5.15) by inducing a profound change in the magnetic topology, similar to the effect previously observed with RMPs [112, 155]. This has been demonstrated to be due to the formation of helical current filaments (HCFs) flowing along field lines in the scrape-off layer induced by LHWs. Magnetic perturbations induced by the currents flowing in these edge helical filament structures have been measured by a set of Mirnov coils during the modulation of LHWs. The change in the magnetic topology has been qualitatively modelled by considering the HCFs in a field-line-tracing code. Because of the geometric effect of the LHW antenna, the perturbation fields induced by the HCFs are dominated by the $n = 1$ components. The magnetic perturbation spectrum calculated based on the experimental parameters indicates a good resonant feature, whereby the plasma edge resonant surfaces are well aligned on the ridge of the spectrum. This is consistent with experimental observations of a weak q_{95} dependence of ELM mitigation with LHWs on EAST [175]. This result offers a new attractive means of optimizing the heat load distribution on the divertor plates and suppressing or mitigating the large transient peak heat and particle loads due to ELMs for the next-step fusion reactors (ITER and DEMO).

5.5 Combination of Different Methods

To gain a higher fusion energy production and to secure the plasma first wall components, a fusion device, i.e. ITER, needs to be operated in a high density (close to the Greenwald limit), high edge radiation and steady state small ELM or ELM-free regime. This may need a combination of several different methods of large ELM control/suppression.

On JET, the preliminary experimental results show that the combined application of an $n = 1$ field and vertical kicks ($n = 0$) leads to a reduction of the threshold perturbation level necessary for ELM mitigation to occur [94]. It could reduce the technical difficulty of accessing further high ELM frequency control using a safe size of vertical kicks.

ELM control with a combination of an $n = 1$ field and pellet injection has been demonstrated on JET. The ELM frequency can be increased without any drop in plasma density. During the application of the $n = 1$ field, the ELM frequency increases, even though some of the ELMs are triggered by pellet injection. Further RMP ELM control experiments with a radiating divertor are planned.

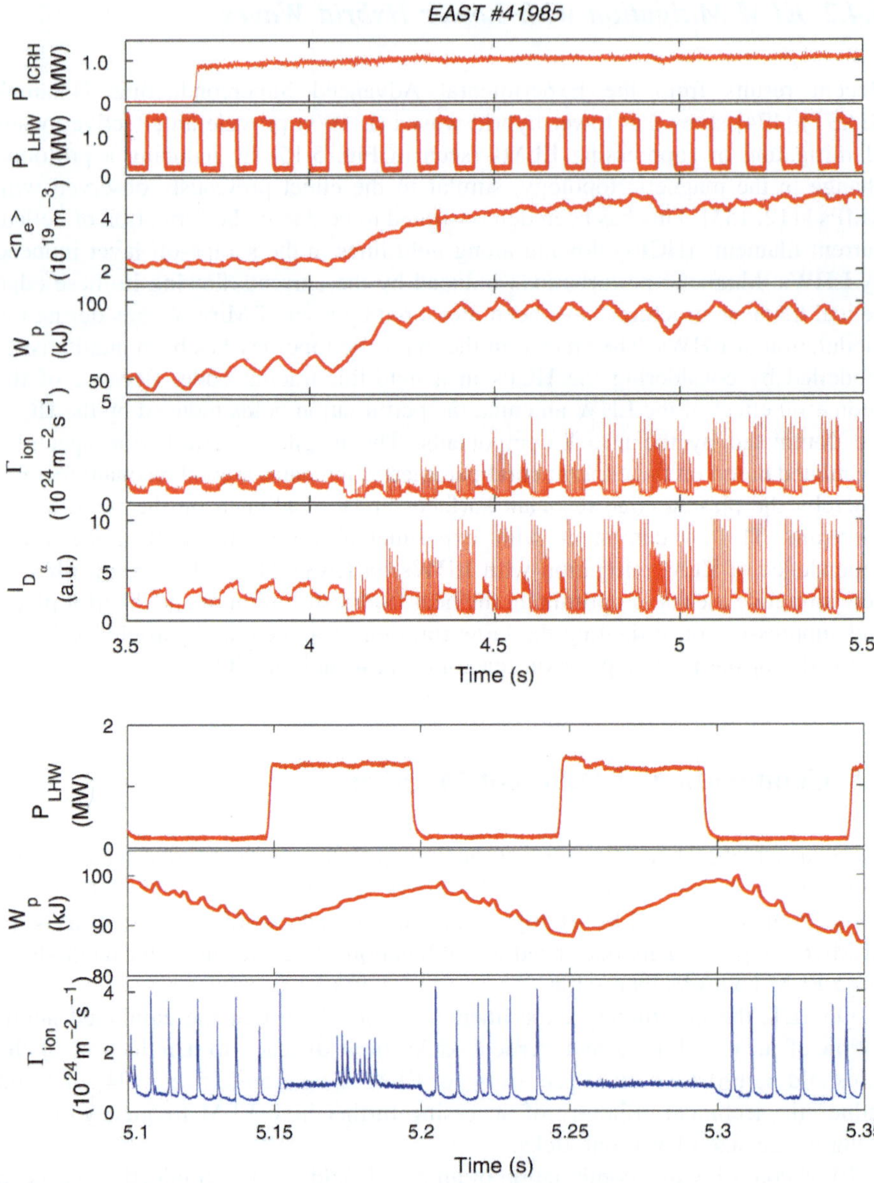

Fig. 5.15 Effect of LHW power modulation on ELMs. The time traces from top to bottom are injected power from ICRH and LHW, central line-averaged density, plasma stored energy, peak particle flux and intensity of D_α emissions in the outer divertor. At the bottom is a focused view of LHW power, stored energy and peak particle flux in the outer divertor. From [174]

5.6 Summary

Erosion and damage caused by large ELMs is one of major hurdles on the route towards achieving magnetic fusion in a reactor scale machine. Scaling predicts that the large ELM energy in ITER will exceed the acceptable level by a factor of $\sim$20. Clearly, a reliable ELM control method is needed.

In the past years, several ELM control/suppression methods have been developed. Some important results are summarized below:

- Both experimental and modelling results show that a reduction of ELM energy by impurity seeding is difficult for large ELMs. A combination of radiating divertor with other active ELM control methods is essential for an application to ITER.
- Vertical kicks need in-vessel coils to reach a high kick frequency. Recent results from JET are promising, and this technique will be used for ITER-like wall experiments on JET, in which case the ELM size need only be reduced by a factor of $\sim$2–3.
- Pellet pacing can typically achieve a factor of two reduction in the energy per ELM. High frequency ELM pacing still needs to be demonstrated on JET. Indeed, pellet pace-making could assist other ELM control techniques, helping to compensate the density pump-out caused by RMP ELM control/suppression.
- RMP ELM suppression/control has shown very promising results up to now, although the physics mechanism is not well understood as yet. Future joint experiments from different devices (DIII-D, JET, MAST, NSTX, AUG, TCV, KSTAR and EAST) will help us to understand ELM suppression physics and provide support for ITER.

ITER may need a combination of different ELM control methods, many of which are still under active investigation.

References

1. F. Wagner, Phys. Rev. Lett. **49**, 1408 (1982)
2. ITER Physics Basis, Nucl. Fusion **39**, 2137 (1999)
3. J.W. Connor, Plasma Phys. Control. Fusion **40**, 531 (1998)
4. J.A. Snipes, Plasma Phys. Control. Fusion **40**, 765 (1998)
5. W. Suttrop, Plasma Phys. Control. Fusion **38**, 1407 (1996)
6. R.J. Buttery et al., in Proceedings of 22nd EPS Conference on Controlled Fusion and Plasma Physics (Bournrmouth), vol. 19C (1995), p. 273
7. P. Gohil, Phys. Rev. Lett. **61**, 14 (1988)
8. M. Jiang, Plasma Phys. Control. Fusion **54**, 095003 (2012)
9. D.C. McDonald, Fusion Sci. Technol. **53**, 891 (2008)
10. K. Hagashima, T. Shoji, Y. Miura, Nucl. Fusion **36**, 335 (1996)
11. N. Oyama, Nucl. Fusion **44**, 582 (2004)
12. N. Oyama, Nucl. Fusion **45**, 871 (2005)

13. A. Kirk, J. Nucl. Mater. **313**, 1081 (2003)
14. R. Maingi, Nucl. Fusion **45**, 1066 (2005)
15. H. Reimerdes, in Proceedings of 24th EPS Conference on Controlled Fusion and Plasma Physics (Berchtesgaden), vol. 21A (1997), p. 533
16. S. Soldatov, Plasma Phys. Control. Fusion **52**, 085001 (2010)
17. W. Suttrop, Plasma Phys. Control. Fusion **42**, A1 (2000)
18. H. Zohm, Plasma Phys. Control. Fusion **38**, 105 (1996)
19. A. Loarte, Plasma Phys. Control. Fusion **44**, 1815 (2002)
20. A. Loarte, J. Nucl. Mater. **313–316**, 962 (2003)
21. T. Ozeki, Nucl. Fusion **30**, 1425 (1990)
22. J. Stober, Nucl. Fusion **41**, 1123 (2001)
23. M. Bècoulet, Plasma Phys. Control. Fusion **45**, A93 (2003)
24. H.R. Wilson, Plasma Phys. Control. Fusion **48**, A71 (2006)
25. J. Ongena, Plasma Phys. Control. Fusion **43**, A11 (2001)
26. G. Saibene, Plasma Phys. Control. Fusion **44**, 1769 (2002)
27. Y. Kamada, Plasma Phys. Control. Fusion **42**, A247 (2000)
28. Y. Kamada, Plasma Phys. Control. Fusion **44**, A279 (2002)
29. G. Saibene, Nucl. Fusion **45**, 297 (2005)
30. J. Stober, Nucl. Fusion **45**, 1213 (2005)
31. R. Maingi, J. Nucl. Mater. **337–339**, 727–731 (2005)
32. K.H. Burrell, Plasma Phys. Control. Fusion **44**, A253 (2002)
33. W. Suttrop, Plasma Phys. Control. Fusion **45**, 1399 (2003)
34. Y. Sakamoto, Plasma Phys. Control. Fusion **46**, A299 (2004)
35. W. Suttrop, Nucl. Fusion **45**, 721 (2005)
36. K.H. Burrell, Nucl. Fusion **49**, 085024 (2009)
37. M. Greenwald, Phys. Plasma **6**, 1943 (1999)
38. K. Kamiya, Nucl. Fusion **43**, 1214 (2003)
39. N. Oyama, Plasma Phys. Control. Fusion **48**, A171 (2006)
40. E.A. Frieman, Phys. Fluids **16**, 1108 (1973)
41. J.W. Connor, Phys. Plasmas **5**, 2687 (1998)
42. J. Manickam, Phys. Fluids B **4**, 1901 (1992)
43. S.M. Kaye, Nucl. Fusion **30**, 2621 (1990)
44. A.J. Webster, C.G. Gimblett, Phys. Rev. Lett. **102**, 035003 (2009)
45. A.J. Webster, C.G. Gimblett, Phys. Plasmas **16**, 082502 (2009)
46. G.T.A. Huysmans, Plasma Phys. Control. Fusion **47**, 2107 (2005)
47. S. Saarelma, Plasma Phys. Control. Fusion **53**, 025011 (2011)
48. H.R. Wilson, S.C. Cowley, Phys. Rev. Lett. **92**, 175006 (2004)
49. A. Kirk, Plasma. Phys. Control. Fusion **46**, 551 (2004)
50. A. Kirk, Phys. Rev. Lett. **92**, 245002 (2004)
51. A. Kirk, Plasma Phys. Control. Fusion **47**, 315 (2005)
52. A. Schmid, Plasma Phys. Control. Fusion **50**, 045007 (2008)
53. C.C. Hegna, Phys. Plasmas **3**, 584 (1996)
54. S. Saarelma, Plasma Phys. Control. Fusion **48**, 31 (2007)
55. J.W. Connor, R.J. Hastie, J.B. Taylor, Proc. Roy. Soc. London Ser. A Math. Phys. Sci. 365, 1720 (1979)
56. R.L. Miller, R.W. Moore, Phys. Rev. Lett. **43**, 11 (1979)
57. R.L. Miller, Phys. Plasmas **5**, 973 (1998)
58. J. Wesson, Tokamaks, 3rd edn. (Clarendon Press, Oxford, 2004)
59. C.E. Kessel, Nucl. Fusion **34**, 1221 (1994)
60. P.B. Snyder, Phys. Plasmas **9**, 2037 (2002)
61. H.R. Wilson, Phys. Plasmas **9**, 1277 (2002)
62. G.T.A. Huysmans, S.E. Sharapov, A.B. Mikhailovskii, W. Kerner, Phys. Plasmas **8**, 4292 (2001)
63. P.B. Snyder, Nucl. Fusion **44**, 320 (2004)

64. P.B. Snyder, H.R. Wilson, Plasma Phys. Controlled Fusion **45**, 1671 (2003)
65. A. Mossessian, Phys. Plasmas **10**, 1720 (2003)
66. S. Saarelma, Nucl. Fusion **43**, 262 (2003)
67. L.L. Lao, Nucl. Fusion **41**, 295 (2001)
68. T.W. Petrie, J. Nucl. Mater. **196–198**, 848 (1992)
69. O. Gruber, Phys. Rev. Lett. **74**, 4217 (1995)
70. A.W. Leonard, J. Nucl. Mater. **290–293**, 1097 (2001)
71. A.W. Degeling, Plasma Phys. Control. Fusion **45**, 1637 (2003)
72. P.T. Lang, Nucl. Fusion **44**, 665 (2004)
73. T.E. Evans, Nat. Phys. **2**, 419 (2006)
74. Y. Liang, Phys. Rev. Lett. **98**, 265004 (2007)
75. M. Shimada, Progress in the ITER Physics Basis: Chap. 1. Overview Summary Nucl. Fusion 47, S1 (2007)
76. K. Tobita, Nucl. Fusion **49**, 075029 (2009)
77. P. Monier-Garbet, Nucl. Fusion **45**, 1404 (2005)
78. R. Sartori, Plasma Phys. Control. Fusion **46**, 723 (2004)
79. J. Rapp, Nucl. Fusion **44**, 312 (2004)
80. J. Rapp, J. Nucl. Mater. **337–339**, 826 (2005)
81. K. Itami, J. Nucl. Mater. **266–269**, 1097 (1999)
82. S. Sakurai et al., J. Nucl. Mater. **290–293**, 1102 (2001)
83. H. Kubo, Nucl. Fusion **41**, 227 (2001)
84. H. Kubo, JT-60 team. Phys. Plasmas 9, 2127 (2002)
85. S. Higashijima, J. Nucl. Mater. **313–316**, 1123 (2003)
86. J. Rapp, Plasma Phys. Control. Fusion **44**, 639 (2002)
87. G.J. Radford, Contrib. Plasma Phys. **36**, 187 (1996)
88. D. Reiter, H. Kever, G.H. Wolf, Plasma Phys. and Contr. Fus. **33**, 1579 (1991)
89. A. Loarte et al., Fusion energy 2000. in Proceedings of 18th International Conference on Sorrentto (Vienna, IAEA, 2000). CD-ROM file ITERP/11R
90. P.T. Lang, Plasma Phys. Control. Fusion **46**, L31–L39 (2004)
91. F. Sartori, in Proceedings of 35th EPS Conference on Plasma Physics, vol 32D (ECA) (Hersonissos, Greece, 2008), P5.045
92. S.P. Gerhardt, Nucl. Fusion **50**, 064015 (2010)
93. E. de la Luna et al., in Proceedings of 36th EPS Conference on Plasma Physics (Sofia) (ECA), vol. 33E (2009), P5.174
94. E. de la Luna et al., 23rd IAEA, EXC/8-4 (2010)
95. P.T. Lang, Nucl. Fusion **43**, 1110 (2003)
96. P.T. Lang, Nucl. Fusion **47**, 754 (2007)
97. L.R. Baylor et al., in Proceedings of 35th EPS Conference on Plasma Physics (Hersonissos, Greece) (ECA), vol. 32D (2008), P4-098
98. A.R. Polevoi, Nucl. Fusion **43**, 1072 (2003)
99. D.K. Mansfield, Nucl. Fusion **53**, 113023 (2013)
100. L.R. Baylor, in Proceedings 37th EPS Conference on Plasma Physics and Controlled Fusion (Dublin, Ireland, 2010), P2.117 http://ocs.ciemat.es/EPS2010PAP/pdf/P2.117.pdf
101. G.T.A. Huysmans, Plasma Phys. Control. Fusion **51**, 124012 (2009)
102. F.M. Poli, Nucl. Fusion **50**, 025004 (2010)
103. G. Kocsis, Nucl. Fusion **47**, 1166 (2007)
104. S.K. Combs, J. Vac. Sci. Tech. A **6**, 1901 (1988)
105. L.R. Baylor, Nucl. Fusion **49**, 085013 (2009)
106. A. Kirk, Nucl. Fusion **50**, 034008 (2010)
107. J.M. Canik, Nucl. Fusion **50**, 034012 (2010)
108. T.E. Evans, Phys. Rev. Lett. **92**, 235003 (2004)
109. T.E. Evans, Nucl. Fusion **48**, 024002 (2008)
110. W. Suttrop, Phys. Rev. Lett. **106**, 225004 (2011)
111. Y.M. Jeon, Phys. Rev. Lett. **109**, 035004 (2012)

112. Y. Liang, Nucl. Fusion **53**, 073036 (2013)
113. W. Suttrop et al., in Proceedings of 39th EPS Conference & 16th International Congress on Plasma Physics (2012), P2.092
114. Y. Liang, Nucl. Fusion **50**, 025013 (2010)
115. I. Barlow, Fusion Eng. Des. **58–59**, 189 (2001)
116. Y. Liang, Plasma Phys. Control. Fusion **49**, B581 (2007)
117. Y. Liang, J. Nucl. Mater. **390–91**, 733–739 (2009)
118. S. Jachmich, J. Nucl. Mater. **390–391**, 768–772 (2009)
119. H.R. Koslowski et al., in 34th EPS Conference on Plasma Physics Warsaw 2-6 July 2007, ECA, vol. 31F (2007), P-5.135
120. M.E. Fenstermacher et al., 23rd IAEA, ITR/P1-30 (2010)
121. J.M. Canik, Nucl. Fusion **50**, 064016 (2010)
122. J.C. Vallet, Phys. Rev. Lett. **67**, 2662 (1991)
123. A. Alfier, Nucl. Fusion **48**, 115006 (2008)
124. S. Saarelma, Plasma Phys. Control. Fusion **51**, 035001 (2009)
125. Y. Liang, Plasma Fusion Res. **5**, S2018 (2010)
126. E.J. Strait, Phys. Rev. Lett. **74**, 2483 (1995)
127. R. Buttery, Phys. Plasmas **15**, 056115 (2008)
128. R. Fitzpatrick, Phys. Plasmas **5**, 3325 (1998)
129. K.C. Shaing, Phys. Plasmas **10**, 1443 (2003)
130. K.C. Shaing, Phys. Fluids **29**, 521 (1986)
131. J.-K. Park, Phys. Rev. Lett. **102**, 065002 (2009)
132. A.J. Cole, Phys. Plasmas **15**, 056102 (2008)
133. W. Zhu, Phys. Rev. Lett. **96**, 225002 (2006)
134. M. Bècoulet, Nucl. Fusion **49**, 085011 (2009)
135. K.C. Shaing, Phys. Plasmas **15**, 082506 (2008)
136. J.-K. Park, Phys. Plasmas **16**, 056115 (2009)
137. A.M. Garofalo, Phys. Rev. Lett. **101**, 195005 (2008)
138. Y. Sun, Plasma Phys. Control. Fusion **52**, 105007 (2010)
139. A. Kirk, Phys. Rev. Lett. **108**, 255003 (2012)
140. A. Kirk, Nucl. Fusion **53**, 043007 (2013)
141. I.T. Chapman, Nucl. Fusion **52**, 123006 (2012)
142. M. Jakubowski, Nucl. Fusion **49**, 095013 (2009)
143. Y. Liang, Phys. Rev. Lett. **105**, 065001 (2010)
144. Y. Liang, Nucl. Fusion **51**, 073001 (2011)
145. C.G. Gimblett, R. J. Hastie, P. Helander, Phys. Rev. Lett. 96, 035006-1-4 (2006)
146. I.T. Chapman, Nucl. Fusion **47**, L36 (2007)
147. R. Moyer, Nucl. Fusion **52**, 123019 (2012)
148. R. Fischer, Plasma Phys. Control. Fusion **54**, 115008 (2012)
149. I.T. Chapman, Plasma Phys. Control. Fusion **54**, 105013 (2012)
150. S.P. Hirshman, D.K. Lee, Comput. Phys. Commun. **43**, 143 (1986)
151. W.A. Cooper, Comput. Phys. Commun. **180**, 1524 (2009)
152. R.J. Sanchez, Comp. Physics **161**, 576 (2000)
153. M.J. Schaffer, Nucl. Fusion **48**, 024004 (2008)
154. M. Bècoulet, Nucl. Fusion **48**, 024003 (2008)
155. M. Jakubowski, PRL **96**, 035004 (2006)
156. O. Schmitz, Plasma Phys. Control. Fusion **50**, 124029 (2008)
157. K. Burrell, Plasma Phys. Control. Fusion **47**, B37 (2005)
158. M. Heyn, Nucl. Fusion **48**, 024005 (2008)
159. R. Fitzpatrick, Nucl. Fusion **33**, 1049 (1993)
160. B.V. Chirikov, Phys. Rep. **52**, 263 (1979)
161. B. Hudson, Nucl. Fusion **50**, 045006 (2010)
162. Park Jong-kyu, Phys. Rev. Lett. **99**, 195003 (2007)
163. M.E. Fenstermacher, Nucl. Fusion **48**, 122001 (2008)

164. M.E. Fenstermacher, J. Nucl. Mater. **390–391**, 793–796 (2009)
165. T.E. Evans, Phys. Plasmas **13**, 056121 (2006)
166. M.E. Fenstermacher, Phys. Plasmas **15**, 056122 (2008)
167. R.A. Moyer, Phys. Plasmas **12**, 056119 (2005)
168. R.A. Moyer, Nucl. Fus. **45**, 595 (2005)
169. C.G. Lowry, Bull. Am. Phys. Soc. **54**, 329 (2009)
170. NSTX Research 5-Year Plan for 2009-13, http://nstx.pppl.gov/fiveyearplan.html
171. W.W. Xiao, Nucl. Fusion **52**, 114027 (2012)
172. Kim Jayhyun, Nucl. Fusion **52**, 114011 (2012)
173. X.L. Zou et al. in Proceedings of 24th IAEA Fusion Energy Conference October 8–13, PD/P8-08 (IAEA, San Diego, 2012)
174. Y. Liang, Phys. Rev. Lett. **110**, 235002 (2013)
175. J.G. Li et al., 17 NOVEMBER 2013. Nat. Phys. DOI:10.1038/NPHYS2795

Chapter 6
Resistive Wall Mode (RWM)

Valentin Igochine

Abstract The advanced tokamak regime is a promising candidate for steady state tokamak operation, desirable for a fusion reactor. This regime is characterized by a high bootstrap current fraction and a flat or reversed safety factor profile, which leads to operation close to the pressure limit (see Chap. 2). At this limit, an external ideal kink mode becomes unstable. This external kink is converted into the slowly growing Resistive Wall Mode (RWM) by the presence of a conducting wall. Reduction of the growth rate allows one to act on the mode and to stabilize it. There are two main factors which determine the stability of the RWM. The first factor comes from external magnetic perturbations (error fields, resistive wall, feedback coils, etc.). This part of RWM physics is the same for tokamaks and reversed field pinch (RFP) configurations. The physics of this interaction is relatively well understood, since it is based on classical electrodynamics, and is used for RWM control with external coils. The second ingredient of RWM physics is the interaction of the mode with plasma flow and fast particles. These interactions are particularly important for tokamaks, which have higher plasma flow and stronger trapped particle effects compared to the present day reversed field pinch device. The influence of the fast particles will also be increasingly more important in ITER and DEMO, which will have a large fraction of fusion born alpha particles. These interactions have kinetic origins that make the computations challenging. Correct prediction of the "plasma-RWM" interaction is an important ingredient which has to be combined with the influence of external fields (resistive wall, error fields and feedback) to make reliable predictions for RWM control in a future reactor.

V. Igochine (✉)
Max Planck Institute for Plasma Physics, Boltzmannstr. 2, 85748 Garching, Germany
e-mail: valentin.igochine@ipp.mpg.de

V. Igochine (ed.), *Active Control of Magneto-hydrodynamic Instabilities in Hot Plasmas*, Springer Series on Atomic, Optical, and Plasma Physics 83,
DOI 10.1007/978-3-662-44222-7_6

6.1 Introduction

The main motivation for fusion research is the development of a reliable and virtually inexhaustible energy source. From the engineering point of view it is attractive to have a stationary solution for this problem. The most advanced present day concept for plasma confinement is the tokamak. Thus, steady state tokamak operation could be an attractive regime for a future fusion power plant based on the tokamak concept. Constant power production in this regime is more convenient for energy conversion and avoids the power storage step required for pulsed operations. Additionally, constant power and mechanical loads on the plasma facing components allow simpler power handling, lower engineering efforts for reactor design, and smaller costs for the fusion power plant. A necessary requirement for tokamak operations is a stationary plasma current, which produces the poloidal component of the magnetic field. This component twists magnetic field lines which ensure plasma confinement in tokamaks. A non-inductive source for the constant plasma current is required for steady state operation because an inductive drive can work only temporarily. There are several different options for such non-inductive current drive in tokamaks: Electron Cyclotron Current Drive (ECCD), Lower Hybrid Current Drive (LHCD), Neutral Beam Current Drive (NBCD) and bootstrap current [1, 2]. From these, the bootstrap current is the most attractive option, as it can produce a sufficiently large amount of the toroidal current in big tokamaks without additional costs for the current drive systems. All other options could be very useful for plasma control, but will play a small role in the total toroidal current balance during the flattop phase of a discharge, since they would consume substantial amount of the produced energy. Thus, all modern concepts for steady state DEMO reactors rely on the bootstrap current as one of the main current sources in the plasma. The amount of bootstrap current is proportional to the pressure gradient [2]. Typically, the maximum pressure gradients are located off-axis which leads to flat or hollow current profiles in the plasma. In this case, the minimum safety factor value is about two, which avoids some dangerous resonant surfaces [(1, 1) kink mode and (3, 2) Neoclassical Tearing Mode]. For even higher safety factor values the most dangerous (2, 1) resonant surface could also be avoided. This is a positive consequence of the elevated safety factor profile in the advanced tokamak scenario. At the same time, high bootstrap current fractions can only be achieved with high pressure gradients (which are unavoidable for advanced tokamak scenarios). It is well known from MHD that current and pressure gradients lead to instabilities. In advanced tokamak scenarios the high pressure gradients and simultaneous broad current profile result in external kink instabilities. In the presence of a resistive wall the kink instability is converted into the slower-growing Resistive Wall Mode (RWM). This instability is the main topic of the present chapter. Here, we present a concise review of resistive wall mode physics, focusing on the mechanisms behind the mode behavior, together with the main control issues (control of this instability was also the subject of a comprehensive review [3]). We start from the basic physics, focus on the recent achievements and discuss possible future steps.

6.2 Stability Boundary of the Resistive Wall Mode

As mentioned before, the resistive wall mode is a pressure gradient driven instability. A typical indicator of the total pressure in a tokamak is the normalized plasma pressure $\beta_t = 2\mu_0\langle p\rangle / B_t^2, \beta_N = \beta_t(aB_t/I_p)$ ($\langle p\rangle$ is the volume average plasma pressure, B_t is the toroidal magnetic field, a is the minor radius and I_p is the plasma current). Using this quantity, it is possible to define two limits for a given external kink mode. Assuming a plasma without an external wall, one obtains an external kink by increasing β_N above a particular threshold value called the "no wall" limit $(\beta_{N,no-wall})$. On the other hand, if a perfectly conducting wall is considered, one could increase β_N further and reach the instability threshold at a higher value, which depends on the profiles shape and is called the "ideal wall" limit $(\beta_{N,ideal-wall})$. In either case, the external kink grows extremely fast after crossing a threshold value (growth rate: $\gamma_0 \sim 1/\tau_{Alfven}$, $\tau_{Alfven} = L/v_{Alfven}$, $v_{Alfven} = B_\phi/\sqrt{\mu_0\rho}$, L is the characteristic length and ρ is the plasma mass density). At this point, control of the mode becomes impossible and confinement is lost. The situation changes if one considers a resistive wall. In this case the external kink is still stable before the "no wall" limit $(\beta_N < \beta_{N,no-wall})$ and highly unstable above the "ideal wall" limit as before $(\beta_N > \beta_{N,ideal-wall})$, but in the range between the "no wall" and the "ideal wall" limits $(\beta_{N,no-wall} < \beta_N < \beta_{N,ideal-wall})$ the mode splits into two branches [4]:

- The fast branch, which is the normal external kink mode, is stable because the rotation of this mode is relatively fast with respect to the wall. At these frequencies the resistive wall acts as an ideal wall and stabilizes this branch.
- The slow branch, on the other hand, has a lower rotation frequency, which enables the mode perturbation to penetrate through the wall and allows this branch to become unstable. But the growth rate of the mode becomes much smaller than the inverse Alfven time (growth rate: $\gamma_0 \sim 2m/\tau_w$, $\tau_w = \mu_0 bd/\eta_{wall}$, where b, d, and η_{wall} are the wall radius, thickness, and resistivity, and m is the dominant poloidal mode number). Here, the wall properties become important because they determine the flux penetration. At such slow time scales control of the mode becomes possible.

In summary, the question of resistive wall mode stabilization in tokamaks is a question of being able to operate in the region between the "no-wall" and "ideal wall" limits. For advanced operation close to the pressure limits the ratio between these two limits is about $\beta_{N,ideal-wall}/\beta_{N,no-wall} \sim 1.2 - 2$ [5], which, when translated to fusion power, is an increase by a factor of 3–4 in the case of RWM stabilization (fusion power scales approximately with β_N^2). This relation between the "no wall" and "ideal wall" can be calculated for a given configuration and position of the resistive wall taking into account possible plasma profiles. This factor is a crucial component which could make the advanced scenario comparable to conventional H-mode discharges in terms of fusion power. In this regime the bootstrap current fraction is close to one ($f_{bs} = I_{bootstrap}/I_{plasma} \approx 1$). Thus, stabilization of the resistive wall mode and operation close to the "ideal wall" limit are extremely important. One should note that in the conventional scenario the gap between the "no-wall" and "ideal wall" limits is small and accessing this region only slightly increases the operational space without strong benefits in fusion performance. Comparisons of the "ideal wall" limit and the "no wall" limit are shown for two different equilibria in Fig. 6.1 [6]. The gap between the limits depends strongly on the toroidal mode number and safety factor profile. For the most unstable n = 1 modes there is a substantial increase in β_N with a reversed shear q-profile and only a moderate increase in the case of a monotonic safety factor profile. Therefore, RWM stabilization is only really worthwhile, in terms of performance gain, in advanced tokamak operation scenarios, which feature reversed or near reversed shear q-profiles.

There is also another measure for the plasma stability with respect to the resistive wall mode which is used in some papers and should be mentioned. One can define the internal inductance, which is the ratio of the average poloidal magnetic field to its value at the plasma boundary, $l_i = \langle B_\theta^2 \rangle / B_{\theta a}^2$. This is an integral measure of the peakedness of the current profile for a particular equilibrium. The internal inductance can be used to estimate the "no-wall" $(\beta_{N,no-wall})$ and "ideal wall" $(\beta_{N,ideal-wall})$ limits, but the relations are not universal. For example, a typical relation for the "no wall" limit in the DIII-D tokamak could be either $\beta_{N,no-wall} \approx 4l_i$ [8] or $\beta_{N,no-wall} \approx 2.4l_i$ [9] depending on the discharge scenario. For this reason, it is better to use the quantity β_N as the measure for the plasma stability against resistive wall modes. The expression of the stability limits in terms of l_i depends strongly on the plasma profiles, plasma shape, properties of the resistive wall and the aspect ratio. The strong aspect ratio dependence can be seen, for example, in a comparison of NSTX (aspect ratio = 1.27) and DIII-D (aspect ratio = 3.1) discharges with similar l_i values. The $\beta_{N,no-wall}$ and $\beta_{N,ideal-wall}$ values are a factor of two larger for the lower aspect ratio case (NSTX) compared to the high aspect ratio case (DIII-D) [10]. Changes in the plasma shape can also influence these limits even if l_i is the same. Thus, all these ingredients have to be taken into account correctly to calculate the "no wall" and the "ideal wall" limits using linear MHD codes. Unfortunately, even this (complex) determination of the plasma stability boundaries does not give a final answer for the boundaries of the RWM unstable region. The actual plasma stability is a very complex subject which depends also on kinetic effects and cannot be described only in terms of β_N, as will be show below.

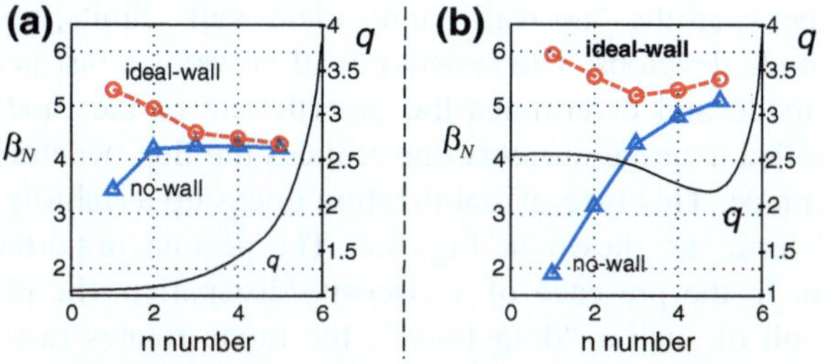

Fig. 6.1 The "ideal wall" and the "no wall" limits are shown for different toroidal mode numbers. **a** The case of the monotonic safety factor profile is considered. **b** The reversed shear safety factor profile is considered. (The figures represent results from reference [6]. The figures are taken from V. Igochine, Nucl. Fusion, 52, 074010 (2012). © IAEA. All rights reserved [7])

6.3 Simple Dispersion Relation for the Resistive Wall Mode

In the previous section, the RWM stability boundaries were defined. In order to understand the behavior of the instability inside of the boundaries, a simple dispersion relation is discussed here. A relatively simple form of such a dispersion relation was derived from a set of linear MHD equations [11, 12]. This simplest form has no kinetic effects and assumes a fluid plasma viscosity[1]:

$$\underbrace{\delta W_p}_{\substack{ideal\\MHD}} + \underbrace{\left(\gamma + in\Omega_{0,pl}\right)D}_{\substack{plasma\\rotation}} + \underbrace{\frac{\delta W_v^b \gamma\tau_w + \delta W_v^\infty}{\gamma\tau_w + 1}}_{\substack{resistive\\wall}} = 0 \tag{6.1}$$

Here $\Omega_{0,pl}$ is the rotation frequency of the plasma, $\left(\gamma + in\Omega_{0,pl}\right)D$ is the energy dissipated in the plasma by rotation of the plasma through the mode, δW_v^b and δW_v^∞ are perturbed energies in vacuum with an ideal wall at b and at ∞, respectively, τ_w is the wall time, $\gamma = \gamma_0 + i\omega_{RWM}$ is the complex growth rate of the mode, and n is the toroidal mode number. The dispersion relation contains parts related to the ideal stability of the mode, the influence of the plasma rotation (interaction of the mode and the plasma) and the influence of the resistive wall. The important consequences of the dispersion relation are shown schematically in Fig. 6.2. The plasma without a wall can be confined up to the "no wall" limit after which a fast external kink instability develops (magenta line). In the presence of an "ideal wall" this limit is shifted to higher β_N values (up to $\beta_{N,ideal-wall}$). In both cases, the control of an ideal instability above these limits is not possible. In the

[1] Other variant of simple RWM representation is nicely given in chapter 3 (subsection 3.3.1.2.1) of the book "Fusion Physics", Ed. M. Kikuchi et al., IAEA VIENNA, 2012, (this book is freely available online http://www-pub.iaea.org/books/iaeabooks/8879/Fusion-Physics).

real situation, between the “no wall” and “ideal wall” limits, a slowly growing resistive wall mode develops. The resistive wall prevents a fast penetration of the flux from the mode and determines the growth rate of the mode. By applying feedback controlled external currents one reduces the flux penetration further and suppresses the mode. This type of stabilization opens up a stability window above the “no wall” limit, as shown in Fig. 6.2. The second important effect is the plasma rotation. In the presence of a viscosity/dissipation, the plasma drags the mode. As a result of such a “drag force”, the mode rotates faster and becomes decoupled from the wall, which prevents penetration of the flux through the wall and again stabilizes the mode. The second situation appears in case of locked RWM, which is static in the lab frame. The plasma flow through the mode structure and interaction of the mode with the plasma leads to energy dissipation and stabilized RWM. The coupling between the plasma and the mode (“drag force”) is stronger for high β_N values because of the higher dissipation at higher β_N.[2] This effectively opens a stability window close to the “ideal wall” limit if the rotation stabilization is dominant. It should be noted that the presented figure gives only a schematic description of RWM behaviour. It contains some simplifications to make the figure clearer. For example, growth rates values γ_0 for “no wall external kink” and “ideal wall external kink” cases are set to zero in the stable region to reduce the number of curves. These values could be strongly negative in reality.

In summary, RWM growth requires flux penetration through the wall. Two different mechanisms can prevent, reduce or suppress this penetration: active feedback control, which is more effective when β_N is larger than but close to the “no-wall” limit, and stabilization via plasma rotation, effective when β_N is smaller than but close to the “ideal wall” limit.

6.4 Structure of the Resistive Wall Mode

Experiments show that both internal (plasma rotation, etc.) and external factors (resistive wall, external coils, etc.) have a strong impact on mode stability [13–16]. The physical reason for these strong effects is the broad eigenfunction of the mode. Resistive wall modes typically have a toroidal mode number $n = 1$ and multiple poloidal mode numbers $m = 2\ldots7$ which are coupled together and provide substantial perturbation amplitudes from the plasma core to the plasma edge. There are also situations in which the higher toroidal mode numbers $n \geq 2$ can be excited simultaneously. These multiple unstable eigenvalues can coexist together in the plasma and were observed, for example, in high beta NSTX plasmas [17].

[2] The dissipation considered here is the due to plasma-mode interaction, described by force $\vec{\nabla} \cdot \vec{\Pi}_1$ in 6.22 and 6.23, which increases with β_N.

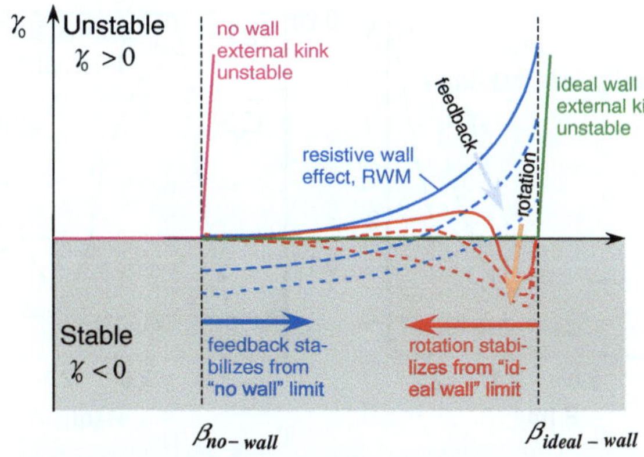

Fig. 6.2 Schematic representation of the RWM behavior between "no-wall" and "ideal wall" limits from the dispersion relation (6.1). The mode is unstable above the "no-wall" if no wall is present (*magenta line*). The mode is stable up to $\beta_{ideal-wall}$ in the case of an ideal wall (*green line*). A slowly growing mode is unstable between the "no-wall" and "ideal wall" limits (*blue curves*). Active feedback pushes this curve down and stabilizes from "no-wall" towards higher β. Inclusion of plasma rotation stabilizes the RWM from the "ideal wall" limit towards lower β (*red curves*). (The figure is from V. Igochine, Nucl. Fusion, 52, 074010 (2012) [7]. © IAEA. All rights reserved.)

It is clear that the external kink part of the RWM eigenfunction creates currents at the plasma boundary [18, 19]. These currents are close to the wall and lead to strong interactions with the currents in the wall and with externally produced perturbations. The influence of the internal currents, located around mid-radius of the plasma, on this interaction is smaller due to the larger distance to the wall. At the same time, the internal part of the eigenfunction is responsible for the strong interaction with the plasma flow. This interaction has a kinetic origin which results in an interaction of the wave (mode perturbations due to RWM) with particles, as will be discussed later. The external kink part of the eigenfunction is less important for the interaction with the plasma, since the amount of plasma particles in the core is much higher than in the plasma boundary region. This conclusion is also confirmed by kinetic calculations, as will be shown later.

Typical displacement eigenfunctions of the resistive wall mode for two slightly different edge safety factors in DIII-D are shown in Fig. 6.3 [20]. Even in the case where the external kink is dominant, the internal kink is non-negligible. Additionally, the calculations for JET indicate that the amplitude of the internal displacement is of the same order or even significantly larger than the external kink part [21]. Only a global character, which includes internal and external parts of the RWM eigenfunction, can explain the simultaneous effects of the resistive wall, external perturbations and plasma rotation on stability, as observed experimentally.

The last important issue concerning the mode structure is the mode "rigidity". It characterizes the rigidity of the mode structure when external actions are

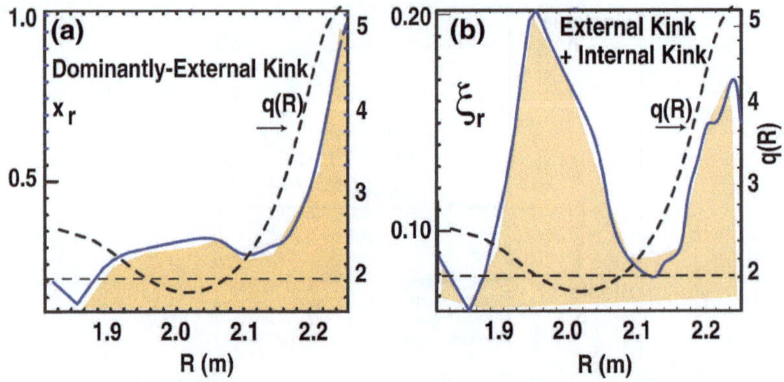

Fig. 6.3 Shape of the radial displacement eigenfunction for two slightly different edge safety factors (*left* $q_a = 5.6$; *right* $q_a = 6$). The figures are adopted from Ref. [20]. The internal part is present in both cases but has a much stronger relative amplitude on the *right*. These figures result from linear MHD calculations which imply that only shapes of the eigenfunctions could be compared. (Non-linear MHD simulations are necessary for the absolute values of the displacements.) (The figures are from M. Okabayashi et al., Nucl. Fusion, 49, 125003 (2009).

applied. This term is defined differently by different authors. Typically one of the following rigidity definitions is used:

(1) Mode "rigidity" in real space assumes that local poloidal (and/or toroidal) suppression of the RWM in one place leads to overall reduction of the mode amplitude [22, 23].
(2) Mode number "rigidity" assumes that suppression of the most unstable toroidal mode number (n = 1) would not destabilize a mode with higher n-numbers [24, 25].
(3) For the "perturbative" calculations discussed in the following sections, mode "rigidity" means that the shape of the eigenfunction is fixed in the analysis of the particle-mode interaction using a kinetic approach (see Sect. 6.6.2).

These definitions are not independent: for example, case which is non-rigid as per definition one could alter the mode spectrum such that it is also non-rigid as per definition two. One should be careful to understand the exact meaning of the word "rigidity" in a particular case. In many cases, "rigid" mode structure of an RWM is a good approximation.

6.5 Physics and Control of Resistive Wall Modes

The physics of the resistive wall mode is quite complicated and is not yet completely understood. Different factors like plasma rotation, fast particles, resistive walls, error fields and feedback impact the mode stability and could be dominant under different circumstances. Thus, resulting stability/instability of an RWM is an integral quantity which contains all of these factors. A change in even one of the ingredients could completely change the picture and drive the instability from a stable to an unstable region or vice versa. Some of these factors are well understood, but others require detailed investigations.

It is useful to separate these factors into two groups, depending on the physics background (see Fig. 6.4). This allows us to gain a better physical understanding of RWM physics:

(1) In the first group we include the effects which could be described purely by electromagnetism. These are the interactions of the mode with any external magnetic fields which could originate from the resistive wall, error fields or feedback coils. In this case only perturbation currents due to the mode inside the plasma enter into the problem. The physical description of the situation could be simplified even more since these currents can be substituted by an appropriate choice of the surface current on the plasma boundary (or boundary conditions at the last closed flux surface in the calculations). The solution of such a problem in real geometry would be complex, but the approach is straightforward. This part of the physics is discussed together with active control in Sect. 6.5.1.

(2) The second group contains the interaction of the RWM with the plasma. This means the influence of the plasma rotation and of fast particles on the mode stability. These effects have kinetic origins and have to be considered as wave-particle interactions. A self-consistent calculation of the mode with a kinetic approach is an extremely challenging task, as will be shown later. Thus, an assumption on this interaction is required which allows a fluid approach in an MHD code to be used. In spite of the significant progress during the last few years, this piece of physics is still not complete. This part of the RWM physics is important because kinetic effects can stabilize an RWM and move the stability boundary well above the "no-wall limit", as will be discussed in detail in Sect. 6.6.2.

The representation in Fig. 6.4 gives the main idea, but the resulting interaction can give opposite effects depending on the particular case. For example, improper actions of external coils can destabilize the mode instead of having the desired stabilization effect.

6.5.1 RWM Interaction with External Magnetic Fields

Many simple models of resistive wall mode stability were established at the beginning of the RWM research to investigate the interaction of the RWM with the wall and with feedback currents. The most widely used models were proposed by Pustovitov [26], Boozer [27], Okabayashi [28], Finn [29] and Fitzpatrick [30]. In the following section, the Pustovitov model is discussed in detail. This model is a

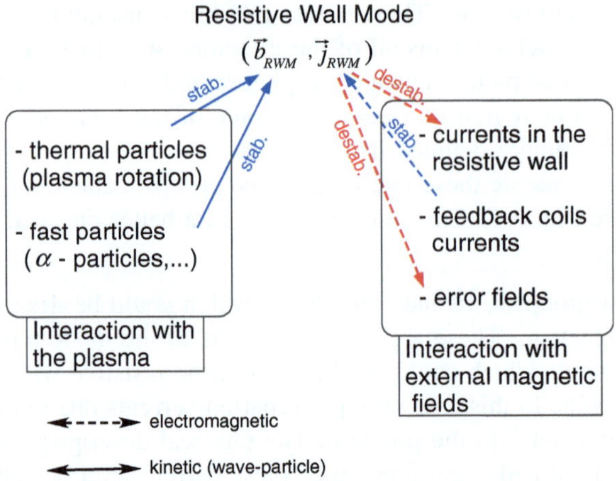

Fig. 6.4 Schematic representation of the resistive wall mode interactions with the plasma and external magnetic fields. Different effects are separated into two groups, depending on the type of interaction. Typical stabilizing and destabilizing influences are marked. The amount of stabilization/destabilization depends strongly on the particular situation (profiles, plasma shape, particles distribution, etc.). (The figure is from V. Igochine, Nucl. Fusion, 52, 074010 (2012).

good example of a typical simple model approach and its limitations. As mentioned before, all currents due to the RWM inside of the plasma volume can be substituted by a single surface current (boundary conditions). This approach is implemented in the Pustovitov model in the most straightforward manner. All information regarding the mode and its interaction with the plasma is hidden in a specially introduced parameter. This parameter is connected to the logarithmic derivative of the perturbed flux from the mode at the boundary. These assumptions are sufficient to derive a dispersion relation for the resistive wall mode and investigate different feedback schemes. In all these models kinetic effects are described by one or two free parameters (representing viscosity and dissipation in the Fitzpatrick model, only dissipation in the Boozer model, etc.). This oversimplification of plasma-mode interaction is the main problem of all simple models. One can determine these free parameters for a given experiment, but scaling to other tokamaks—or even scaling to substantially different regimes in the same tokamak—is impossible. At the same time, these simple approaches are extremely useful for getting a basic understanding of RWM physics and control. They also become powerful tools in the case of weak interaction between the resistive wall mode and the plasma (when kinetic effects are small). Such a situation arises in reversed field pinches (RFPs) without neutral beam injection, which also have resistive wall modes.

Resistive wall mode physics in RFPs is not identical to that in tokamaks. The RWM in RFPs is a current driven instability, in contrast to the pressure driven RWM in tokamaks, which is the primary concern in advanced scenarios. In addition, the interaction between the plasma and the mode play an important role in tokamaks and is negligible in RFPs [31, 32]. This difference actually makes RFPs a good test device for the electromagnetic part of RWM physics. Additional advantages for these tests come from the large number of control coils which, for example, completely cover the surface of the RFX-mod device. In the tokamak case, only a high field side set of coils is typically installed. It should be noted that the current driven RWMs are also investigated in tokamaks, utilizing their high reproducibility to study the RWM control issues. The structure of the current driven RWM at the plasma surface is similar to what is expected for pressure driven RWMs and thus could be used for investigation of the RWM interaction with external currents and walls [33, 34]. At the same time the pressure driven RWM could have quite a different eigenfunction in the plasma core with higher amplitudes of the internal components and a much richer poloidal spectrum [35]. Thus, the plasma-RWM interaction could be quite different for current driven and pressure driven RWMs in tokamaks.

6.5.1.1 Main Idea of RWM Stabilization with External Coils

An external kink mode is, by definition, stable up to its "ideal wall" limit, $\beta_{N,ideal-wall}$, if the plasma is surrounded by a perfectly conducting wall. The perturbed flux from the mode does not penetrate through the wall in this case. Real walls have finite resistivity. It allows slow penetration of the perturbed flux and leads to an unstable RWM. This definition of the RWM contains the key idea of the active RWM stabilization. Any actions which prevent penetration of the flux through the vessel, and thus mimic the action of a perfectly conducting wall on the mode, will stabilize RWMs. In reality, two situations provide this effect depending on the mode rotation:

(a) Fast rotation of the mode: $\omega \gg 1/\tau_{wall}$. Fast rotation of the mode in the laboratory frame produces fast oscillations of the magnetic field at the wall and skin effect prevents flux penetration. In this case, the rotation frequency should be much higher than the inverse wall time. This stabilization mechanism provides *automatic* stabilization up to the ideal wall limit for *all sufficiently fast rotating external kink modes*, but it obviously fails for slow rotating or locked modes.
(b) Slow rotation of the mode, $\omega \sim 1/\tau_{wall}$, or locked mode case. Use of external currents to compensate flux penetration for a particular mode (m, n) is used to handle the slowly rotating case. The external currents mimic the "ideal wall" acting either on the helical component of the mode only, or on all measured field perturbations. The external currents are used by all active RWM stabilization algorithms with external coil currents, and exactly this case is discussed below.

In the slowly rotating case, all stabilizing and destabilizing actions on the mode can be considered as an interaction of different currents in the plasma, in the wall and in the active control coils. The problem can be further simplified if all plasma currents are represented as a surface current at the plasma surface (or as boundary conditions for the magnetic potential or magnetic field at the plasma-vacuum interface). This consideration removes the principal differences between pressure driven RWMs, which are the main concern for advanced scenarios, and current driven RWMs, which can be generated during the plasma current ramp up in tokamaks. It also removes differences between the RFP and tokamak situations. Also the different safety factor profiles lead to different mode numbers in these devices. Based on these points, joint discussion of RWM stabilization in RFPs and tokamaks is given in this section.

6.5.1.2 Algorithms for RWM Stabilization

An overall scheme of resistive wall mode stabilization is shown in Fig. 6.5.The main components of this system are: magnetic sensors, a controller, power amplifiers and active control coils. Magnetic sensors provide information about the unstable mode, in particular the mode amplitude, mode numbers and phase in the laboratory frame. The controller identifies the mode structure and provides suggestions for the required current amplitudes in the control coils. Power amplifiers supply the coils with the suggested currents. Finally, coil currents mimic ideal wall behavior for a given unstable mode or perform other actions requested by the control algorithm. It is important to note that multiple choices are possible for all components of the system and the actual choice influences stability and robustness of the control scheme. In the following, the basic possible options for these components are discussed.

(i) Magnetic Sensors

Typically, magnetic sensors in a tokamak measure either the poloidal (b_θ) or the radial (b_r) component of the magnetic field. Both types of magnetic sensors, as well as a combination thereof, can be used for RWM control. From practical point of view, it is better to use control schemes with a minimal value of the feedback gain G, which relates the flux produced by the control coils to the sensor measurements. A simple analysis with an analytical model using the relationship between the fluxes at the plasma and the resistive wall helps to identify the effectiveness of different types of sensors [36]. It can be shown that the b_θ sensor is equivalent to an ideal sensor with an additional high-pass filtered term. This improves the sensitivity at high frequencies, which is important for the fast reaction of the control scheme. Poloidal sensors have two main advantages over radial sensors:

(i) faster time response as mentioned above, and
(ii) these sensors are decoupled from the control coils, which produce mainly radial magnetic fields and do not mix strongly with poloidal mode components.

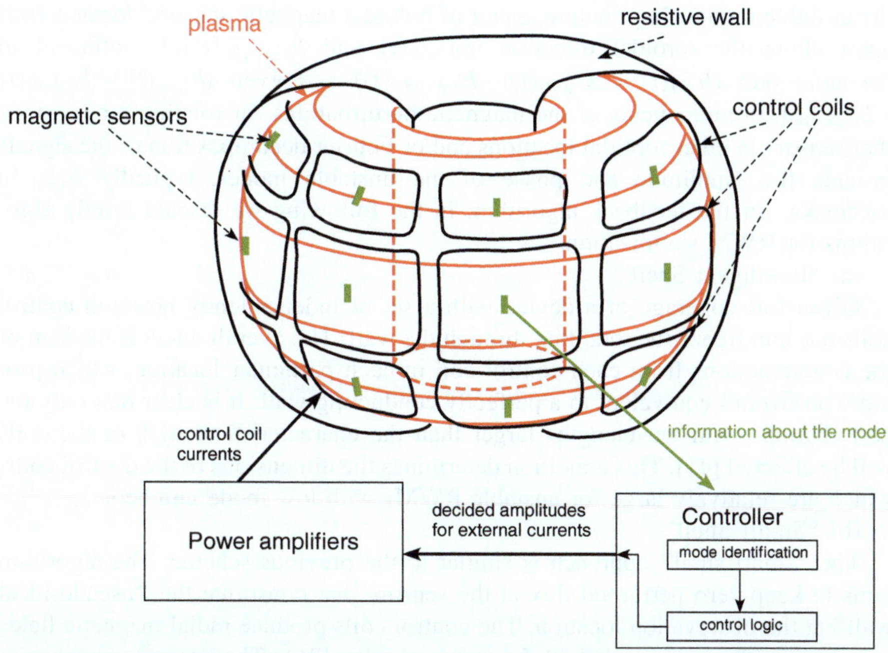

Fig. 6.5 An overall scheme of the resistive wall mode stabilization is shown. Information from the magnetic sensors is used by the controller for mode identification. Control logic proposes actions and power amplifiers provide required currents in the control coils

Advantages of the poloidal sensors are also demonstrated by linear MHD calculations with integrated feedback control taking into account the actual geometry of sensors and active coils (see for example Ref. [37–39]). The combination of poloidal and radial sensors is an even better choice for control [40]. The sensors must be placed inside the conducting wall, otherwise the wall will screen high frequencies due to skin current effect, which would add a substantial time delay into feedback algorithm. Experimental results from different devices are consistent with these predictions [36]. The practical choice of sensors is an optimization process within control theory, which includes optimization of control algorithms, takes into account time delays of different components and assumes particular sensors. Engineering restrictions define possible sensor positions and their maximum number as well as minimal possible time delays for all feedback components.

(ii) Controller

The controller is the core component which proposes actions on the mode, based on the control algorithm. It is important to remember that the optimum choice for an algorithm depends on the available sensors and amplifiers. The controller requires input variables, which are typically the amplitude and phase of the unstable modes. In principle, a pair of poloidal magnetic sensors[3] located 180° apart along the toroidal direction ($b_{\theta,\phi=0}(t)$ and $b_{\theta,\phi=180}(t)$) is sufficient to decouple odd ($b_{odd}(t) = b_{\theta,\phi=0}(t) - b_{\theta,\phi=180}(t)$) and even ($b_{even}(t) = b_{\theta,\phi=0}(t) + b_{\theta,\phi=180}(t)$) components of the magnetic perturbations for rotating modes. Further sensors in other toroidal locations and/or Fourier decomposition of the signals provide the amplitude and phase of the unstable modes, typically $b_{n=1}$ in tokamaks, for the feedback algorithm. In the following we discuss briefly algorithms for RWM stabilization.

(a) "Intelligent Shell"

Given full coverage of a device with a set of independently powered control coils one can freeze the total flux through the wall. The overall effect is the sum of the counteractions from each control coil in each particular location, which provides an overall equivalent to a perfectly conducting wall. It is clear that only the perturbations with wavelengths larger than the characteristic length of the coils will be affected [41]. This condition determines the dimensions of the control coils, which are relatively large for unstable RWMs with low mode numbers.

(b) "Smart Shell"

The "smart shell" approach is similar to the previous scheme. The algorithm aims to keep zero perturbed flux at the sensors that constitute the "pseudo-ideal wall" at the observation location. The control coils produce radial magnetic fields opposite to the measured field from the plasma [36]. The plasma response is obtained by subtracting the "vacuum magnetic field" from the raw signals. The "vacuum magnetic field" is the field produced by the coils in the absence of the plasma, but in the presence of all passive conducting structures. Ideally, the dynamic response of a 3D wall to the external magnetic field perturbations would be subtracted from the raw signals to make cancellation of the perturbed flux from the unstable mode as exact as possible [42].

(c) "Mode Control"

Implementation of a filtering technique adds flexibility to the control schemes. Typically, real time Fourier transform of the signal is used to decompose different harmonics of the signal; in other words, to distinguish modes with different poloidal (m) and toroidal (n)mode numbers: $b_{measured} = \sum_{m,n} b_{m,n} e^{i(m\theta+n\phi)}$. Finally, the control algorithm is applied in Fourier space for each mode $b_{m,n}$. This increases flexibility of the control scheme and allows control of different modes independently. For example, it is possible to suppress all modes except a particular single harmonic, which is allowed to grow, or to simultaneously control multiple modes [43, 44]. One can clearly see that mode control strongly extends the duration of the plasma discharge and is able to keep the amplitudes of unstable modes at a low level (see Fig. 6.6).

[3] Radial magnetic sensors can be used in the same way.

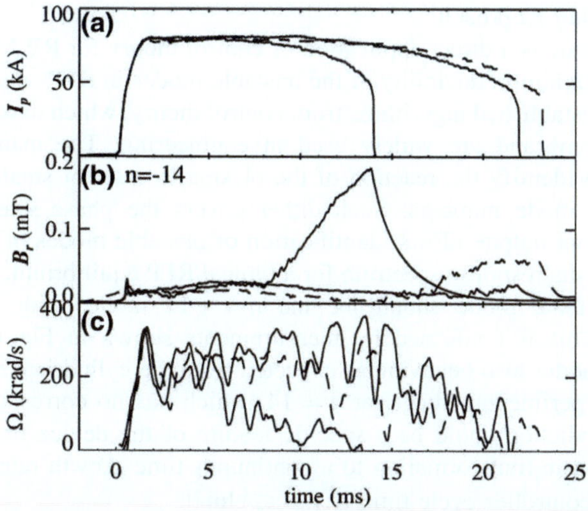

Fig. 6.6 Time evolution of n = −14 resonant mode in the EXTRAP T2R reverse field pinch. **a** Plasma current, **b** radial field amplitude, **c** tearing mode rotation angular velocity. Three discharges are compared: shot 15 863, without feedback (*dotted-dashed line*), shot 15 867 with feedback on all modes except $-2 \leq n \leq 2$ (*dashed line*) and shot 16 369 with feedback on only the main internal RWMs $-11 \leq n \leq -8$ (*solid line*). (Reprinted figure with permission from P.R. Brunsel et al., Phys. Rev. Lett., 93, 225001 (2004) [43] Copyright (2004) by American Physical Society.)

(d) "Selective Virtual Shell"

This is a mode control variant which cancels the radial field for all modes except for a limited set of selected ones (passing modes). Cancellation can be achieved either at the measuring sensor radius or at any radius between sensor and the plasma boundary. The latter variant computes the field distribution in the region external to the plasma and applies conditions of the virtual shell at the prescribed location. Passing modes are either controlled differently or free to grow [45].

(e) "Clean Mode Control"

This variant is a further development of "mode control" and "virtual shell" techniques. The discrete nature of the active control coils produces not only the Fourier harmonics requested by the feedback algorithms but also an infinite sequence of sideband harmonics in the magnetic field. The perfect cancellation of the sensor flux corresponds to the cancellation of the plasma harmonics at the sensor and the aliased sideband harmonics from the coils. Thus, the plasma part is not perfectly canceled in this case. Removing the sideband effect by taking into account the actual coil geometry is the idea of "clean mode control" [46]. The approach uses identical periodicity of the sensors and coils, and for practical computation within the feedback loop truncates the sideband series keeping only the dominant harmonics.

(f) "Dithering Approach"

This technique is a direct application of control theory for RWM identification and control. High reproducibility of the unstable modes in RFPs makes it possible to apply well established algorithms from control theory, which describe control of arbitrary systems and are widely used in engineering. The main idea of this approach is to identify the reaction of the plasma to a set of small perturbations with different mode numbers. Such dither covers the phase space of possible modes. The final output allows identification of unstable modes [47]. Figure 6.7a shows the plasma response spectrum for a typical RFP equilibrium, indicating that the most unstable mode should be the $n = -11$ mode, with an eigenvalue $|\lambda| \approx 1.025$. This is confirmed by measurements shown in Fig. 6.7b, with the neighboring modes also behaving as expected from the cylindrical model. The hot spot in the experimental image, at $n = 14$, which has no correspondence in the theoretical prediction could be a specific feature of the device or of the control system. The basic transformation to a continuous-time growth rate γ depends on the real-time controller cycle time τ_s:$\gamma = \tau_s^{-1} \ln|\lambda|$.

This result is an independent confirmation of the observation that RWM behavior in RFPs can be well described by a circular cylindrical model [44, 48]. The important advantage of the application of control theory is its ability to synthesize a control system directly from measured plasma response data, without any additional models for the plasma behavior [49]. This is a promising approach

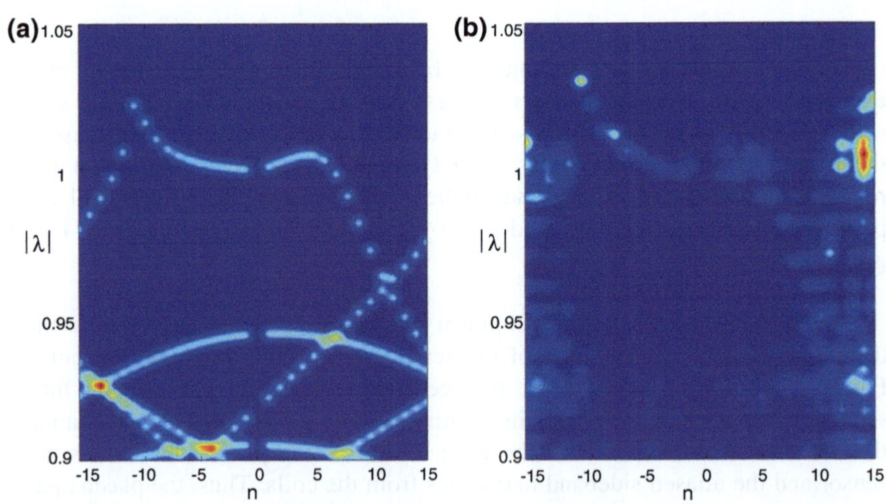

Fig. 6.7 Spectral representation of the plasma response for EXTRAP T2R reversed field pinch is shown. **a** Cylindrical ideal MHD resistive shell modes in theory; as seen through the discrete sensor array. **b** The same matrix from the experimental measurements is shown. (The figures are from K.E.J.Olofsson, et al., Plasma Phys. Control. Fusion 53, 084003 (2011). © IOP Publishing. Reproduced by permission of IOP Publishing. All rights reserved.)

for MHD mode control in the future. The basic assumptions used for such a control scheme are mode rigidity, that the MHD mode structure remains constant and all external actions *affect only mode amplitudes and phases.*

(g) "Active Mode Rotation"

Flexible feedback in RFPs also allows investigation of the case with partially suppressed RWMs and phase shifted feedback. The principal schemes of the feedback action are shown in Fig. 6.8.

Full suppression of the mode amplitude is shown in Fig. 6.8a. Here the plasma mode is cancelled by the feedback field, resulting in the mode amplitude going to zero (this is exactly the case discussed up to now). Partial suppression is obtained with a smaller B_{ext} (reference case $\Delta\phi = 0$), which keeps the mode amplitude at a constant level (Fig. 6.8b). In this case, "partial suppression" means the *growth rate* is forced to zero, but the *mode amplitude* has a constant "finite" value. This allows one to investigate the influence of phase shifted external perturbations, as shown below. For cases (a) and (b), the imaginary part of the gain is zero ($\Delta\phi = 0$). In Fig. 6.8c partial suppression of the mode amplitude is obtained by applying a complex gain with a non-zero imaginary part. This corresponds to introducing an angle ($\Delta\phi \neq 0$) between the plasma and external components and a controllable rotation of the RWM (Fig. 6.9). The mode rotation frequency is independent of the plasma current and has a $f \sim tg(\Delta\phi)$ dependence [50, 51]. This result can be obtained in the frame of cylindrical theory (see next section). Controlled rotation of the mode is not the aim for the RWM stabilization, but rather a proof of principle for phase shifted feedback and exploration of the physical

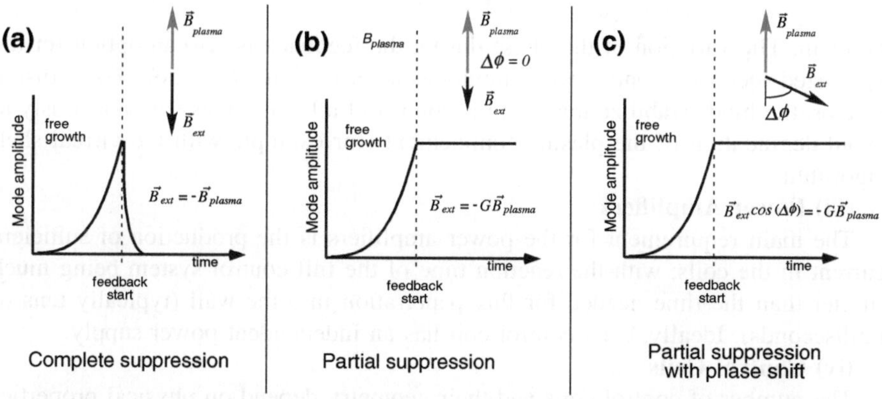

Fig. 6.8 Principal schemes of the closed-loop feedback experiments and their influence on the mode amplitude. **a** The mode amplitude is suppressed completely after feedback start; $\vec{B}_{pl} = -\vec{B}_{ext}$. **b** Partial suppression of the mode amplitude. Feedback keeps the mode amplitude at a constant level with no phase shift, $\Delta\phi = 0$, this is the starting point for our experiments. **c** Partial suppression with phase shift. Feedback keeps the mode amplitude at a constant level and rotates by an angle $\Delta\phi \neq 0$. Note that $|B_{ext}|\cos(\Delta\phi)$ is the same for (**b**) and (**c**). (The figures are from V.Igochine et al., Plasma Phys. Control. Fusion 51, 055008 (2009). © IOP Publishing. Reproduced by permission of IOP Publishing. All rights reserved.)

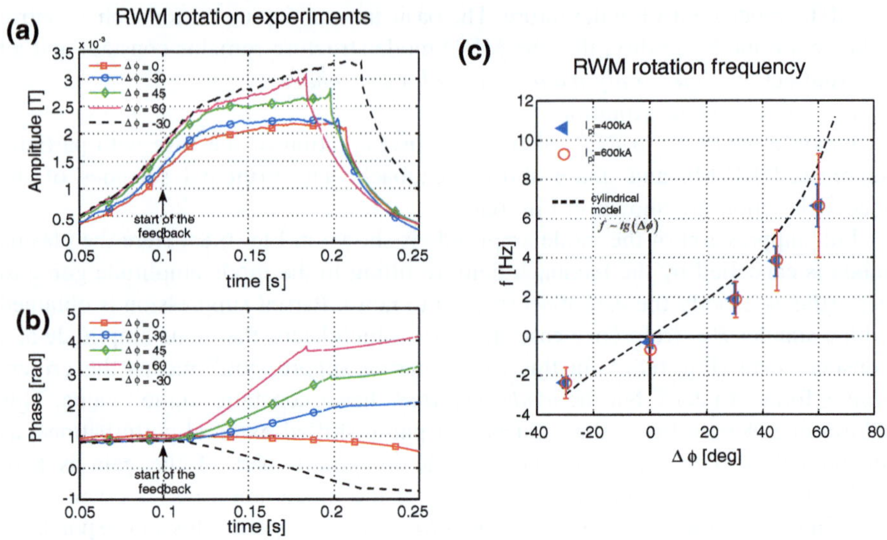

Fig. 6.9 Results of RWM rotation experiments in RFX-mod reversed field pinch with closed-loop feedback are shown. The feedback starting point is indicated by an arrow. **a** Amplitude of the $(m = 1, n = -6)$ mode for different discharges plotted for different $\Delta\phi$; **b** the corresponding phase of the $(m = 1, n = -6)$ mode; **c** dependence of the mode rotation frequency on the phase shift $\Delta\phi$ between the mode and external perturbations for two different plasma currents. The model is discussed in the next section. (The figures are from V.Igochine et al., Plasma Phys. Control. Fusion 51, 055008 (2009). © IOP Publishing. Reproduced by permission of IOP Publishing. All rights reserved.)

problem. The variation of the phase during the feedback is also an option for the optimized feedback control with simultaneous error field correction. For a fusion reactor, the mode stabilization requires control of all harmonics simultaneously to avoid degradation of the plasma confinement (for example with the virtual shell algorithm).

(iii) Power Amplifiers

The main requirement for the power amplifiers is the production of sufficient current in the coils, with the reaction time of the full control system being much shorter than the time needed for flux penetration into the wall (typically tens of milliseconds). Ideally, each control coil has an independent power supply.

(iv) Control Coils

The number of control coils and their geometry depend on physical properties of the mode and on engineering restrictions. In tokamaks, the ballooning effect leads to larger amplitude of the RWM perturbations at the low field side of the tokamak. Control coils located at this position couple well to the unstable kink mode. Fortunately, this coil position is also favorable from the engineering point of view and the control coils in tokamaks are typically located at the low field side. The number of the coils in the toroidal direction is between 6 (DIII-D tokamak) and 9 (ITER tokamak), depending on the vacuum vessel design. It has been shown numerically that a single row of control coils is able to stabilize RWMs [52]. Locating the coils inside the vacuum vessel is clearly more favorable for the same reasons as internal location of the sensors. Finally, control of edge localized modes (ELMs) strongly influences design and number of the internal control coils. It was found in several tokamaks that the upper and lower coils are able to suppress or to mitigate ELM instability (see for example [53, 54] and references therein). Integration of these coils into RWM control provides clear beneficial effects for RWM stabilization, because of better coverage of the plasma surface with control coils. The ITER design foresees three toroidal arrays of 9 coils each for RWM and ELM controls (27 coils in total) which is expected to be sufficient for RWM stabilization. At the same time, the plasma surface in RFPs is fully covered by the control coils (for example: RFX-mod: 192 coils, EXTRAP T2R: 128 coils) which makes these devices valuable "test beds" for different control techniques as one can see from previous results.

6.5.1.3 Cylindrical Model of RWM and Feedback Control

There are several simple models that allow one to implement different feedback schemes and sensor types [36, 55]. In this chapter, we discuss the model proposed by Pustovitov [56] which is rather flexible and allows the implementation of different sensors and control algorithms. This approximation is even able to provide good descriptions of real experiments, in particular circular high aspect ratio plasmas in RFPs. It assumes circular cylindrical geometry of the system, which implies independent treatment of each (m, n) mode and neglects toroidal coupling. The derivation here is given for single helicity, i.e., assuming all currents in the plasma, in the wall and in the control coils have the same helicity as the unstable mode. In the case of several modes this approach can be generalized to a system of equations. In this case the total radial perturbation is the sum of all considered modes:

$$B_r \equiv \vec{B} \cdot \nabla r = \sum_{m,n} b_{r,mn}(r) e^{(im\theta + in\phi)} e^{(\gamma_{mn} + i\omega_{mn})t} \tag{6.2}$$

where $b_{r,mn}$ is the mode amplitude and (m, n) is omitted in the following. The linear growth rate, γ, and the mode frequency, ω, describe the time behavior of the perturbations. If a linear plasma response is assumed, the current and pressure profiles in the plasma are not affected by the feedback system and all information regarding the mode and its interaction with plasma is contained in a specially introduced complex parameter Γ_m (defined in equation 6.5), independent on the perturbation amplitude.

The main regions of the model are shown in Fig. 6.10. The diffusion equation for the radial component of the amplitude of the perturbed magnetic field, b_r, in the resistive wall can be written in the following form:

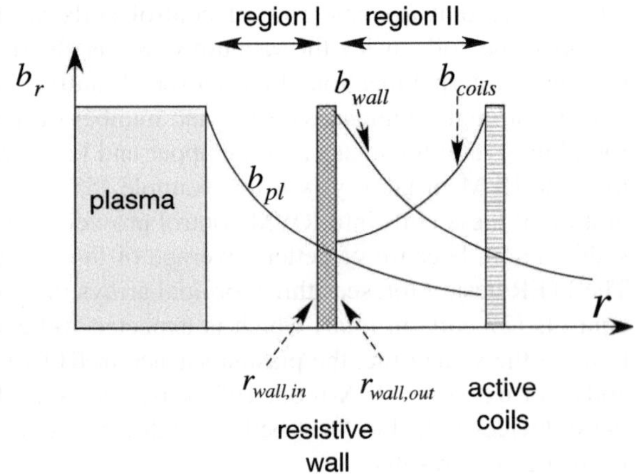

Fig. 6.10 Representation of Pustovitov's model is shown. External coils are located outside the resistive wall

$$\frac{\partial b_r}{\partial t} = \frac{1}{\mu_0 \sigma} \frac{\partial^2 b_r}{\partial r^2} \approx \frac{1}{\mu_0 \sigma} \left(\left.\frac{\partial b_r}{\partial r}\right|_{r=rw,out} - \left.\frac{\partial b_r}{\partial r}\right|_{r=rw,in} \right) \frac{1}{\delta} \tag{6.3}$$

where σ and δ are resistivity and thickness of the wall respectively. In the case of a locked mode, $\omega = 0$, the jump of the derivative at the wall can be written in the following form:

$$\left.\frac{\frac{\partial b_r}{\partial r} r}{b_r}\right|_{r=rw,out} - \left.\frac{\frac{\partial b_r}{\partial r} r}{b_r}\right|_{r=rw,in} = (\gamma + i\omega)\mu_0 \sigma \delta r_w = (\gamma + i\omega)\tau_w, \tag{6.4}$$

here, all plasma effects as well as the drive of the mode are contained in:

$$\left. r\frac{\frac{\partial b_r}{\partial r}}{b_r}\right|_{r=rw,in} \equiv \underbrace{-(m+1)}_{\substack{vacuum\\ part}} - \Gamma_m \tag{6.5}$$

Outside the resistive wall, region II, the magnetic field can be represented as a solution of the Laplace equation in vacuum:

$$b_r = \left(b_{pl} + b_{rw}\right)\left(\frac{r}{r_w}\right)^{-m-1} + b_{coils}\left(\frac{r}{r_w}\right)^{m-1} \tag{6.6}$$

The derivative jump at the outer side of the wall follows directly from this representation of the magnetic field and calculations at the wall position:

$$\left. r\frac{\frac{\partial b_r}{\partial r}}{b_r} \right|_{r=rw,out} = -(m+1) + 2m\frac{b_{coils}}{b_{sum}} \tag{6.7}$$

where $b_{sum} = b_{pl} + b_{rw} + b_{coils}$. Combination of (6.4), (6.5) and (6.7) gives dispersion relation for the RWM:

$$\gamma\tau_w = \Gamma_m + 2m\frac{b_{coils}}{b_{sum}} \tag{6.8}$$

A more general derivation of the model, starting from toroidal equations, is given in Ref. [57]. The mode is stable if the growth rate of the mode is negative, $\gamma<0$; otherwise the mode is unstable, $\gamma > 0$. Without any external currents, $b_{coils} = 0$, the growth rate of the mode is $\gamma\tau_w = \Gamma_m$. Feedback actions can be introduced via the b_{coils} value, which includes magnetic field from all external currents. For our analysis, it is necessary to prescribe a dependence of b_{coils} from the input signal: $b_{coils} = -Gb_{input}$, where G is the feedback gain. In case of radial field sensors this expression reduces to $b_{coils} = -Gb_{sum}$, because these sensors see the radial magnetic field from all components of the system. The final dispersion relation is

$$\gamma\tau_w = \Gamma_m + 2m\frac{-Gb_{sum}}{b_{sum}} = 2m(G_0 - G) \tag{6.9}$$

The system is stable only if $\gamma<0$, giving the minimal required gain G_0 as

$$G > G_0 = \frac{\Gamma_m}{2m} \tag{6.10}$$

The same procedure can be applied for the poloidal field sensors [40]. Using $div\vec{B} = 0$ and neglecting the small toroidal component, the poloidal field inside the resistive wall, in region I, can be expressed in terms of the total magnetic field:

$$b_{\theta,in} = \frac{m+\Gamma_m}{m} b_{sum}. \tag{6.11}$$

The gain required for stabilization follows directly from the dispersion equation:

$$G > G_0\frac{m}{m+\Gamma_m}. \tag{6.12}$$

The RWM feedback suppression with poloidal sensors inside the resistive wall requires a lower gain factor compared to the radial field sensor case. Application of the model to the internally located control coils is discussed in a separate work by the same author [25].

The active mode rotation experiments discussed above can be also analyzed with this model [58]. In the thin wall approximation, the diffusion equation for the magnetic field in the resistive wall reduces to :

$$\tau_w \frac{\partial b_r}{\partial t} = \Gamma_m b_r - \Gamma_m^0 b_{coils} \tag{6.13}$$

with $\Gamma_m^0 = -2m$. The feedback in the experimental situation can be represented in the following form:

$$b_{coils} + \tau_f \frac{\partial b_{coils}}{\partial t} = -G b_r, \tag{6.14}$$

where $\tau_f = L_f / R_f$is the characteristic time of the feedback system with effective inductance L_f and effective resistivity R_f of the feedback circuit. The feedback system in the experiment has a reaction time much shorter than the wall time: $\tau_w \gg \tau_f$. Taking this into account one can combine these equations into the dispersion relation for a RWM:

$$\tau_w \frac{\partial b_r}{\partial t} = b_r \left(\Gamma_m - G \Gamma_m^0 \right). \tag{6.15}$$

The mode rotation is not zero in this case, $\omega \neq 0$, which gives the following relation:

$$n \omega \tau_w = \mathrm{Im}\left(\Gamma_m - G \Gamma_m^0 \right). \tag{6.16}$$

The feedback algorithm applies a phase shift (see Fig. 6.9). Thus, feedback gain can be expressed in the form

$$G = |G| e^{i\Delta\phi}. \tag{6.17}$$

This gives for the mode rotation frequency

$$\omega = \omega_0 + \frac{\Gamma_m^0}{n \tau_w} |G| \sin(\Delta\phi). \tag{6.18}$$

The feedback scheme also keeps the mode amplitude constant, which implies zero growth rate, $\gamma = 0$, and provides the required value for Γ_m^0.

$$\gamma \tau_w = \mathrm{Re}\left(\Gamma_m - G \Gamma_m^0 \right) \tag{6.19}$$

$$\gamma_0 \tau_w = -\Gamma_m^0 |G| \cos(\Delta\phi) \tag{6.20}$$

The final expression for the mode rotation frequency becomes

$$\omega = \omega_0 + \frac{\gamma_0}{n}\tan(\Delta\phi). \tag{6.21}$$

This is exactly the experimentally observed behavior in Fig. 6.9.

6.5.1.4 3D Effects for RWM Stability and Control

The simple model discussed above is a useful tool for the basic understanding of the control processes and can even give correct dependencies, but realistic 3D simulation is clearly required to predict the actual growth rates. In this case, realistic current distributions in the conducting structures and in the control coils have to be considered. Since the physical background is just an electromagnetic interaction, the problem is complex but solvable. This is shown by the implementation of the real wall in different codes: VALEN [59], STARWALL [60, 61] and CarMa [62, 63]. Modeling of the RWM shows that realistic 3D structures are required for growth rate predictions even in case of circular RFP plasma [64]. All relatively large gaps or holes in the conducting wall result in detrimental effects on the mode stability and reduce the beta limit. At the same time, additional conducting structures within the wall could provide a substantial stabilizing effect. In ITER, blanket modules will have a beneficial effect on RWM growth rates, which will largely compensate the detrimental effect due to the presence of ports. The final calculated RWM growth rates in ITER are close to the toroidally symmetric 2D case (without ports and blanket modules) [65].

The next important effect is the mode coupling via the wall. Unstable RWMs with different toroidal mode numbers (n) become coupled through the 3D wall, which modifies the resulting stability of the plasma. The coupling pattern depends on the number of holes, their periodicity and shapes. Moreover, multiple possible locations for mode locking become possible and these locations have slightly different growth rates for the same mode (the same toroidal mode number, n) [61]. The reduced set of coils, compared to the full coverage situation, requires higher currents in the coils for mode control, and the position of these coils also becomes important. Dedicated experiments in RFX-mod with a downgraded set of control coils show that equally distributed coils are not able to suppress a mode which has a helicity similar to the number of coils. The mode moves toroidally and grows in the gaps between the coils. The same number of unequally distributed coils is able to control the mode because no gaps for mode control are present and the mode behaves rigidly. Thus, the possibility of the mode to grow in the gaps depends on the mode helicity and the number of control coils.

The finite thickness of the wall provides additional effects on the RWM dynamics. The standard thin-wall approximation, which is used in most of the formulas in this chapter, assumes constant amplitude of the perturbed magnetic field across the wall (for examples 6.1 and 6.3). This is identical to the assumption of a skin depth larger than the wall thickness, which is not always the case. The inclusion of the skin effect always makes the growth rates larger than in the thin-wall approximation. The difference increases with increase of the growth rates, but it is also noticeable even for the conventional slow RWMs [66]. Simulations with CARMA code, considering volumetric structure of the wall, confirm this result [67]. In the case of the static RWM, the finite thickness of the wall does not influence the stability boundary. The situation changes for relatively fast rotating modes. Mode rotation can open stability window above some critical level [68, 69].

The next step in the direction of the full RWM modeling is the implementation of the real tokamak environment (3D wall + feedback) into the codes together with a good model for the plasma-RWM interaction and a feedback algorithm. This is currently the subject of intensive research. The available theoretical works provide the basis for coupling the full 3D electromagnetic treatment of the wall, plasma calculations and linear MHD codes which includes the influence of the plasma on the mode (e.g. Ref. [70]). The first results demonstrating the feasibility of the code coupling already exist [71] but further improvement of the codes is needed. Finally, such tools will be used to predict RWM stability in ITER and DEMO. One of the most crucial points in this modeling is the interaction of the RWM with the plasma, which is taken into account with a linear MHD code. This is discussed in the next section.

6.5.2 RWM Interaction with Plasma

Experimental evidence for the importance of plasma—mode interactions was already obtained from the beginning of RWM research in tokamaks [13–15]. This evidence immediately posed questions about the physical mechanism behind the interaction and the possibility of implementing this interaction in linear MHD descriptions of the plasma. It was soon realized that Alfvén continuum damping alone is not sufficient to explain the experimental observations (e.g. a stable RWM region at very low plasma rotation which will be discussed later). The interaction is stronger than predicted by the continuum damping alone and a more detailed analysis of the mode-plasma interaction is necessary. The physical mechanism which could provide this additional damping is an interaction of the mode (wave) with plasma particles. The main problem is to describe this interaction (energy transfer between the RWM and plasma particles), which includes an interplay with both thermal particles as well as fast particles. The difficulties arise when attempting to formulate a set of assumptions that enable the kinetic problem (wave-particle) to be reduced to an MHD formulation (fluid description). If such a description were found, it would be possible to solve the problem self-consistently within the linear MHD approach. The simplest assumption which can be made is to represent this interaction as a fluid viscous force $\vec{\nabla} \cdot \overleftrightarrow{\Pi}_1$ in the linearized MHD momentum balance equation [11]:

$$\rho_0(\gamma + in\Omega_{0,pl})\vec{v}_1 = -\vec{\nabla} \cdot \vec{p}_1 + \vec{j}_1 \times \vec{b}_0 + \vec{j}_0 \times \vec{b}_1 - \vec{\nabla} \cdot \overleftrightarrow{\Pi}_1 - \rho_0 \vec{U}(\vec{v}_1) \quad (6.22)$$

where ρ is the density, v is the fluid velocity, b is the magnetic field, j is the current, p is the fluid pressure, n is the toroidal mode number, $\gamma = \gamma_0 + i\omega_{RWM}$ is the complex growth rate, and $\vec{U} = \vec{v}_1 \times \left(\frac{\vec{v}_0}{R} \times \hat{\vec{R}}\right) + \left(\vec{v}_1 \cdot \vec{\nabla}\right)\vec{v}_0$. Equilibrium quantities and perturbed quantities are denoted by subscript 0 and 1, respectively. The rotation frequency of the plasma in equilibrium $\Omega_{0,pl}$ is non-uniform and depends only on the equilibrium flux function.

The first model of this kind was proposed by Hammet and Perkins in 1990 [72]. It assumes an approximation for the Landau damping in the sound wave range. This model is often called the "sound wave damping" model. The force part has the following representation in this model:

$$\vec{\nabla} \cdot \overleftrightarrow{\Pi}_1 = \kappa_{\parallel} \sqrt{\pi} \left| k_{\parallel} v_{th_i} \right| \rho_0 \vec{v}_1 \cdot \hat{b}_1 \hat{b}_1, \quad (6.23)$$

where $k_{\parallel} = (n - m/q)/R$ is the parallel wave vector and v_{th_i} is the ion thermal velocity. The model contains a free "geometrical factor", $\kappa_{\parallel}$, which varies from 0.1 to 1.5. The model, developed by A. Bondeson and M.S.Chu a few years later, has no free parameters and takes into account mode coupling [73]. Thus, the "drag force" for a particular (m, n) mode depends on the toroidally coupled $(m \pm 1, n)$ modes and the modes are no longer independent. Predictions using this model were in relatively good agreement with experimental measurements at that time, as obtained with the MARS-F code for the DIII-D case shown in Fig. 6.11 [74].[4]

New problems come into play with reduction of the plasma rotation. It was clear from first estimations that plasma rotation in ITER will be significantly weaker than that in the present day devices, since the relative input torque from neutral beam injection will be smaller (taking into account the plasma volume in ITER). Thus, predictions at low plasma rotation were identified as very important for ITER. In order to address this issue, one neutral beam source was redirected in DIII-D to make balanced injection possible and to reduce the resulting torque on the plasma [74, 75]. JT-60U [76] and NSTX [77] were also operating in this regime and a strong reduction of the rotation threshold was observed. In spite of the relatively good agreement with the first results in DIII-D, it was soon realized that further reduction of the plasma rotation is possible, which does not agree with the Bondeson model

[4] Measurements of RWM stability in the presence of a strong neutral beam torque required some form of "magnetic braking" to reduce the plasma rotation to the critical value. Two braking methods were used: (1) reduction of the current in the error correction coils, allowing the uncorrected part of the intrinsic error field to create a drag on the plasma rotation; (2) application of an additional nonaxisymmetric field with an external set of coils.

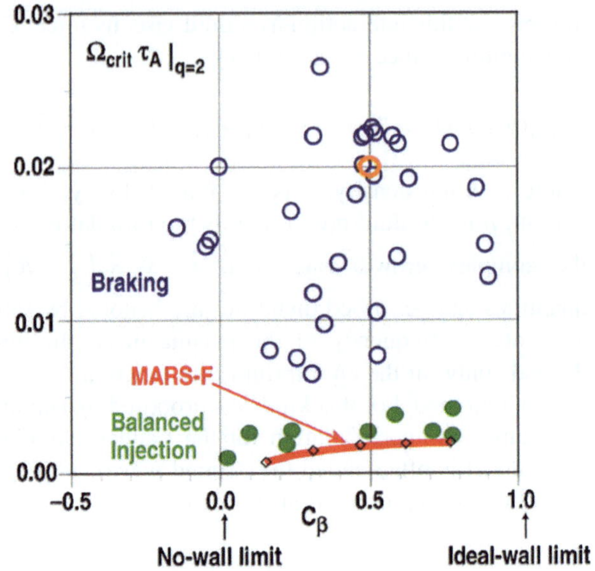

Fig. 6.11 Comparison of the critical plasma rotation at q = 2 resonant surface for RWM onset with MARS-F results (*open diamonds and red curve respectively*) from Ref [74]. Experimental points for balanced NBI cases are shown in *green* (*filled circles*). Results of the braking experiments are shown in *blue* (*open circles*)

predictions shown in Fig. 6.11. Here, the plasma rotation frequency required for RWM stabilization is plotted versus normalized distance between the "no-wall" and "ideal wall" limits, $C_\beta = \left(\beta_N - \beta_N^{no-wall}\right)/\left(\beta_N^{ideal-wall} - \beta_N^{no-wall}\right)$. Stable operation is not possible below a critical rotation value, which is in contradiction with stable operation at very low rotation frequency. Thus, some assumptions of the Bondeson model are no longer valid in these low rotation plasmas and additional stabilization mechanisms have to be taken into account. In the model, several low frequency resonances were assumed to be zero, in particular the magnetic drift frequency and the diamagnetic drift frequency. To understand the problem including these resonances, it is useful to discuss another approach for investigation of plasma-RWM interaction. Up to now, only so called "self-consistent" MHD calculations have been discussed. In these calculations the eigenfunction is changed "self-consistently" according to linear MHD equations, but the interaction with the plasma is calculated with several assumptions. Another way to address the same problem is the so called "perturbative" approach.

The growth rate of the mode, γ, can be obtained from the perturbed MHD energy. In the RWM case we obtain the following expression [78]:

$$\gamma\tau_w^* \simeq -\frac{\delta W_{MHD}^{\infty}}{\delta W_{MHD}^{b}}, \tag{6.24}$$

where, δW_{MHD}^{∞} and δW_{MHD}^{b} are the fluid potential energy with a conducting wall at ∞ and b, respectively. The normalized wall time is $\tau_w^* = \tau_w\left(1 - r_a^{2|m|}\big/r_b^{2|m|}\right)\big/(2|m|)$, where r_a and r_b are the plasma minor radius and the wall minor radius respectively.

The fluid energy includes the plasma and vacuum contributions. This formulation can be extended to include also the influence of the kinetic effects on mode stability. The stability of the RWM is then determined by a dispersion relation derived from the kinetic-MHD energy principle for low frequency modes [79, 80].

$$\gamma\tau_w^* \simeq -\frac{\delta W_{MHD}^{\infty} + \delta W_k}{\delta W_{MHD}^{b} + \delta W_k} \tag{6.25}$$

where δW_k is the kinetic part of the potential energy. In this approach, kinetic codes use fixed eigenfunctions from linear MHD to evaluate the interaction between the plasma and the mode for a given particle distribution (in another words to obtain δW_k).

The "perturbative" approach differs from the "self-consistent" approach in the following aspects:

(1) Kinetic modifications of the RWM structure are not taken into account. The mode is assumed to be "rigid", which means that the shape of the mode eigenfunction is not affected by the plasma.
(2) Kinetic effects are calculated accurately by evaluating kinetic integrals.
(3) The fluid continuum damping (due to the mode resonance with the Alfven waves) is not present in the perturbative calculations.

A comparison of the models is presented also in Table 6.1, where advantages and disadvantages of the approaches are marked with "+" and "−", respectively. Both perturbative and self-consistent approaches are realized in the kinetic-MHD code MARS-K [81].

Note that this kinetic—MHD energy principle treats the bulk plasma as an ideal one. Therefore, it is implicitly assumed that no island can develop at the rational surfaces $q = m/n$. Additionally, (6.24) and (6.25) allow an accurate evaluation of the RWM growth rate only near the point of marginal stability of the plasma. In spite of all the restrictions, very useful results can be obtained from the "perturbative" approach, which clarifies the plasma—mode interaction. In what follows we discuss two questions to understand the physics of the resistive wall mode:

(1) Which resonances/particles have a strong impact on mode stability?
(2) Where do the plasma-mode interactions mainly occur?

Qualitative answers to the two questions can be obtained from the kinetic approach. Formally, δW_k is a change of the potential energy, which can be calculated for the kink mode as follows [21, 82]:

$$\delta W_k = \frac{1}{2}\int d\Gamma\left(mv_{\parallel}^2 + \mu B\right)\delta f \sum_m \vec{\kappa}\cdot\vec{\xi}^{(m)*}(r,t)e^{-i(n\phi - m\theta)} \tag{6.26}$$

Here, ϕ is the toroidal angle, θ is the poloidal angle, $\vec{\kappa}$ is the curvature vector ($\vec{\kappa} = \vec{b}\cdot\nabla\vec{b}$, $\vec{b} = \vec{B}/B$), and $\delta f = f - f_0$ is the perturbed part of the distribution

Table 6.1 Comparison of the "self-consistent" and "perturbative" approaches

« Self-consistent » approach (*codes* MARS-K, CASTOR-FLOW, etc.)	« Perturbative » approach (*codes* HAGIS, MISK, MARS-K, etc.)
System of linear MHD equations is solved	Kinetic integrals are calculated
(−) Approximation for kinetic effects	(+) Correct calculation of the kinetic effects
(+) Plasma influences on the mode structure	(−) The mode structure is fixed
(+) Interactions with Alfven continuum are present	(−) No interactions with Alfven and sound wave continuum are present

Advantages and disadvantages are marked by "+" and "−", respectively. Both approaches are implemented in MARS-K code

function in the phase space volume element ($\Gamma = d^3xd^3v$). This integral can be simplified for circular plasmas in the large aspect ratio approximation. It is possible to show analytically that the resulting expression for δW_k contains resonant conditions:

$$\delta W_k \sim \sum_{l=-\infty}^{\infty} \frac{(\omega_{RWM} + i\gamma_0 - n\omega_{E\times B})\frac{\partial f_j}{\partial \varepsilon} - \frac{1}{eZ_j}\frac{\partial f_j}{\partial \Psi}}{\langle \omega_D^j \rangle + l\omega_b^j - i\nu_{eff}^j + n\omega_{E\times B} - \omega_{RWM} - i\gamma_0}. \tag{6.27}$$

Here, f_j is the distribution function of the particles j, ε is the pitch angle, ω_{RWM} is the RWM frequency in the plasma frame, Ψ is the magnetic flux, Z_j is the effective charge, and $\nu_{eff}^j$is the collision frequency.[5] The first four frequencies in the denominator are the precession drift, bounce, collision, and $E \times B$ frequencies, respectively. The $E \times B$ frequency is $\omega_{E\times B} = \omega_\phi - \omega_{*i}$, where ω_ϕ is the toroidal rotation frequency and ω_{*i} is the ion diamagnetic frequency. For thermal particles the Doppler shifted mode frequency, $\omega_{E\times B} - \omega_{RWM}$, can resonate with the following frequencies:

- the transit frequency of the passing particles, $\omega_t \sim v_{th}/R$;
- the bounce frequency of the trapped particles, $\omega_b \sim \sqrt{r/R}(v_{th}/R)$;
- the precession drift motion, $\omega_d \sim \rho/r(v_{th}/R)$;

leading to large and complex δW_k. Here, $\rho = mv_\perp/eB$ is the Larmor radius and v_{th} is the thermal velocity. These frequencies are quite different: $\omega_d << \omega_b < \omega_t$. Thus, a scan of the plasma rotation should identify several resonances. A scan of δW_k as a function of plasma rotation as calculated for a JET discharge is shown in Fig. 6.12 [21]. The thermal particle distribution function is assumed. One can see significant changes in the kinetic part of the potential energy with changes of the plasma rotation. An increase of the δW_k corresponds to an increase of the mode stability. Such an increase is seen close to the precession drift frequency and bounce frequency where the mode resonates with the motion of the particles.

[5] In practice, only the first bounce harmonics l are important. For $l > |4|$ the resulting integral provides negligible changes of δW_k.

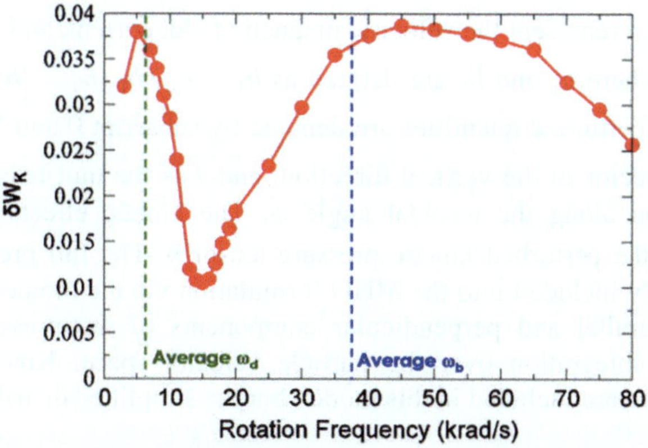

Fig. 6.12 (*color online*) The real part of the change in the potential energy of the kink mode as a function of the plasma toroidal rotation. The influence of the kinetic damping changes significantly as the rotation varies. (The figures are from I.T.Chapman et al., Plasma Phys. Control. Fusion 51, 055015 (2009). © IOP Publishing. Reproduced by permission of IOP Publishing. All rights reserved.)

Using the populations of marker particles, which are evenly distributed in phase space, it is possible to identify the place of dominant plasma—mode interaction with a kinetic code [21]. While the eigenfunction is large near the edge, calculations show that the strongest interaction can be occur at mid-radius. It should be noted that, in these calculations, a particle is considered as lost when its orbit reaches the separatrix. This means that the role of particles near the plasma edge may be underestimated. The influence of faster particles could also modify the result, but the importance of the internal part of the eigenfunction is clear.

It has been shown that the low-frequency RWM can resonate with both the precession drift frequency and bounce frequencies of thermal ions (depending on the actual distribution of plasma particles). Thus, the assumption in the Bondeson model [73], $\omega_d = \omega_b = 0$, is not valid anymore and these resonances have to be treated carefully. A recently developed model by Liu et al. [83] takes into account all of these resonances. The magnetic drift frequency and diamagnetic drift frequency are no longer zero. The model treats the problem in full toroidal geometry where the kinetic integrals are evaluated. The kinetic effects are introduced into the MHD equations in a way different from (6.22). The equation now is written in the following form [81]:

$$\rho_0(\gamma + in\Omega_{0,pl})\vec{v}_1 = -\vec{\nabla}\cdot\overset{\leftrightarrow}{p} + \vec{j}_1 \times \vec{b}_0 + \vec{j}_0 \times \vec{b}_1 - \rho_0\left[2\Omega_{0,pl}\hat{\vec{Z}} \times \vec{v}_1 + (\vec{v}_1 \cdot \nabla\Omega_{0,pl})R^2\nabla\phi\right] \tag{6.28}$$

$$\overset{\leftrightarrow}{p} = p\overset{\leftrightarrow}{I} + p_{\parallel}\hat{\vec{b}}_1\hat{\vec{b}}_1 + p_{\perp}\left(\overset{\leftrightarrow}{I} - \hat{\vec{b}}_1\hat{\vec{b}}_1\right) \tag{6.29}$$

Here $\vec{v}, \vec{b}, \vec{j}, \overleftrightarrow{p}$ represent the velocity, magnetic field, current, and pressure tensor respectively, where $\hat{\vec{b}}_1$ and b_0 are defined as $\hat{\vec{b}}_1 = \vec{b}_1 / b_0$, $b_0 = \left|\vec{b}_0\right|$. Equilibrium quantities and perturbed quantities are denoted by subscript 0 and 1, respectively. $\hat{\vec{Z}}$ is the unit vector in the vertical direction, and $\overleftrightarrow{I}$ is the unit tensor. $\Omega_{0,pl}$ is the plasma rotation along the toroidal angle ϕ. The kinetic effects enter into the equations via the perturbed kinetic pressure tensor $\overleftrightarrow{p}$. The full pressure tensor is self-consistently included into the MHD formulation via the momentum equation, (6.28). The parallel and perpendicular components of the pressure tensor are calculated by integration over the particle velocity space. Kinetic resonances discussed before are included in this model, but the simplified distribution function is assumed.

As was pointed out, any inclusion of the kinetic effects into the MHD description requires a set of approximations which will work only if all of the important physical effects are taken into account. In spite of the fact that the Liu model includes a much more careful treatment of the problem, some assumptions were made. It neglects the perturbed electrostatic potential, assumes zero banana width for trapped particles and no finite Larmor radius (FLR) corrections to the particle orbits, etc. Thus, there is no guarantee that all of the important effects are present, especially if one keeps in mind the history of the model development. At the same time, "perturbative" calculations with kinetic codes could be extremely useful for addressing the same problem from the other side. Although it is not possible now to exactly benchmark the two approaches, one could get an idea about the possible influence of the plasma on the mode from the "perturbative approach" and about changes of the eigenfunction from "self-consistent" calculations. Comparison of the results from these two approaches gives the best overview of the problem. The most natural way is to merge the two approaches in the future. Such merging could be done, for example, by extending kinetic codes to perform self-consistent calculations. These calculations will be probably very time-consuming, but they are necessary to benchmark the plasma-RWM interaction model in several cases. This model would finally be included into a linear MHD code which will be coupled to 3D external structures, as discussed in Sect. 6.5.1.4.

At present, a simultaneous calculation with the "perturbative" and "self-consistent" approaches is the best way to address the RWM stability problem. In this way, estimates of kinetic effects and changes of the eigenfunction due to interaction with the plasma can be obtained. Results of such calculations for ITER are discussed in Sect. 6.6.

6.5.3 Experimental Evidence of Kinetic Effects on RWM Stability

The complex problem of the interaction between plasma particles and the RWM was discussed in the previous section. The RWM stability encompasses many effects. It is a challenging task to separate the kinetic effects in experiments and to measure them. In spite of these difficulties, some measurements show the importance of the kinetic effects. As discussed before, the stability of the RWM should vary with changes in the plasma rotation. The active "MHD spectroscopy" technique could be used to characterize the stability of the RWM. In these experiments the plasma response to an externally applied quasistatic $n = 1$ magnetic field was measured in DIII-D for various plasma rotations [84]. The main result was that the changes in the RWM stability depend on the plasma rotation and are in qualitative agreement with predictions of kinetic calculations. The features at low rotation seem to be directly related to the resonance with the precession frequency of the thermally trapped ions. This effect is expected from kinetic calculations.

Analysis of the plasma stability in NSTX with and without feedback at low plasma rotation also points out the importance of kinetic effects, since these effects could explain the observed variation of the RWM stability with evolution of the rotation profile [85].

These results are the first experimental evidence for the importance of kinetic effects. The results are promising, but further investigations are required to prove the impact of kinetic effects in experiments and ideally to measure the magnitudes of these effects.

6.6 RWM Stability in ITER

The ITER project [86–88] will be one of the most important experiments in fusion research for decades to come. Thus, accurate predictions for ITER operation in advanced scenarios are of particular interest. The determination of the RWM stability boundary in ITER, and verification of that boundary during ITER operation are necessary points on the way to DEMO [89], which will be the first prototype fusion energy reactor. Accurate evaluation of the stability boundary will be especially important for advanced tokamak operation in DEMO [90–93]. In this case, these predictions will influence strongly even the size of the device. In the next two sections the influence of the thermal and fast particles in ITER on RWM stability are discussed. The new model [83] was used to study RWM stability in ITER together with perturbative calculations. Both types of calculations were carried out with the MARS-K code, which includes perturbative and self-consistent approaches.

6.6.1 Influence of the Plasma Rotation on the RWM Stability

Research into the stabilization effects of precession resonances at slow rotation velocities in ITER is presented in ref [94]. The calculations were based on the steady state 9 MA scenario in ITER with high pressure and high bootstrap current fraction, $\beta_{N,no-wall} \approx 2.7$, $\beta_{N,ideal-wall} \approx 3.8$ (scenario 4) [95]. The predicted rotation at the q = 3 surface is less than 0.25 % $\omega_{Alfve'n}$ [94]. (The q = 2 surface, which is typically used for critical rotation value (e.g. Fig. 6.11), is absent for these ITER plasmas.) The predicted plasma central rotation, ω_0, is less than 2 % $\omega_{Alfve'n}$. One should note that there is significant uncertainty in the predicted rotation amplitude and rotation profile in ITER, and that the real rotation values could be much smaller.

For steady-state advanced scenarios, the self-consistent calculations predict full stabilization of the RWM at very slow plasma rotation (less than 0.2 % of the Alfvén speed at the plasma centre) and moderately high plasma pressures $\left(C_\beta = \left(\beta_N - \beta_N^{no-wall}\right)/\left(\beta_N^{ideal-wall} - \beta_N^{no-wall}\right) = 0.4\right)$. It is important that no lower rotation threshold, below which the mode becomes unstable, was found in the calculations (see Fig. 6.13, right). Instead, an upper stability threshold was found. This contradicts the previously discussed Bondeson model predictions, in which a slow rotation threshold is present, see Fig. 6.11. Existence of an upper threshold could be due to the inclusion of the precessional resonances in the simulations. Further increase of plasma rotation could again stabilize the mode above the upper limit due to resonance damping, continuum damping, and inclusion of fast particle effects or any combination of these processes. As for ITER, the results show the mode to be stable well above the "no wall" limit.

Even more optimistic results were obtained by the "perturbative" approach, where the fixed eigenfunction of the ideal kink mode (from fluid RWM calculations) is used to evaluate the kinetic integrals (see Fig. 6.13, left). The difference in the results of the perturbative and non-perturbative calculations is partially due to the modification of the RWM eigenfunction, and self-consistent determination of the mode eigenvalue in the latter approach as discussed in the previous sections.

In summary, for a plasma toroidal rotation speed up to the predicted value for ITER, the kinetic damping of the RWM is mainly provided by precessional drift resonances of trapped thermal particles. This damping stabilizes RWM up to β_N values half way between the "no wall" and "ideal wall" limits $\left(C_\beta = 0.4\right)$.

6.6.2 Influence of the Alpha Particles on RWM Stability

One of the main differences between ITER plasmas and the plasmas in present day tokamaks is the existence of a large fraction of fast alpha particles resulting from fusion reactions. To understand the whole picture in ITER, the influence of these particles on the RWM stability must be taken into account. Such modeling was

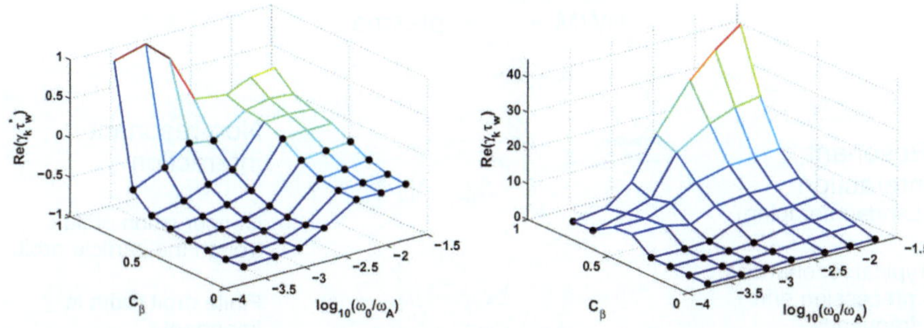

Fig. 6.13 Growth/damping rate of the RWM for ITER advanced tokamak plasmas, predicted by perturbative kinetic calculations (*left*) and self-consistent kinetic calculations (*right*). The precessional resonance damping is included. Black dots indicate a stable RWM. (Figures are from Ref. [96]. © IAEA)

published in the paper by Y. Liu [97]. The calculations were made based on the same steady state scenario as in the previous Sect. 6.6.1. The α contribution is found to be generally stabilizing, compared with the thermal particle kinetic contribution alone (which was discussed before). This conclusion is based on the two approaches. The "perturbative" approach generally predicts higher stabilization than the "self-consistent" approach, which is a typical result for such comparisons. The stabilizing effect from α particles is more pronounced at fast enough RWM mode rotation, when the rotation frequency roughly matches the α precession frequency, which is of the order of a few percent of the toroidal Alfven frequency for ITER. Thus, this stabilization is expected to be significant only if the central plasma rotation will be close to predicted upper level in ITER or higher. One should note that the discussed modeling neglects non-resonant interaction and the effect of the finite drift orbit widths of trapped α particles, which is expected to be reasonably large compared with the plasma minor radius in ITER. These effects can be taken into account in perturbative simulations, by assuming distribution functions expected in ITER for all species. In this case, non-resonant interaction can also be taken into account [98]. Fast particles, circulating toroidally, encompass a volume in the tokamak within which the magnetic flux associated with this movement is adiabatically conserved (a similar effect works in the mirror system as discussed in chap. 2). Any slow perturbation from an RWM tries to change this flux and works against flux conservation. As a result, the presence of fast particles produces a magnetic flux which counteracts the perturbation from the RWM and provides a stabilizing effect. The frequency resonance is not required in this case and the mode could have a frequency much lower than the particle rotation frequency (see Fig. 6.14). For finite orbit width, another effect could be important. Changes in the potential energy depend on the displacement amplitude and magnetic field curvature (see 6.26). For the same amplitude of the displacement in the region of positive and negative curvature, non-resonant parts of the δW_K integral cancel. If the displacement amplitudes are different, cancellation of the

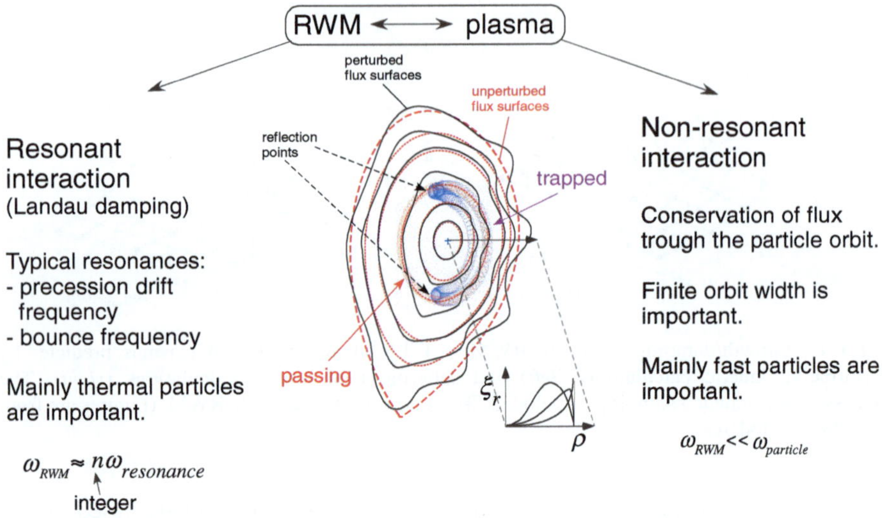

Fig. 6.14 Resonant and non-resonant interaction of RWM with different species. Equilibrium flux surfaces are shown with *dashed lines*. Perturbed surfaces are shown with *solid lines*. Perturbation amplitudes are strongly amplified to make them visible. Typical trapped and passing particles trajectories are shown together with radial displacement eigenfunction for different poloidal mode numbers

integral is not complete and the remaining part influences mode stability. The effect of all energetic species are relatively unaffected by the plasma rotation frequency, because it arises from orbit effects rather than the resonant effects which dominate the thermal species contribution.

Finally, having calculated changes in potential energy due to resonant and non-resonant interactions in the perturbative approach, it is possible to clarify the effect of different species. Calculations for the ITER case show that while the maximum contribution to δW_k comes from thermal particles; the other species are also important and give an integral contribution higher than the thermal particles alone (Fig. 6.15) [98]. Stability of the resistive wall mode in ITER is dominantly determined by the trapped particle contribution, but the passing particles also play a significant role, in particular for the ICRH population, where they have a more stabilizing contribution than the trapped particles. The thermal particles alone account for a significant increase of the stability limit into the wall-stabilized region up to $C_\beta = 0.3$. The additional non-resonant damping from NBI, ICRH and α particles extends this region further to more than the target value for ITER advanced scenario: $C_\beta = 0.45$. These results show again the importance of a correct description of the RWM-plasma interaction, which is provided by the perturbative approach, and the need for self-consistent treatment of RWM in high beta tokamak plasmas. The distribution functions used for different species provide an additional source of uncertainties for the result. Finally, the overall conclusion

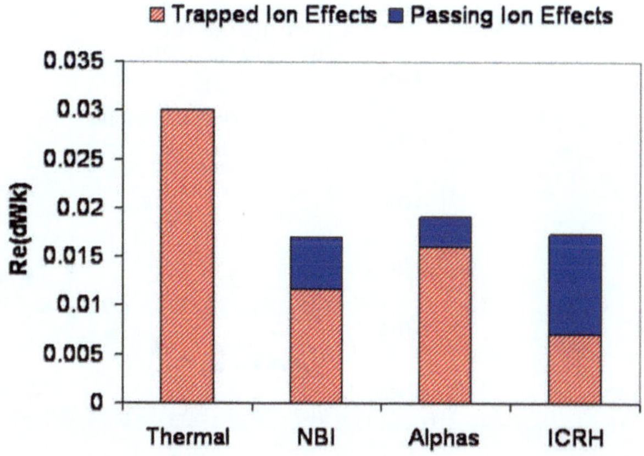

Fig. 6.15 The contributions to the real part of δW_k from both passing and trapped particles for the thermal, α, NBI and ICRH distributions as calculated by HAGIS at $\nu = 0.01\nu_{Alfven}$ and $C_\beta = 0.5$. (The figure is from [98].)

from current research remains positive for ITER and predicts passive stabilization of the RWM just from the foreseen heating and current drive actuators up to the target value $\beta_N = 2.95$.

6.7 Resonant Field Amplification

Small external magnetic perturbations, which have components and frequencies which are resonant with the marginally stable mode, can be amplified by the plasma even below no-wall limit [99–101]. This process is called resonant field amplification (RFA) and characterized by RFA amplitude:

$$A = \frac{B_r - B_r^{vac}}{B_r^{vac}}, \tag{6.30}$$

where B_r^{vac} is the externally applied radial field with the same helicity as the marginally stable mode; B_r is the total helical radial field which consists of the applied field and the field generated by the plasma in response to the applied field. The RFA amplitude is significantly enhanced when the plasma exceeds the no-wall stability limit. This quantity is also used in the experiments as an indication of the pressure-driven limit for RWM (no-wall beta limit); it could also be an indication for the proximity of the other limits, for example current limit [102]. This non-monotonic RFA increase is shown for a JET case in Fig. 6.16, which shows the first peak in RFA at low β_N due to the marginal edge stability as the first ELM occurs. The RWM starts to give a dominant contribution to the plasma response at higher β_N.

This makes it a more general measure of the plasma stability, but in the case of RWMs the other factors have to be considered for the analysis. For example, the existence of unstable ideal modes in the same discharge if the no-wall limit is crossed. In this case, it is important to identify the ideal character of the triggered mode, which is characterized by a constant phase of the perturbation, as shown in

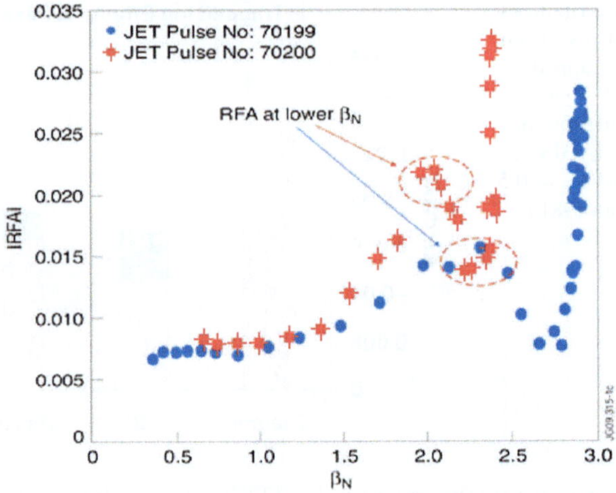

Fig. 6.16 Non-monotonic n = 1 RFA response is observed in JET at low β_N, well below the RFA threshold. The RFA threshold manifests itself as much stronger increase at a later phase with higher β_N. (The figure is from M.P. Gryaznevich et al., Nucl. Fusion, 52, 083018 (2012).

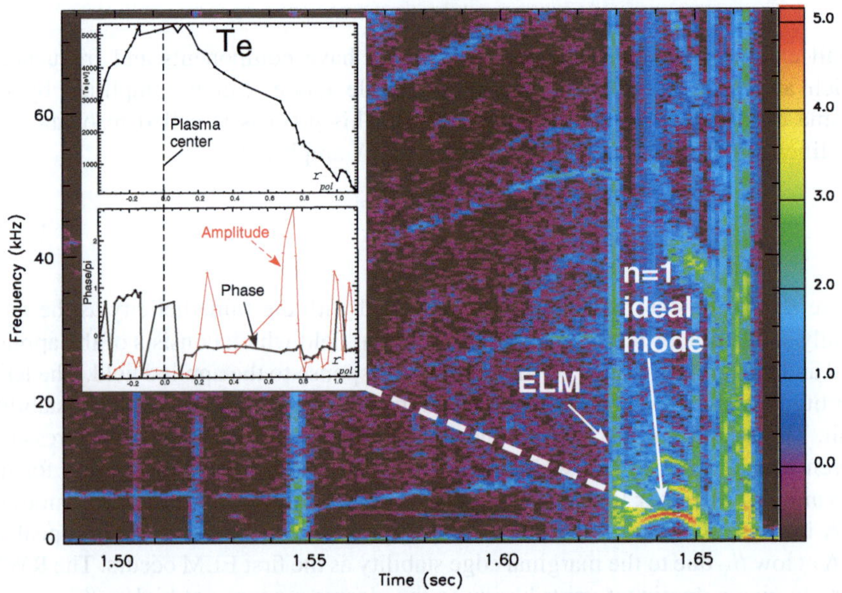

Fig. 6.17 Spectrogram of the ASDEX Upgrade discharge 29100 is shown. The ideal mode is triggered by the edge localized mode (ELM) at about 1.63 s. The electron temperature profile together with the amplitude and phase of the unstable mode are shown in the insert. Constant phase of the perturbation demonstrates the ideal mode character [103]

Fig. 6.17 (in contrast to the phase jump for the resistive instability). In this case, an ELM triggers the ideal instability just above the no-wall limit.

Further verification can be made by numerical modeling of the experimental equilibrium and identification of the no-wall limits for this equilibrium. Such dedicated studies allow a better formulation for the no-wall limit identification to be found. For example, in the JET tokamak, the maximum of the logarithmic derivative of the RFA, $d(\ln A)/d\beta_N$, gives a better estimation for the no-wall limit compared to the simple RFA maximum [104]. The measurements of RFA response are widely used for comparison with different theoretical predictions [105–107].

6.8 Triggering of RWM at Low Plasma Rotation

There is one important point which has to be mentioned together with RWM physics in order to complete the RWM physics discussion. It is not directly responsible for the physics of the RWM, but determines the stability of the discharge with respect to the RWM. Several years ago, it was shown that the RWM could be stabilized at low plasma rotation [75, 76]. However, stable operation without RWM in slowly rotating plasmas is not guaranteed [20]. Various MHD instabilities can trigger the RWM near the no-wall limit. These include: (i) Edge Localized Modes (ELMs), (ii) off-axis fishbones, and (iii) the energetic particle-driven wall mode (EWM) [108]. It was shown recently that the latter two modes have the same origin [109].

A global MHD event can interact with the RWM branch and can trigger the RWM if the following conditions are fulfilled:

- The MHD event produces distortions in the same place where the potential RWM could be present.
- The MHD event has a component with the same toroidal mode number n (typically $n = 1$ for $n = 1$ RWM).
- The MHD event and the potential RWM have comparable frequencies.

This forced excitation of RWMs takes place within a fraction of the time scale of the driving MHD event, which is typically much faster than the wall time constant. It is interesting that the magnitude ratio between the excited mode and the driving force is similar in all these cases. This indicates similarities for the triggering process.

The existence of the different trigger mechanisms gives additional confirmation of the global nature of the RWM eigenfunction (see Sect. 6.4). In case of ELM triggering, the *external part* of RWM is coupled to an ELM perturbation, which has components only at the plasma edge. Fishbone-driven RWMs are triggered differently. The energetic ions injected by neutral beams are expelled by the off-axis fishbone bursts with the maximum perturbation occurring at the plasma mid-radius ($\rho \approx 0.5$). Thus, the coupling occurs between the off-axis fishbone and the *internal part* of the RWM eigenfunction.

The existence of these coupling phenomena moves the problem of RWM stability and control from a single mode consideration into the complex space of global MHD stability control in the plasma. This is, in some sense, similar to the challenges which one faces in trying to sustain the advanced scenario in a tokamak. As it was shown by Luce [110] one must simultaneously meet the correct point in current drive, transport, and MHD stability, which requires integrated control for advanced scenario regimes. The same is true for control of the RWM in this regime. It must be integrated into the global MHD control approach to avoid triggering of RWM or to suppress any triggered modes by proper feedback action. It has to be mentioned that resistive instabilities can also be destabilized with increase of β_N [111]. A possible solution is to avoid the most dangerous rational surfaces or stabilize the modes via external current drive.

6.9 Achievement of high β_N plasmas in tokamaks

Maximization of the achieved beta is the practical aim for plasma operation in advanced scenarios. As was pointed out before, non-uniformity of the external structures and/or magnetic perturbations from other instabilities can trigger RWMs. Avoiding these factors by static correction of the error fields and control of other instabilities allows the operational window without triggered RWMs to be extended. A further increase of beta brings the plasma close to the stability threshold and feedback correction of the magnetic perturbations from RWMs is required. If such a correction is applied from the beginning, the average duration of the discharge is increased, as well as the average value of β_N. It was shown in the NSTX tokamak that the combination of static error field corrections (for $n = 3$) and feedback control (for $n = 1$) greatly increases the probability of sustaining high β_N over long pulses [112]. A similar approach with simultaneous correction has been applied in the DIII-D tokamak. Here, the dynamic error field correction is applied together with feedback control of the mode. The most important component of the error field correction is to eliminate the resonant component of the stable kink mode and to avoid its amplification in a marginally stable situation. On the other hand, in the unstable plasma, with self-generated RWMs, the main aim is to suppress the mode with feedback. The resulting external magnetic field action is the sum of preprogrammed error field correction, fine-tuned dynamic error field correction and fast feedback correction [20]. Experimental mode stabilization in high beta plasmas is a complex problem which naturally includes all the ingredients of complex plasma-mode interaction, as well as interaction with other modes and with external conductors. It is not surprising that these operations at high β_N are possible but not unconditionally guaranteed in present tokamaks and feedback suppression is required if the mode becomes unstable or triggered above the no-wall limit.

In summary, one can formulate the current situation with achievement of the high β_N in present devices as follows:

- Correction of the error fields helps to archive higher β_N.
- RWM stabilization up to $\beta_{N,ideal-wall}$ is archived, but not in a stationary regime.
- Other instabilities can trigger RWM, which requires special attention.
- Particle effects are important for RWM stability and have to be treated carefully.

All these problems are subjects of the intensive research and have to be solved for reliable and stationary plasma operations close to $\beta_{N,ideal-wall}$.

6.10 Conclusions and Discussion

The advanced tokamak regime is a promising candidate for steady state tokamak operation, which is preferable for a fusion reactor. This regime is fully non-inductive and the plasma current is mainly sustained by the bootstrap current. The bootstrap current value is proportional to the pressure gradient. Thus, high pressure is unavoidable in the advanced regime. The resistive wall mode is a global kink instability appearing in tokamaks at high pressure. It typically has a single toroidal mode number ($n = 1$) and multiple significant poloidal mode numbers ($m = 2\ldots6$). Due to their global nature, RWMs interact with plasma particles and resistive walls and can be triggered by several other instabilities. The electromagnetic part of the resistive wall mode interaction is relatively well understood. The interaction of the resistive wall mode with the plasma is a more challenging problem and is still open. Currently, there are two modelling approaches employed to address this interaction: the "self-consistent" and the "perturbative" methods. The first approach contains an approximation for the kinetic effects, but calculates RWM stability self-consistently. The second approach treats the kinetic interaction correctly, but does not treat the problem self-consistently. These two approaches are complementary and have to be used simultaneously. For steady-state advanced scenarios in ITER, the self-consistent calculations predict a full stabilization of the RWM at very slow plasma rotation and moderately high plasma pressures $\left(C_\beta = 0.4\right)$. "Perturbative" calculations predict even higher stabilization. Both results are very positive for ITER and show that a substantial increase in fusion performance is possible even without active feedback control.

At the same time, self-consistent modeling with a correct kinetic description is required to make correct predictions. Even if such modeling requires large computer resources, it must be performed to verify approximate models for linear MHD codes. This verification must also be done experimentally. The absence of a verified model for linear MHD is presently one of the main unknowns in RWM physics. A linear MHD code with a verified model could be coupled with an electromagnetic code (3D external vessel + feedback coils + error fields + feedback control) to model the RWM in a realistic environment. The first example of such a coupling already exists [62, 63, 65, 66], which shows the feasibility of the task. This set of codes allows us to predict the spectrum and linear growth rates of the RWM in future machines such as ITER and DEMO. The importance of non-linear RWM evolution is also the subject of the intensive research, which becomes possible by coupling of non-linear MHD codes with a 3D wall description [113].

The other important point is the development of integrated scenario control for advanced tokamak operation. This development includes integrated control and avoidance of RWMs and other MHD instabilities which could act as a trigger. One of the possible routes is to avoid RWM triggering by flexible feedback correction with a wide bandwidth feedback system. In this scheme, the feedback corrects slow (error field correction) and fast (RWM) dynamics simultaneously [34]. A possible variant of an integrated control strategy is the "three step strategy":

1. **Avoid RWM if possible** This implies correction of error fields and mitigation of ELMs. (Only relatively small perturbations from other instabilities are allowed).
2. **Prevent coupling to the wall and suppress small RWM** by active feedback ***if* (1) *fails***.
3. **Decouple from the wall, rotate and suppress** by feedback ***if* (1) *and* (2) *fail***.

We define the proposed strategy in such a way that the efforts are minimal at the beginning. The second step, and especially the last one, are more challenging and should be done only if absolutely necessary. This will keep the amount of possible actions and efforts at the minimum level. One can see that this strategy has to be integrated into an overall MHD control. It is also clear that some points have to be addressed experimentally (for example, decoupling of RWMs from the wall has been demonstrated for RFPs [50, 51, 114] but has to be tested for tokamaks).

The other important issue is the stability boundaries $\beta_{N,no-wall}$ and $\beta_{N,ideal-wall}$. There are three important points which one has to keep in mind when working with these terms.

(1) The stability boundary is different for different mode numbers (see Fig. 6.1).
(2) These two limits were defined on the basis of an ideal MHD approximation. Kinetic effects lead to changes in the mode stability and can considerably change the stability boundaries, as was shown in the previous sections. In other words, it is not possible to keep the plasma stable in the presence of an ideal instability. This ideal instability limit is a hard limit for the plasma confinement. At the same time, stability boundaries for the RWM are strongly affected by kinetic effects.
3) β_N represents the average pressure in the plasma. The real drive for the mode comes from the pressure profile gradient, not from the pressure itself. Also, an increase of the gradients is unavoidable for a strong increase of the total plasma pressure. Total pressure β_N can be moderately increased without increasing the gradient in the mode region. Thus, different stability situations are possible for the same β_N even without inclusion of the kinetic effects.

Identification of the "true" stability boundary requires modeling with accurate profile gradients and an accurate account of the particle influence. Such modeling is challenging because profile measurements tend to be of insufficient quality (e.g. current profile, rotation profile, etc.), and because of the lack of a "final" model for RWM-plasma interaction. Thus, rough estimates in terms of well measured quantities, such as β_N, are required to understand the position of the plasma in operational space and to compare the results from different experiments. Studying the plasma response to external perturbations gives experimental identification of the RWM stability boundary. The reaction of the plasma increases after crossing the "no-wall" limit. The typical estimate for the "ideal wall" limit is a factor of two above the "no-wall" limit for advanced scenario. But this is a rule of thumb which could be wrong under certain circumstances. The value of β_N is the best indicator available but the factors affecting the stability limits have to be kept in mind.

The next important issue is mode "rigidity". This term is defined differently by different authors as discussed in Sect. 6.4. In many cases, a "rigid" mode structure is assumed using one of these definitions, which is a good approach. At the same time, there is also evidence for "non-rigidity" of the mode (for example in Ref. [77]). The question of mode "rigidity" has to be addressed carefully in future experiments because "non-rigidity" could change requirements for the mode stabilization and for the design of the RWM feedback control. In this situation, experiments on RFPs are of particular interest. The large number of control coils allow the configuration to be varied in a very flexible way and to explore mode rigidity in terms of local toroidal/poloidal suppression as well as in terms of unstable modes with different toroidal mode numbers [24, 115–117].

We would like to note that RWM physics could be important for a future power plant. This power plant must be robust and stable in its operational window. At the same time, its operational space does not necessarily need to be as wide as in modern tokamaks or as in ITER. One has to decide in which scenario the plant will operate before it is built. The cost of the electricity becomes also one of the major criteria for such a power plant. The cost could be reduced by:

- reduction of the reactor size, which arises from advanced tokamak operations (with higher β_N and higher bootstrap current fraction),
- use of a minimal set of diagnostic and control tools, which are feasible for the reactor,
- operation at the maximum achievable normalized beta (costs $\propto \beta_N^{-0.4}$) [118].

It is also clear that power plants will be either ITER size or bigger than ITER ($R = 5.2 - 9.5\,m$, see for example comparison table for different DEMO designs in reference [93]). Thus, plasma rotation is expected to be smaller compared to ITER, which could open a stable operational space due to resonance of the particles with the RWMs. Thus, it could be possible to design a future power plant for the operation space where no RWM control is required and high β_N values are achieved by kinetic stabilization of RWMs. This will combine resonant stabilization from the slow particles and non-resonant stabilization from the fast alpha particles. In this case no special control coils are required, which will simplify the device and reduce the costs. This shows the importance of the problems discussed in this chapter, and, in particular, for a correct model for the plasma-RWM interaction. A feasibility study for such a scenario requires a lot of work both to develop self consistent plasma scenarios and for a quantitative understanding of the kinetic effects on the RWM physics. It is clear that, at present, the power plant issues are much more uncertain than for ITER and thus further research will be required. Both RWM physics and RWM control are far from complete and remain areas of active research. The present chapter provides a general description of the RWM issues; particular details can be found in the referenced papers or in references therein.

References

1. N.J. Fisch, Rev. Mod. Phys. **59**, 175 (1987)
2. A.G. Peeters, Plasma Phys. Control. Fusion **42**, B231 (2000)
3. M.S. Chu, M. Okabayashi, Plasma Phys. Control. Fusion **52**, 123001 (2010)
4. A. Bondeson, D.J. Ward, Phys. Rev. Let. **72**, 2709 (1994)
5. H. Reimedes et al., Phys. Plasmas **13**, 056107 (2006)
6. J. Manickam et al., Phys. Plasmas **1**, 1601 (1994)
7. V. Igochine, Nucl. Fusion **52**, 074010 (2012)
8. A.M. Garofalo, Nucl. Fusion **40**, 1491 (2000)
9. E.J. Strait et al., in *Fusion Energy* 2002 (Proc. 19th Int. Conf. Lyon, 2002) (Vienna: IAEA) CD-ROM file EX/S2-1 and, http://www-pub.iaea.org/MTCD/publications/PDF/csp_019c/html/node137.htm
10. H. Reimerdes, Phys. Plasmas **13**, 056107 (2006)
11. M.S. Chu et al., Phys. Plasmas **2**, 2236 (1995)
12. R. Betti, J.P. Freidberg, Phys. Rev. Lett. **74**, 2949 (1995)
13. A.M. Garofalo et al., Phys. Rev. Lett. **82**, 3811 (1999)
14. A.M. Garofalo et al., Phys. Plasmas **10**, 4776 (2003)
15. M. Okabayashi, Plasma Phys. Control. Fusion **44**, B339 (2002)
16. A.M. Garofalo et al., Nucl. Fusion **42**, 1335 (2002)
17. S. Sabbagh et al., Nucl. Fusion **46**, 635 (2006)
18. A.J. Webster, Phys. Plasmas **17**, 110708 (2010)
19. L.E. Zakharov, Phys. Plasmas **15**, 062507 (2008)
20. M. Okabayashi, Nucl. Fusion **49**, 125003 (2009)
21. I.T. Chapman et al., Plasma Phys. Control. Fusion **51**, 055015 (2009)
22. A. Bondeson et al., Plasma Phys. Control. Fusion **45**, A253 (2003)
23. Y. Liu, Plasma Phys. Control. Fusion **51**, 115006 (2009)
24. M. Takechi et al., 37th EPS Conference on Plasma Physics, Dublin, Ireland, P2.192, http://ocs.ciemat.es/EPS2010PAP/html/author.html. Accessed 21–25 June 2010
25. V.D. Pustovitov, M.S. Mayorova, Plasma Phys. Control. Fusion **48**, 51 (2006)
26. V.D. Pustovitov, Plasma Phys. Rep. **30**, 187 (2004)
27. A.H. Boozer, Phys. Rev. Let. **86**, 5059 (2001)
28. M. Okabayshi, N. Pomphrey, R.E. Hatcher, Nucl. Fusion **38**, 1607 (1998)
29. J.M. Finn, Phys. Plasmas **2**, 198 (1995)
30. R. Fitzpatrick, Phys. Plasmas **9**, 3459 (2002)
31. S.C. Guo, Phys. Plasmas **6**, 3868 (1999)
32. D. Yadykin et al., Plasma Phys. Control. Fusion **53**, 085024 (2011)

33. M. Okabayashi et al., 36th EPS Conference on Plasma Physics, Sofia, Bulgaria, ECA Vol.33E, P-4.141 (2009) or CD-ROM file P4.141 http://epsppd.epfl.ch/Sofia/start.htm. Accessed 29 June–3 July 2009
34. Y. In et al., Plasma Phys. Control. Fusion **52**, 104004 (2010)
35. Y. Liu et al., Phys. Plasmas **17**, 072510 (2010)
36. E.J. Strait et al., Nucl. Fusion **43**, 430 (2003)
37. Liu et al., Phys. Plasmas **7**, 3681 (2000)
38. A. Bondeson et al., Nucl. Fusion **41**, 455 (2001)
39. M. Okabayashi et al., Phys. Plasmas **8**, 2071 (2001)
40. V.D. Pustovitov, Plasma Phys. Control. Fusion **44**, 295 (2002)
41. C.M. Bishop, Plasma Phys. Control. Fusion **31**, 1179 (1989)
42. L. Piron et al., Plasma Phys. Control. Fusion **53**, 084004 (2011)
43. P.R. Brunsell et al., Phys. Rev. Lett. **93**, 225001 (2004)
44. P.R. Brunsell et al., Plasma Phys. Control. Fusion **47**, B25 (2005)
45. M. Cavinato et al., IEEE Transaction of nuclear science **53**, 1015 (2006)
46. P. Zanca et al., Nucl. Fusion **47**, 1425 (2007)
47. K.E.J. Olofsson et al., Plasma Phys. Control. Fusion **53**, 084003 (2011)
48. J.R. Drake, Nucl. Fusion **45**, 557 (2005)
49. K.E.J. Olofsson et al., Plasma Phys. Control. Fusion **52**, 104005 (2010)
50. V. Igochine, Plasma Phys. Control. Fusion **51**, 055008 (2009)
51. T. Bolzonella, Phys. Rev. Lett. **101**, 165003 (2008)
52. Y.Q. Liu, A. Bondeson, Phys. Rev. Lett. **84**, 907 (2000)
53. T.E. Evans et al., Phys. Rev. Lett. **92**, 235003 (2004)
54. W. Suttrop et al., Phys. Rev. Lett. **106**, 225004 (2011)
55. M. Okabayshi, et al., Nucl. Fusion **38**, 1607 (1998)
56. V.D. Pustovitov, Plasma Phys. Rep. **27**, 195 (2001)
57. V.D. Pustovitov, Phys. Plasmas. **15**, 072501 (2008)
58. V.D. Pustovitov, Plasma Phys. Rep. **37**, 35 (2011)
59. J. Bialek et al., Phys. Plasmas **8**, 2170 (2001)
60. P.Merkel et al., 2006 21st IAEA Fusion Energy Conf. 2006 (Chengdu, China)(Vienna: IAEA) TH/P3-8
61. E. Strumberger et al., Phys. Plasmas **15**, 056110 (2008)
62. R. Albanese et al., IEEE Trans. Magn. **44**, 1654 (2008)
63. A. Portone et al., Plasma Phys. Control. Fusion **50**, 085004 (2008)
64. F. Villone et al., Phys. Rev. Lett. **100**, 255005 (2008)
65. F. Villone et al., Nucl. Fusion **50**, 125011 (2010)
66. V.D. Pustovitov, Pl. Phys. Rep. **38**, 697 (2012)
67. F. Villone, V.D. Pustovitov, Phys. Lett. A **377**, 2780 (2013)
68. V.D. Pustovitov, Nucl. Fusion **53**, 033001 (2013)
69. V.D. Pustovitov, Phys. Plasmas **19**, 062503 (2012)
70. V.D. Pustovitov, Plasma Phys. Control. Fusion **50**, 105001 (2008)
71. Y. Liu, F. Villone, Plasma Phys. Control. Fusion **51**, 115008 (2009)
72. G.W. Hammet, F.W. Perkins, Phys. Rev. Lett. **64**, 3019 (1990)
73. A. Bondeson, M.S. Chu, Phys. Plasmas **3**, 3013 (1996)
74. E.J. Strait et al., Phys. Plasmas **14**, 056101 (2007)
75. H. Reimerders et. al., Phys. Rev. Lett. **98**, 055001 (2007)
76. M. Takechi et al., Phys. Rev. Lett. **98**, 055002 (2007)
77. S.A. Sabbagh et al., Phys. Rev. Lett. **97**, 045004 (2006)
78. S.W. Haney, J.P. Friedberg, Phys. Fluids B **1**, 1637 (1989)
79. B. Hu, R. Betti, Phys. Rev. Lett. **93**, 105002 (2004)
80. B. Hu et al., Phys. Plasmas **12**, 057301 (2005)
81. Y. Liu et al., Phys. Plasmas **15**, 112503 (2008)
82. B.N. Breizman et al., Phys. Plasmas **5**, 2326 (1998)
83. Y. Liu et al., Phys. Plasmas **15**, 092505 (2008)

84. H. Reimerdes et al., Phys. Rev. Lett. **106**, 215002 (2011)
85. S.A. Sabbagh et al., Nucl. Fusion **50**, 025020 (2010)
86. M. Shimada et al., Nucl. Fusion **47**, S1 (2007)
87. T.C. Hender et al., Nucl. Fusion **47**, S128 (2007)
88. C. Gormezano et al., Nucl. Fusion **47**, S285 (2007)
89. V. Mukhovatov et al., Nucl. Fusion **47**, S404–S413 (2007)
90. J. Garcia et al., Nucl. Fusion **48**, 075007 (2008)
91. W.M Stacey. in *Front Matter, in Fusion: An Introduction to the Physics and Technology of Magnetic Confinement Fusion*, 2nd edn. (Wiley, New York, 2010)
92. I.T. Chapman et al., Fusion Engin. and Design **86**, 141 (2011)
93. H. Zohm, Fusion Sci. Technol. **58**, 613 (2010)
94. Y. Liu, Nucl. Fusion **49**, 035004 (2009)
95. A. Polevoi et all, in *Fusion Energy 2002* (Proc. 19th Int. Conf. Lyon, 2002) (Vienna: IAEA) CD-ROM file and http://www.iaea.org/programmes/ripc/physics/fec2002/html/fec2002.htm
96. Y. Liu et all, in *Fusion Energy 2010* (Proc. 23rd Int. Conf. Daejeon, 2010) (Vienna: IAEA) CD-ROM file [TH/P9-26] and, http://www-naweb.iaea.org/napc/physics/FEC/FEC2010/html/index.htm
97. Y. Liu, Nucl. Fusion **50**, 095008 (2010)
98. I.T. Chapman et al., Phys. Plasmas **19**, 052502 (2012)
99. A.H. Reiman, D. Monticello, Phys. Fluids B **3**, 2230 (1991)
100. M.P. Gryaznevich et al., Plasma Phys. Control. Fusion **50**, 124030 (2008)
101. A.M. Garofalo, R.J. La Haye, J.T. Scoville, Nucl. Fusion **42**, 1335 (2002)
102. M.P. Gryaznevich et al., Nucl. Fusion **52**, 083018 (2012)
103. V. Igochine et al., 40th EPS Conference on Plasma Physics, 2013, P5.147
104. Y. Liu et al., Plasma Phys. Control. Fusion **51**, 115005 (2009)
105. Y. Liu, Phys. Plasmas **17**, 072510 (2010)
106. H. Reimerdes, Nucl. Fusion **45**, 368 (2005)
107. H. Reimerdes et al., Phys. Rev. Lett. **93**, 135002 (2004)
108. G. Matsunaga et al., Phys. Rev. Lett. **103**, 045001 (2009)
109. M. Okabayashi et al., Phys. Plasmas **18**, 056112 (2011)
110. T.C. Luce, Phys. Plasmas **18**, 030501 (2011)
111. D.P. Brennan et al., Phys. Plasmas **10**, 1643 (2003)
112. S.A. Sabbagh et al., Nucl. Fusion **50**, 025020 (2010)
113. M. Holzl et al., Journal of physics. Conference Series **401**, 012010 (2012)
114. V. Igochine et al., in Proceedings of the 35th EPS Conference on Plasma Physics, Hersonissos, 2008, ECA 32D, P2.066 and, http://epsppd.epfl.ch/Hersonissos/html/i_index.htm
115. T. Boltzonella et al., in *Fusion Energy* 2010 (Proc. 23rd Int. Conf. Daejeon, 2010) (Vienna: IAEA) CD-ROM file [EXS/P5-01] and, http://www-pub.iaea.org/mtcd/meetings/cn180_papers.asp
116. T. Boltzonella et al., 37th EPS Conference on Plasma Physics, Dublin, Ireland, 21 - 25 June, 2010, P2.176, http://ocs.ciemat.es/EPS2010PAP/html/author.html
117. D. Yadikin, P.R. Brunsell, J.R. Drake, Plasma Phys. Control. Fusion **48**, 1 (2006)
118. D.J. Ward et al., in *Fusion Energy 2000* (Proc. 18th Int. Conf. Sorrento, 2000) (Vienna: IAEA) CD-ROM file FT/P2-20 and, http://www.iaea.org/programmes/ripc/physics/fec2000/html/node1.htm

Chapter 7
Disruptions

Tim C. Hender

Abstract This chapter discusses disruptions in tokamaks. Disruptions are a rapid loss of the confined plasma and its current. Their consequences include large heat and mechanical loads on structures surrounding the plasma, and in some cases the generation of a very high energy runaway electron beam carrying a substantial fraction of the original plasma current. These consequences become worse in larger tokamaks, which have higher currents and magnetic fields, and are therefore a major design constraint. After outlining the physics of disruptions, this chapter discusses the methods used to detect impending disruptions, means to avoid the disruptions and ways to mitigate the consequences if a disruption is unavoidable. This is an intensely studied topic and means to detect and mitigate a very high percentage of disruptions are becoming increasingly well developed.

7.1 Introduction

Disruptions in the tokamak are a rapid loss of the confined plasma and its current, with consequent heat loads to the plasma facing walls and electromagnetic forces on the device structure. The energy stored in a tokamak plasma rises roughly as L^5 (where $L =$ the linear dimension) [1], so the energy dissipated at the wall in a disruption rises approximately as L^3. Therefore doubling the size of the tokamak (e.g. the step from JET to ITER) increases the wall load at disruption by an order of magnitude (if it is spread uniformly). Likewise, for a given twist of the edge magnetic field (its safety factor), the electromagnetic loads at disruption rise as $\sim LB_{\phi}I_p$ (where $B_{\phi} =$ Toroidal magnetic field and I_p is the plasma current). Since

T.C. Hender (✉)
Directorate, CCFE, Culham Science Centre, Abingdon,
Oxon OX14 3DB, UK
e-mail: tim.hender@ccfe.ac.uk

V. Igochine (ed.), *Active Control of Magneto-hydrodynamic Instabilities in Hot Plasmas*, Springer Series on Atomic, Optical, and Plasma Physics 83,
DOI 10.1007/978-3-662-44222-7_7

$I_p \sim LB_\phi$ (at a given edge safety factor) the EM load $\sim L^2 {B_\phi}^2$ and so typically increases approximately as $\sim L^3$ (since empirically in tokamaks the toroidal field in general rises slightly slower than linearly with size). The result is that disruptions are a major constraint in the design of large tokamaks such as ITER and will be for fusion power plants based on the tokamak. Further, means to avoid, control and mitigate the consequences of disruption are essential in tokamaks of the scale of JET and larger.

Present day (and future) tokamak plasmas have a D-shaped cross section, which has been found to improve plasma confinement and stability (for a given edge safety factor). The vertical elongation of the plasma means it is unstable to vertical perturbations and must be held in-place by a control system. Failure of this control system due a large and rapid change of plasma parameters, hardware failure or programming error results in the plasma moving upwards or downwards—this is known as a Vertical Displacement Event (VDE). As the plasma moves and hits the wall eventually a disruption is triggered and the plasma energy and current are rapidly lost. Alternatively a disruption may occur and the rapid change of plasma parameters then triggers a vertical instability. Disruptions (as opposed to VDEs) most usually result from some form of macroscopic plasma instability, which may be initiated by operating too close to a stability limit, by a control system error or by some foreign material falling into the plasma. The accepted sequence of events in a canonical disruption is shown in Fig. 7.1.

There are of course many exceptions to this picture. A disruption in JET is shown in Fig. 7.2, and illustrates a case with a collapse of the thermal energy on O(1) ms and the quenching of the current on O(10) ms. The plasma centroid moves upwards as the vertical control of the plasma is lost, following the thermal quench.

The key issues arise from the resultant effects of the disruption:

- A rapid heat loss to the walls—in a high performance ITER plasma the 350 MJ of plasma thermal energy will be lost in around 1 ms [2].
- Large forces on the conducting structure containing plasma, due to induced eddy currents as the plasma current decays in the disruption and also due to currents with part of their circuit in the plasma, and part in the wall (known as halo currents). These currents can result in forces of approaching 8,000 tonnes on the ITER vacuum vessel [3].
- Runaway electrons induced by the large toroidal electric field due to the rapid decay of the plasma current in a disruption—in ITER it is predicted around 10 MA might be carried by 10–20 MeV electrons [4], with consequent potential for damage as these high energy electrons hit the wall.

In the remainder of this chapter the causes and consequences of disruptions are reviewed in more detail, then methods to detect impending disruptions are discussed which can be used to avoid the disruption or apply mitigating actions, as examined in the latter sections of this chapter.

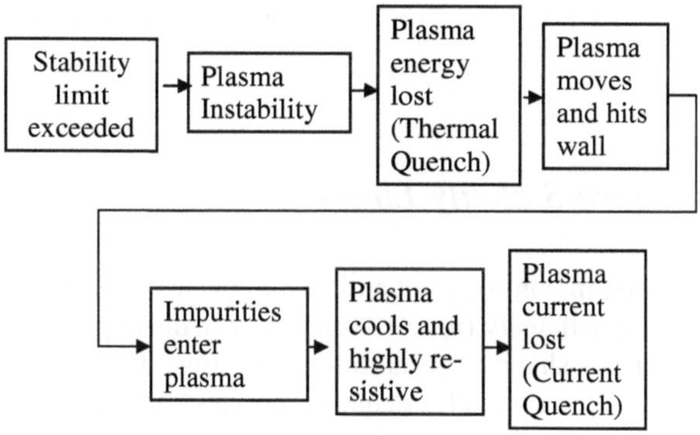

Fig. 7.1 Sequence of events leading to a canonical disruption

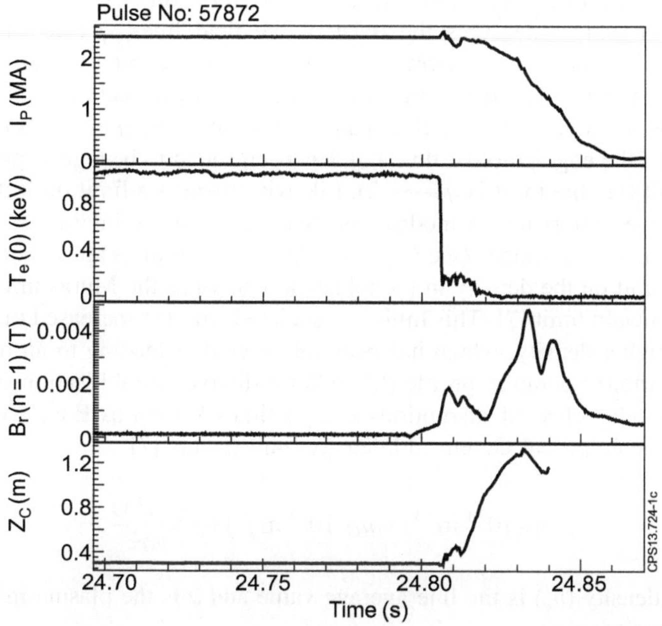

Fig. 7.2 The plasma current decay (I_p) in a JET disruption follows a more rapid loss of the plasma energy, illustrated by the central electron temperature ($T_e(0)$) behaviour. A growing helical distortion of the plasma measured by $B_r(n = 1)$ is followed by the energy loss, with resulting loss of control and upward movement of the plasma current centroid (Z_c)

7.2 Causes of Disruptions

As discussed above disruptions may be due to operating too close to a stability limit, or may be due to loss of control, or because some material falls into the plasma. In this section the main stability limits, which when exceeded can lead to disruptive plasma instabilities are discussed. Also the VDE is examined in more detail in this section.

7.2.1 Performance Stability Limits

In the tokamak the main controlled quantities are the toroidal field (B_ϕ), the plasma current (I_p), the plasma density (n_e, electron density), the applied additional heating power and the plasma shape. The additional heating may be from high energy injected neutral particles (typically deuterium) or radio frequency waves that resonate with the ion or electron gyro-frequencies. The additional heating directly controls the plasma temperature (and thus the kinetic pressure of the plasma), and the heating mix can influence the pressure profile within the plasma. MHD stability imposes limits on normalised quantities—an MHD equilibrium is defined by the safety factor which measures the twist of the field lines, q (see Chap. 2), the pressure profile, the ratio of thermal to magnetic pressure, β (see Chap. 2), and the plasma shape. There is a hard limit imposed by ideal external kink modes, which are unstable for edge-$q < 2$. In diverted tokamaks the edge-q is ∞ but the value of q at 95 % of the edge poloidal flux, q_{95}, has been found to be a good proxy for the edge-q limit (i.e. the limit is $q_{95} \sim 2$). Likewise there is a limit on β due to ideal pressure driven external kink modes (see Chap. 2), or due to internal $m = 2, n = 1$ neo-classical tearing modes (see Chap. 8). As discussed in Chap. 2 there is also an empirical limit on the density in the tokamak, known as the Murakami [5], Hugill [6] or Greenwald limit [7]. This limit is associated with the increased radiation that occurs at higher density, which has been interpreted as leading to an inward contraction of the temperature profile [8] and/or radiative destabilisation of magnetic islands [9]—this class of disruptions is sometimes known as Edge Cooling Disruptions. The empirical Greenwald density limit (n_G) is [7]:

$$n_e(10^{20}\ \mathrm{m}^{-3}) < n_G(10^{20}\ \mathrm{m}^{-3}) \equiv \frac{I_p(MA)}{\pi a^2} \tag{7.1}$$

where the density (n_e) is the line average value and a is the plasma minor radius. Since approximately

$$q_{95} = \frac{2\pi a^2 B_t}{\mu_0 R I_p} \frac{1 + \kappa^2}{2} \tag{7.2}$$

where quantities are in MKS units and κ is plasma elongation. Combining (7.1) and (7.2) gives

$$\frac{1}{q_{95}} = \frac{R n_G(10^{20}\ \mathrm{m}^{-3})}{B_t(T)} \frac{0.4\pi}{1 + \kappa^2} \tag{7.3}$$

which is the form of the density limit proposed by Murakami and by Hugill. An examination of JET disruptions shows substantially increased disruptivity at, and beyond, the density limit and at $q_{95} = 2$ (Fig. 7.3)

In practice the hard limit at $q_{95} \sim 2$, where a very fast growing instability on the Alfven timescale occurs, manifests itself as a slower growing $m = 2$, $n = 1$ tearing mode at slightly higher edge-q (see Chap. 2 for details on tearing modes). Likewise the density limit also tends to manifest itself in the destabilisation of an $m = 2$, $n = 1$ tearing mode. These tearing modes rotate with the plasma electron fluid, but in larger tokamaks tend to stop rotating as they grow towards the disruption event. This cessation of rotation is termed *mode locking* [11] and the primary force slowing the plasma (and mode) arises from eddy currents driven in the conducting structures surrounding the plasma by the rotating instability.

The manifestation of the β-limit is more varied. Typically in the ELMy H-mode, the envisaged baseline operating mode for ITER, $m = 2$, $n = 1$ neo-classical tearing modes (NTMs) are triggered at sufficiently high β (see Chap. 8). These NTMs may, or may not, grow to sufficient amplitude to cause a disruption. In more advanced operational scenarios an internal transport barrier [12] may occur, with strong peaking of the pressure profile which can drive Alfvénic growth rate instabilities, which very rapidly lead to disruption (see for example [13]). Finally at high plasma pressure resistive wall modes may occur (see Chap. 6) that can cause disruptions.

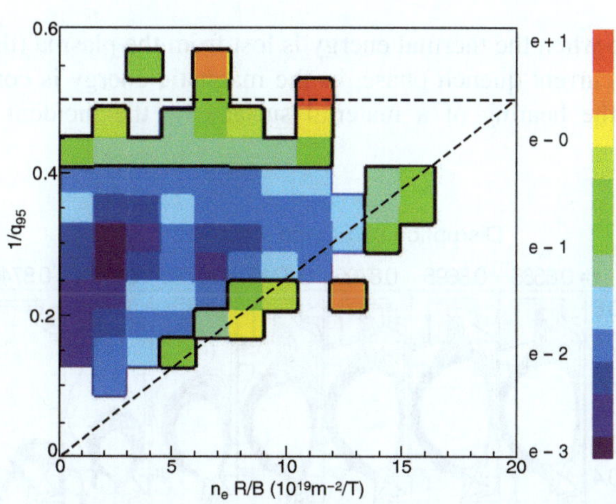

Fig. 7.3 Statistics from JET for the operational period from 2000 to 2007. The contour plot shows the disruptivity (s^{-1}) as function of the inverse edge safety factor ($1/q_{95}$) and the normalised plasma density (n_e). The inverse of the disruptivity is the time (statistically) to disruption for the given parameters (q_{95} and n_e) and is derived by sampling the plasma state every 250 ms (when the plasma current is above 1 MA). In total, more than 15,000 plasma discharges were sampled. The broken lines are the expected stability limits (see text) and *solid black lines* enclose regions where the disruptivity >0.04 s^{-1} From [10]

7.2.2 VDEs

The elongation of present day tokamak plasmas for stability and confinement reasons means that they are vertically unstable and must be position controlled by a feedback system (see Sect. 7.3.2). Passively the surrounding conducting structures slow the motion of plasma to a timescale related to their resistive skin time, making feedback control possible. A disruption results in rapid large changes in plasma parameters, which almost invariably cause the vertical feedback control to be lost. Alternatively other large disturbances (e.g. large edge instabilities know as ELMs) may cause loss of vertical control, or some hardware failure or programming error may cause loss of control. Figure 7.4 shows an example from Alcator C-Mod in which a disruption leads a vertically unstable plasma that moves downwards. At later times halo currents indicated by the arrows flow (see next section).

In general the non-symmetries in the single x-point magnetic geometry (as shown in Fig. 7.4) mean the plasma has a higher probability of moving upwards or downwards, depending on details of the magnetic configuration, in a VDE.

7.3 Consequences of Disruptions

7.3.1 Heat Loads

Heat loads arise when the thermal energy is lost from the plasma (thermal quench) and during the current quench phase, as the magnetic energy is converted to heat energy. Since the heating of a material surface by the incident energy (*U*) is

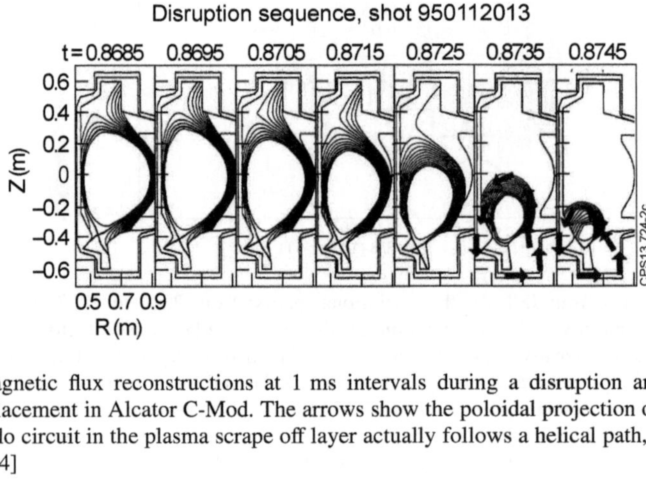

Fig. 7.4 Magnetic flux reconstructions at 1 ms intervals during a disruption and subsequent vertical displacement in Alcator C-Mod. The arrows show the poloidal projection of halo current flow. The halo circuit in the plasma scrape off layer actually follows a helical path, in order to be force free [14]

thermally conducted into the structure, the governing parameter for the surface heating is $U/(A \times \tau^{0.5})$, where A is the area and τ is the timescale of deposit. If all the thermal energy of a high performance ITER plasma is conducted to the divertor during a disruption then this would result in a very significant thermal load to the divertor [3]. However, there are several mitigating factors:

- In most tokamak disruptions the thermal energy content of the plasma, at the time of the thermal quench, is substantially lower than its peak value [15]. However there are exceptions such as disruptions at high-β where there is essentially no loss of confinement before the disruption.
- Some of the thermal quench energy is lost by radiation; exactly how much depends on the wall materials and thus the radiating impurities in the plasma. For example JET with a carbon wall lost around 75 % of the thermal energy radiatively, but with the new tungsten/beryllium wall this reduces to <50 % [16].
- The width of the peripheral region of the plasma (known as the scrape-off layer) in which the heat is conducted to the divertor broadens during the thermal quench; this broadening is in the range 5 to ~20 [2].

Taking these mitigating factors into account significantly lessens the divertor disruptive heat loads, but they remain high enough that further active mitigation (see Sect. 7.6) is essential.

The magnetic energy associated with the plasma current is $1/2LI_p{}^2$ (where L is the plasma inductance). During the current quench phase the majority of this magnetic energy is converted into heat by Ohmic heating in the highly resistive plasma. Some the original magnetic energy is inductively coupled into the vacuum vessel and coils. Typically in JET around 80 % of the magnetic energy is consumed in Ohmically heating the plasma [17]. The majority of this thermal energy is then radiated to the walls—since radiation gives a fairly uniform spreading of the heat loads it is the most desirable means of dissipating the plasma energy.

7.3.2 Eddy Currents, Halo Currents and Forces

Vertical instability occurs because an upwards (or downwards) displacement of the plasma (ξ) from its equilibrium position produces a Lorentz force

$$F_Z \approx -2\pi R I_p \xi \frac{dB_{R,ext}}{dz} \equiv 2\pi I_p \xi n_v B_{Z,ext} \tag{7.4}$$

where z is the coordinate in the vertical direction, $B_{R,ext}$ and $B_{Z,ext}$ are the radial and vertical magnetic field produced by the coils external to the plasma, and n_v (known as the field index) is defined as

$$n_v = -\frac{R}{B_{z,ext}}\frac{dB_{z,ext}}{dR} \tag{7.5}$$

Here the fact has been used that the external fields are in vacuum ($\vec{J} = 0$) and so $dB_{R,ext}/dz = dB_{z,ext}/d$R. Since the radial force on the plasma must be inwards to counter the force that wants to expand the plasma 'hoop', we have $I_p B_{z,ext} < 0$. Therefore the condition for vertical instability is $n_v < 0$ (see Fig. 7.5), since in that case an upwards motion of the plasma ($\xi > 0$) gives $F_z > 0$ and the motion is reinforced (i.e. instability). The neutrally stable ($n_v = 0$) plasma has a 'natural' elongation that increases as the aspect ratio decreases [18]. For $n_v < 0$ the plasma elongation is increased because the external field corresponds to coil currents above/below the plasma 'pulling' on it vertically and/or by coils nearer the mid-plane 'squeezing' the plasma horizontally.

As a vertically unstable plasma moves it must be instantaneously in force balance—normally the inertia of the plasma plays no role in this force balance and the balancing force arises from currents flowing in vacuum vessel, and other structures, surrounding the plasma. These vessel currents may be eddy currents driven by the movement of the plasma (or changes in the plasma current magnitude) and/or halo currents that flow partly in the plasma periphery and partly in the vacuum vessel.

An example of halo current flow is shown in Fig. 7.4. Halo currents may have a toroidally asymmetric component as observed in many tokamaks (see for example [2, 19–21]). The toroidally symmetric component of the halo current is driven through conservation of poloidal and toroidal magnetic flux. In the plasma, which generally has low β by the time halo currents are established, the halo current flow will be parallel to the magnetic field. Between 2 footpoints where a halo field line intercepts the vacuum vessel structure the conservation of poloidal flux drives the halo current in the I_p direction and toroidal flux conservation drives a poloidal current whose toroidal field is in the same direction as the existing field. Such descriptions of halo currents have been very successful in describing the toroidally symmetric results [22].

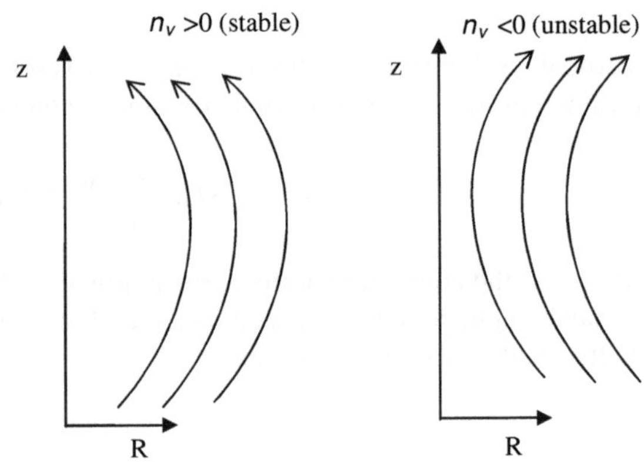

Fig. 7.5 Shape of the external magnetic field for positive and negative field indices

The asymmetric halo current has a predominantly $n = 1$ character (where n is the toroidal mode number); this is most easily seen if the halo current rotates, but can also be observed by measurements at several toroidal angles (see for example [19]). Empirically the asymmetric halo current is quantified by the Toroidal Peaking Factor

$$\mathrm{TPF} = \frac{I_{h,\max}}{I_{h,average}} \tag{7.6}$$

Figure 7.6 shows the relationship between TPF and halo fraction for a wide range of tokamaks. The data are bounded by TPF × $(I_{h,max}/I_{p0}) = 0.7$, where I_{p0} is the plasma current just before disruption

The origin of the asymmetric halo currents is not fully resolved. In most models the asymmetry originates due to an external kink mode triggered by the reduction in edge-q as the plasma area shrinks faster than the plasma current decreases ($q_{edge} \propto \mathrm{Area}/I_p$). In one model it is noted that helical edge currents flow in a kink mode [24], and since the plasma intercepts the wall they will have part of their path in wall (these currents have been termed *Hiro* currents). Alternatively 3D nonlinear modelling results attribute the halo current asymmetry to the vertical plasma position asymmetry [25].

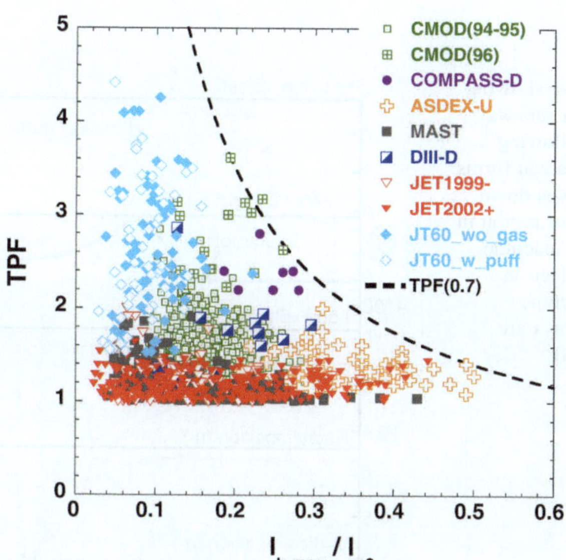

Fig. 7.6 Data on the halo current asymmetry, quantified by the parameter TPF, for the tokamaks Alcator C-Mod, COMPASS-D, ASDEX Upgrade, MAST, DIII-D, JET and JT60-U [23]. (Copyright IAEA, reproduced with permission from the proceedings of the International Atomic Energy Agency, 20th IAEA Fusion Energy Conference 2004, 1st–6th November 2004, Vilamoura, Portugal, IAEA, Vienna (2005))

7.3.3 Runaway Electrons

At disruption the low plasma electron temperature (T_e) and large influx of impurities which can increase the effective charge, Z_{eff} (though at such low temperatures some impurities are only partially ionised), leads to a high plasma resistivity (η)

$$\eta \propto \frac{Z_{eff}}{T_e^{3/2}} \tag{7.6}$$

and so from Ohm's law a large electric field occurs, $E = \eta J$, where J is the plasma current density. When $E > E_D$, the so called Driecer electric field [26], then the thermalised electrons will accelerate to reach runaway energies of 10–20 MeV. Also if the thermal decay of the plasma is sufficiently fast that the electrons cannot relax to a thermal distribution, then the 'hot-tail' can give rise to enhanced runaway electron generation [27]. It is observed that up to around two-thirds of the original plasma current can be carried by runaway electrons during a disruption. Such behaviour is sometimes observed in JET (Fig. 7.7).

In large tokamaks (around JET size and beyond) the formation of secondary runaways by close angle collisions in which the runaway electrons transfer a substantial fraction of their momentum to a slow electron is a key effect [29]. This

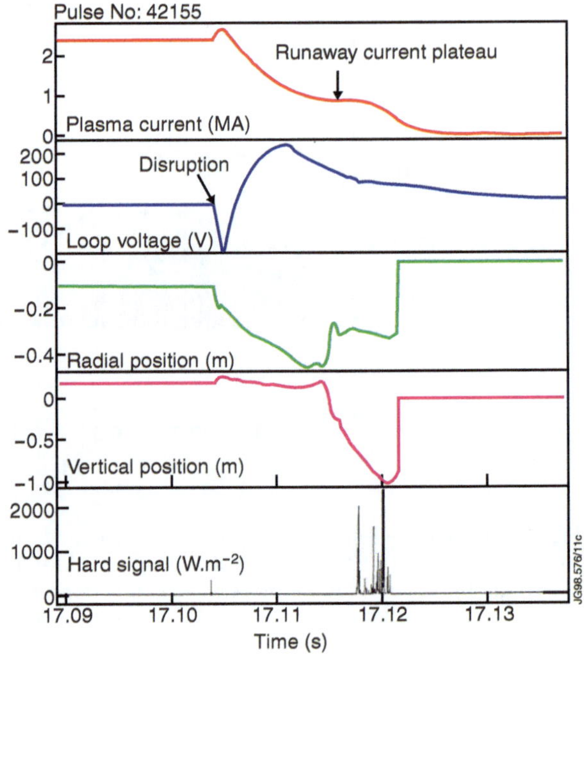

Fig. 7.7 JET pulse showing the formation of a runaway current plateau following a disruption. The plateau forms as the plasma moves down towards the divertor region in the bottom of the machine. As the runaway electron beam hits the machine structure hard X-rays are emitted. From [28]

process leads to exponential growth of the runaway population and in ITER for example many e-foldings are possible.

Disruption generated runaways have caused damage in tokamaks including Tore Supra [30, 31], and JET [31]. Not only is the thermal energy of the runaway electrons deposited in the material surface they strike, but also a fraction of the magnetic energy associated with the runway beam can be converted and deposited as thermal energy [32]. Runaway electrons are thus a significant issue for future large tokamaks and methods for their control will be discussed in Sect. 7.6.

7.4 Detection of Disruptions

The detection of an impending disruption with adequate warning time is obviously key to any disruption control or avoidance scheme. In ITER detection and mitigation of a very high percentage of disruptions is needed. Most present larger tokamaks use an empirical mixture of signals to indicate an impending disruption, allowing appropriate action to be taken. In JET for example a detection threshold is routinely set on the $n = 1$ locked mode amplitude—the control action taken when the threshold locked mode amplitude is exceeded is to reduce the plasma current and plasma elongation, both of which reduce the disruption forces. Also in JET the ITER-like wall, with beryllium plasma facing layers in the main chamber, has led to the use of real time protection based on infra-red imaging of surfaces, and a safe pulse termination system [33]. In ASDEX Upgrade the present system is based on detection of locked modes and loss of plasma vertical position control—either of which trigger a massive gas injection valve (see Sect. 7.6).

Given the underlying complexity and variety of disruptions it is unlikely that they can be predicted with high reliability by tracking a small number of quantities. As discussed in the remainder of this section more sophisticated learning based methods have been developed to predict disruptions, which to some extent provide a more systematic basis for determining the key measured inputs to reliably predict disruptions. As will be seen further developments are still needed to predict disruptions with adequate reliability and robustness (to changing operational regimes).

7.4.1 *Artificial Neural Networks*

Artificial neural networks (NNs) have been used extensively to predict disruptions in tokamaks. In most applications a multi-layer perceptron network has been used (Fig. 7.8). This consists of an input layer where the signals (often normalised) are taken from the tokamak, one, or more, layers of hidden neurons and an output layer. In most applications of disruption prediction there is a single output, which might for example be the predicted time to disruption.

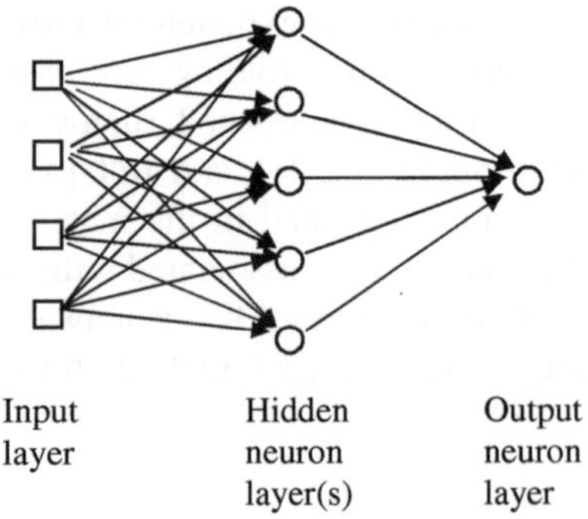

Fig. 7.8 Schematic of a multi-layer perceptron neural network

The network is trained on a set of data to find the connection weights that minimise the prediction error—further details on neural networks can be found for example in [34].

One of the first applications of NNs to the prediction of a disruption was on the TEXT tokamak [35]. In this the measured poloidal magnetic field (Mirnov) fluctuations were used to their predict future behaviour more than 1 ms in advance. In later work it was found that superior results were obtained by predicting the future behaviour of a soft X-ray line passing through the plasma core [36]. This approach was extended on ADITYA tokamak [37] using 4 Mirnov signals, a central SXR signal and a Balmer α signal (H_α). The results showed it was possible to predict behaviour up to 8 ms in advance.

On the DIII-D tokamak 33 input signals were used to train a NN to predict the normalised β_N (=$\beta/(I_p(\mathrm{MA})/a(\mathrm{m})\ B_\phi(\mathrm{T}))$) at disruption [38]. This network was trained on 56 β-limit disruptions and tested against 28 other disruptive pulses. A measure of the success for disruption prediction is the probability of true positive detections of a disruptions versus the probability of false positive detections—obviously the aim is maximise the true positive detections and minimise the number of false detections. From Fig. 7.9 it can be seen that the around 90 % of disruptions can be correctly predicted with an accompanying false positive rate of about 20 % by the NN in this DIII-D example.

In the JT-60U tokamak NNs have been used to predict disruptions caused by the density limit, ramp down of the plasma current, locked modes caused by low density and β-limit disruptions [39]. The JT-60U NN predicted the *stability level*, defined as 1 for a highly stable plasma and 0 at disruption, from 9 input parameters. The NN is trained in 2 steps—in the first step a guess is used for the stability level as the disruption is approached, and the output stability level is used in the second step to retrain the NN. In addition further training data is introduced in the second step and iteratively adjusted to maximise the NN performance. Excluding β-limit disruptions the NN achieved 97–98 % prediction rate 10 ms before the disruption (and >90 % 30 ms before) with a false positive rate of just 2.1 % for non-disruptive pulses; this testing was done on data from 9 years of operation (1991–1999). However, β-limit disruptions, which generally lacked clear precursors, were less

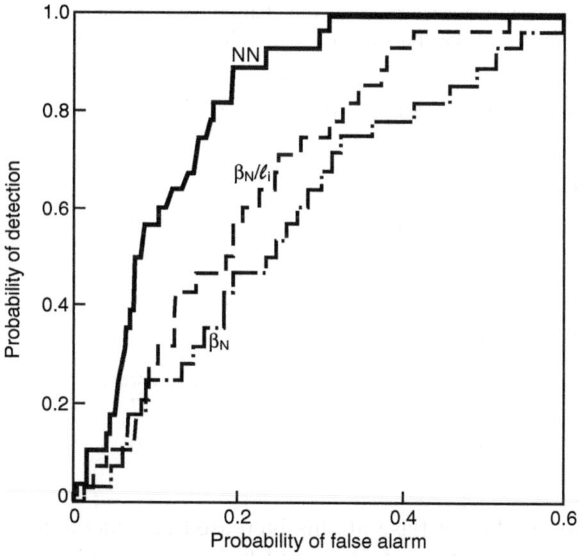

Fig. 7.9 Variation of the probability of a correct detection of disruption versus the probability of false alarm for the test set of data. The implicit variable is the detection β_N threshold which as it is lowered increases the probability of a correct detection but also increases the false positives. Results are shown from the neural network (NN), and also from theoretically based formulae for the limit (labelled β_N and β_N/l_i). See [38] for details

successfully predicted—this was improved by training a sub-NN to predict the β_N-limit, which is then treated as an input for main stability level NN [40].

In JET there has been a lot of work on disruption prediction with NNs starting in the 1990s [41]. More recently a multi-layer perceptron NN has been trained on 86 disruptive pulses and 400 successful pulses [42]. For the disruptive pulses the training was on the 400 ms preceding the disruption and for successful pulses for a randomly selected 400 ms. Nine inputs of global plasma and machine parameters were used at input to the network; a saliency analysis showed the most important input parameters are I_p, the total input power, the plasma internal inductance (l_i), q_{95} and the poloidal beta (β_p). The network target for disruptive pulses was a function that varied from a value of 0 at 400 ms before the disruption and a value of 1 at the disruption. For successful pulses the network target is 0. Table 7.1 shows the performance of this network in terms of missed alarms (MA) in disruptive pulses and false alarms (FA) in successful pulses. In addition to the training set of pulses, a validation set, to ensure the NN does not over-train, and a test set of pulses are reported in Table 7.1.

The threshold level is set with a bias to avoiding false alarms, which accounts for their low %. It can be seen that within the test set 84 % of disruptions are successfully predicted 100 ms in advance. A deficiency of the NN approach is that when it is applied outside the domain of its training that it rapidly deteriorates in performance [43]. In the case of the JET NN discussed above, when applied to the

Table 7.1 JET NN performances. Δt_p is the time before the disruption. The numbers in brackets indicate the total number of the shots for training, validation and performance test [42]

Δt_p *(ms)*									
	40	60	80	100	120	140	160	180	200
Training set									
MA (86)	2	3	4	3	4	5	8	9	17
FA (400)	0	0	0	0	0	0	0	0	0
Validation set									
MA (35)	0	0	1	1	2	3	4	5	5
FA (246)	0	0	0	1	1	1	1	1	2
Test set									
MA (62)	9	11	13	10	13	17	18	21	22
FA (132)	0	0	0	0	0	0	0	0	0

whole pulse the number of false alarms increased considerably; rather than train the NN on the whole pulse, which would be computionally costly, a clustering procedure based on Self Organising Maps (SOM) [44] was adopted, to select significant representative samples for the training [45]. A SOM is a form of NN that maps muti-dimensional spaces to 2-D space, such that it clusters similar behaviours. In the application discussed in [45] the training data is divided into samples near disruption (*disrupted samples*), *non-disruptive samples* well away from disruption, and intermediate *transition* samples. Since the transition region is not well defined these samples are not used in the training of the disruption prediction NN. A 100 ms before the disruption this NN successfully predicted 77 % of disruptions and had 1 % of false alarms; these results were on a test set data from a JET campaign that occurred 15 months before the training set data was obtained.

Neural Networks have also been extensively developed on the ASDEX Upgrade tokamak. A NN was invoked in real time to trigger injection of frozen deuterium pellets [46]; so called *killer pellets* that mitigate the effect of the disruption (this is discussed further in Sect. 7.6). This NN was trained on data from 99 disruptive discharges and 386 non-disruptive discharges. The input to the network was either binary quantities such locked mode (input = 1) or no locked mode (input = 0), or normalised physics quantities such as q_{95}, and their time derivatives. Figure 7.10 shows an off-line validation of the network for a density limit disruption, where the network output is the time to disruption, which in training is taken to have a maximum value of 0.8 s.

In online use 79 % of disruptions (of the types in the training set) were recognised. The network was set to trigger the pellet injector if a disruption was detected in 88 discharges, and in 10 of these the killer pellet was injected to ameliorate the disruption effects.

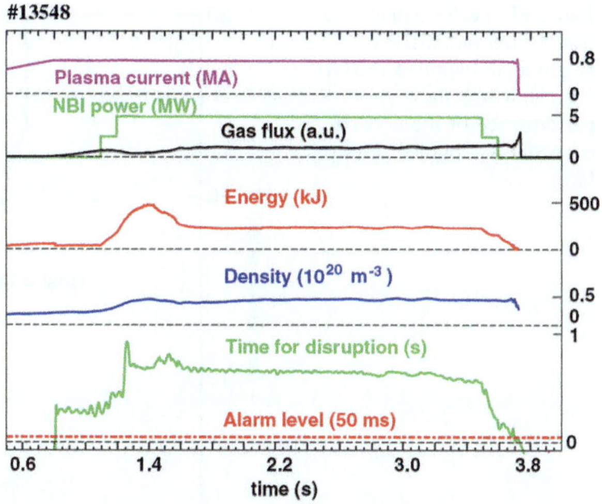

Fig. 7.10 Network prediction for a density limit disruption in ASDEX Upgrade. The alarm threshold is reached 20 ms before the disruption. From [46], copyright 2003 by the American Nuclear Society, La Grange Park, Illinois

In subsequent studies on ASDEX Upgrade a self organising map (as discussed above for JET) has been used to select the NN training samples [47]. This network successfully detected about 82 % of disruptions in a test set belonging to the same experimental campaign as the training set. However, when a later set of campaigns is used the success rate drops to 64 %. The use of an adaptive predictor in which missed disruptions are added to the training set, and the NN retrained, restored the success rate to over 80 %.

The above examples highlight one deficiency of NNs—they perform poorly when applied outside their training domain. A second problem is that they require a training set of disruptions, which are increasingly undesirable on larger tokamaks. A possibility is to train using data from a different tokamak and this was explored in [48] for training and application between ASDEX Upgrade and JET. To achieve this cross-training normalised plasma variables must be used; the chosen parameters were q_{95}, β_N, l_i, n_e/n_G, $\tau_E/< \tau_E >$, and locked mode indicator that switches rapidly from 0 (no locked mode) to 1 (locked mode)—τ_E is the energy confinement time and $\langle\rangle$ denotes an average over the dataset (separately for JET and ASDEX Upgrade). To apply the training data on ASDEX Upgrade to JET (or vice versa) a scaling for time is needed. A minor radius squared, a^2, scaling of time was explored on the basis that this is how the resistive diffusion time scales. Figure 7.11 shows results for a single layer network trained on ASDEX Upgrade and applied to JET. The NN predicts the time to disruption and implicitly the threshold time to disruption is varied along the curves in Fig. 7.11. In JET an early failure was defined as where a disruption was predicted to occur >0.1 s in advance of the actual event and a late failure was when it is predicted <0.01 s in advance (deemed to be too little time to take mitigating actions). Figure 7.11 also shows, as would be expected, that the NN does substantially better on ASDEX Upgrade data.

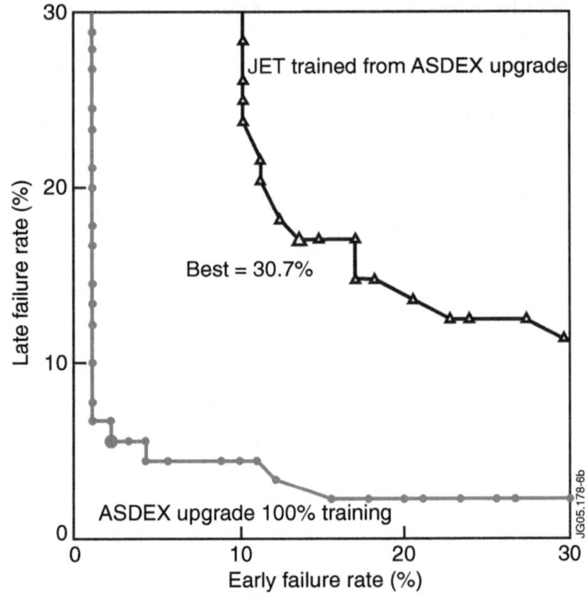

Fig. 7.11 Performance of a NN trained on ASDEX Upgrade and applied to JET. Also shown is the performance of the network on ASDEX Upgrade. From [48]

In summary artificial neural networks are the most thoroughly explored option for disruption detection, but in general their performance is not sufficient to capture nearly all disruptions, as will be required on future large tokamaks. They also have the deficiencies of needing training data and not working well outside their training domain. Hence attention is increasingly focussing on other methods of disruption prediction.

7.4.2 Other Methods

Although neural networks are the most explored method of disruption prediction, other methods have been explored including fuzzy logic [49], classifiers known as Support Vector Machines (SVM) [50], discriminant analysis [51] and multiple threshold tests [52].

The fuzzy logic differs from NNs in that the rules for classifying a disruptive pulse are explicit and can be used to develop additional physics insight. The implementation on JET [49] predicted the probability of a disruption, using 12 inputs and 36 logic based rules. The input variables are classified into between 3 and 5 membership classes—so for example locked mode amplitude is classified as low, medium or high. Likewise the output (disruption probability) is classified into 5 categories between low and high. 36 logic rules were then applied—an example is ‘If I_p is low and locked mode amplitude is high then the output (disruption probability) is high’. These rules can reflect empirical understanding, but the rules and the categorisation of the input variables are optimised using training data of disruptive and non-disruptive pulses. Taking a threshold probability of 0.45 for a disruption alarm and testing on 292 disruptive pulses it was found that 7.9 % of disruptions were missed, while testing of 385 non-disruptive pulses showed 17.7 % of false alarms—the number of false alarms is reduced by increasing the threshold but at the expense of increasing the number of missed disruptions.

The SVM classifier method has been applied on JET [50, 53]. The particular implementation is called APODIS (Advanced Predictor Of DISruptions). APODIS is used to classify whether a pulse state is disruptive or non-disruptive and does this through the entire duration of the pulse. APODIS has two main novelties in relation to other predictors. The first one reflects the fact that APODIS uses a special combination of SVM classifiers with the classifiers distributed into a two layer architecture. The first layer of classifiers (3 in the present version) operate in parallel on consecutive time windows that are 32 ms long. Their respective predictions are combined by the second layer classifier, whose output is the APODIS decision function and establishes whether or not to trigger a disruption alarm. The second main novelty of APODIS is the use of features from both the time and frequency domains. In the time domain the mean values of the signals every 32 ms are used, while in the frequency domain the standard deviation of the power spectrum (after removing the DC component) is used every 32 ms [54]. APODIS has been installed in the JET real-time network [55]. The training was based on 100 non-disruptive discharges and 125 disruptive discharges from JET in carbon wall operation, and required the use of multicore high performance computers. The best predictor was selected from several sets of 100 non-disruptive discharges and 125 disruptive discharges. These sets were randomly chosen from a total number of 4070 non-disruptive and 246 disruptive discharges. The utility of the classifier method was demonstrated by testing on JET ITER-like Wall pulses and was very successful (Table 7.2). An important point is an average warning time for an impending disruption of 426 ms was achieved.

An assessment of how many disruptive discharges are necessary to train the APODIS predictor from scratch has been made; this is an important issue because only a small number of disruptions are acceptable for ITER or DEMO, and so assembling a large training data set is not possible. An adaptive APODIS predictor has been trained in an adaptive way from the first disruption of JET operating with its new metallic wall. The results show that 42 disruptions are necessary to achieve success rates of 93.8 % and false alarm rates of 2.8 % [56]; these rates have been determined from 1036 non-disruptive discharges and 201 disruptive discharges.

Allied to the APODIS approach a clustering method to identify the type of disruption (density limit, neo-classical tearing mode, auxiliary power shut-down etc.) is being developed [57]. This method is based on the geodesic distance on a probabilistic Gaussian manifold, that allows error bars on measurement to be taken into account. The overall success in classifying 795 JET disruptions is 85 % (it has to be remembered that the classification by human experts is not always unambiguous). Once a disruption is predicted then such a classification of the disruption type will allow targeted mitigating actions to be taken.

Table 7.2 Performance of JET APODIS code in classifying disruptive and non-disruptive pulses

	Correctly classified	Incorrectly classified (false/missed alarm)
Non-disruptive pulses	645	6
Disruptive pulses	300	5

The discriminant analysis technique was applied to edge cooling disruptions (see Sect. 7.2.1) in ASDEX Upgrade [51]. A separation of disruptive and non-disruptive pulses is given by

$$0.091 \times l_i^{0.95} \times (n_e/n_G)^{0.46} \times U_{loop}^{\prime 1.24} \times P_{frac}^{0.21} \times \beta_N^{-0.05} = 1.04 \qquad (7.7)$$

where P_{frac} is the fraction of radiated to total input power and U'_{loop} is the loop voltage (with 4.8 added to avoid negative values). When the left-hand side of 7.7 exceeds 1.04 a disruption alarm is generated. This equation gives insight into the causes of edge cooling disruptions, so for example during the edge cooling the current profile contracts raising l_i. In practice a probability function is used to give the likelihood that a disruption will occur [51]. Testing of this discriminant method showed about 80 % of disruptions can be detected 20 ms in advance, allowing adequate time for mitigating actions.

On NSTX a disruption detection system has been developed by combining thresholds for direct instability detection (vertical motion, stationary and rotating $n = 1$ instabilities), equilibrium features (profile shapes, wall gaps), and transport indicators (e.g. plasma rotation, energy and particle confinement). Based on an analysis of the NSTX disruption database the coefficients used in the detection algorithm are chosen to minimise the total failure rate (defined as the sum of false positives, late warnings (<10 ms before the disruption), and missed warnings). Tests show the total failure rate can be as low as 6.5 % when applied to a set of around 2000 disruptions [52]. This method has the advantage that an underlying physics cause, for the disruption, can be attributed, unlike with neural networks.

7.4.3 Remaining Issues

All the methods of disruption prediction that have been discussed rely on a training dataset, and with the exception of the study in [48], this training dataset has come from the tokamak for which the prediction is being made. Since in tokamaks of ITER scale disruptions are highly undesirable and have to be predicted with a high level accuracy, assembling a training set is challenging. It is desirable that further exploration of machine independent disruption predictors occurs and that means to minimise the size of the disruption training set are further explored. An option that could be explored is real-time prediction of the plasma stability—this would either be a direct calculation from first principles of the plasma stability (as discussed in [58]) or could be an MHD spectroscopy technique [59]. The MHD spectroscopy technique has been most applied to high frequency Alfven Eigenmodes (see Chap. 9 and [60]), but also to diagnosing the proximity to resistive wall mode instability (see Chap. 6 and [61]). In this technique an external magnetic field which has components of the MHD eigenmode whose stability is being examined is applied and typically the frequency is scanned across the expected frequency of the natural instability. If the plasma is close to instability then the applied field is amplified and the proximity to instability can be determined by the degree of amplification. This approach has been used to monitor the stability of a disruption causing instability [62], but much more development is warranted. Overall, significant progress in methods to predict disruptions has been achieved but the rigorous demands for future large tokamaks to be able to achieve a very high success rate in predicting disruption and the difficulties of producing training data mean this area needs further attention.

7.5 Disruption Avoidance and Control

As always prevention is better than cure, and so methods to steer the plasma discharge away from a disruption or applying a control actuator to avoid a disruption are highly desirable.

Various ways to steer away from impending disruptions have been developed, including:

- On TEXTOR a real-time detection system was used to detect growing long wavelength instabilities and apply torque with a neutral beam to prevent mode locking [63]
- Historically on ASDEX Upgrade detection of an impending radiative (detached) disruption was controlled by cessation of gas fuelling and an increase in applied neutral beam heating power [64]. Presently detachment is monitored through the divertor temperature and radiation [65]
- Control of the plasma pressure using the applied heating power as an actuator has been practiced on several tokamaks (e.g. on JT-60U [66] and DIII-D [67]), as a means to avoid pressure limit disruptions.

Direct control of the instabilities that cause the disruption has been tried. On the DITE tokamak the magnetic signatures of the instabilities causing density limit disruptions were detected and used in a feedback control loop to apply external fields to reduce their amplitudes. In the case with feedback a significantly higher plasma density could be achieved before the disruption was triggered (Fig. 7.12).

In another example of disruption avoidance electron cyclotron resonance heating (ECRH, whose frequency resonates with the electron gryo-frequency) was applied in the FTU [69] and ASDEX [70] tokamaks near the primary instability

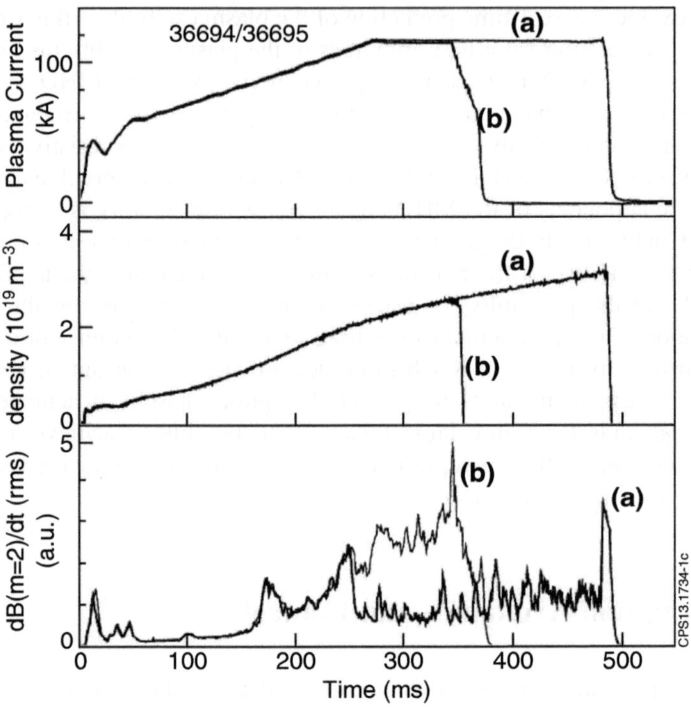

Fig. 7.12 Two nominally identical pulses with **a** magnetic feedback applied from 240 ms and **b** no feedback. The feedback lowers the magnetic fluctuation amplitude (*dB/dt*) and allows a higher density to be achieved. From [68], copyright 1990 by the American Physical Society

location ($q = 2$) and permitted the avoidance of disruptions in otherwise identical discharges. Similarly disruptions were avoided near an edge-q of 3 in JFT-2 M by ECRH Upgrade [71], and such results have been observed in other tokamaks.

7.6 Disruption Mitigation

If disruption avoidance or control methods fail then methods are needed to mitigate the consequences of the disruption. Primarily the peak heat loads on material surfaces and the EM disruption forces must be reduced, and a means to inhibit or safely reduce post-disruption plasma current carried by runaway electrons is needed. In order to meet design criteria in ITER the mitigation methods need to be both highly effective in reducing loads and highly reliable [72].

The most developed mitigation method involves injecting large amounts of gas (at least an order of magnitude more than the plasma content) using fast opening valves—this method is termed massive gas injection (MGI). The other primary method that has been developed is to inject cryogenic pellets of gas, which again substantially increase the plasma density. For runaway electrons other methods of mitigation have been investigated and are discussed in sub-section c below.

a. MGI

In MGI experiments, the gas is delivered to the plasma by one or more fast-valves, with opening times of order 1 ms, mounted outside or inside the vessel. The distance of the valves to the plasma can vary from a few centimetres (e.g. in ASDEX Upgrade [73]) to a few metres with long delivery tubes (4 m in JET, [74]). The time-of-flight of the gas to the plasma depends on its sound speed. Often a light carrier gas (e.g. deuterium) is used to entrain the higher-Z impurity gas (e.g. Argon or Neon), which is more efficient in radiating the plasma energy. A typical example of MGI quenching an ASDEX Upgrade plasma [73] is shown in Fig. 7.13.

The neutral gas propagates from the valve, or flight tube, to the edge of the plasma surface and where it starts to ionise and the cooling phase commences. The neutral gas is observed to penetrate to the $q = 2$ surface in several tokamaks (e.g. TEXTOR [75] and Tore Supra [76]). An example of light from Ar_{II} emission for MGI in TEXTOR is shown in Fig. 7.14, where it can be seen that before the thermal quench the neutrals reach the $q = 2$ surface. Further studies on TEXTOR with different $q = 2$ radii confirmed the extent of the pre-thermal quench neutral penetration [77].

Studies on Tore Supra [76] also show the cooling front from the MGI reaches $q = 2$ before the thermal quench—it is conjectured that observed plasma instabilities eject core energy outside $q = 2$, thereby ionising the injected gas and preventing further penetration into the core. Data from MAST [78] indicates a pre-thermal quench build-up of electron density primarily in the vicinity of $q = 2$ (Fig. 7.15), though the physics underlying this is not fully understood.

The duration of this cooling phase tends to decrease with an increase in injected gas (for JET [74] and for Tore Supra [76]), but can saturate as shown in ASDEX Upgrade for larger amounts of injected gas [73]. The edge-q also affects the duration of the cooling and for example JET [74] and DIII-D [79] data show it increases with increasing edge-q.

As the cooling phase proceeds the level of macroscopic core instabilities increase, and this triggers a rapid loss of energy during the thermal quench phase. Tomographic reconstructions of radiated power show it rises rapidly in the core in the thermal quench phase [74, 80] as shown for a JET case in Fig. 7.16.

3D plasma stability computations using the NIMROD code [81] coupled to a radiation package have helped confirm the interpretation of the thermal quench phase during MGI [82]. In these MHD simulations a symmetric neutral source distributed at the plasma edge simulated the MGI. The modelling reproduces and explains the sequence of events observed experimentally: the perpendicular diffusion of the impurities into the plasma, the propagation of the cold front into the plasma and consequent current diffusion, with the resulting development of tearing modes and magnetic field stochastisation, coinciding with the TQ.

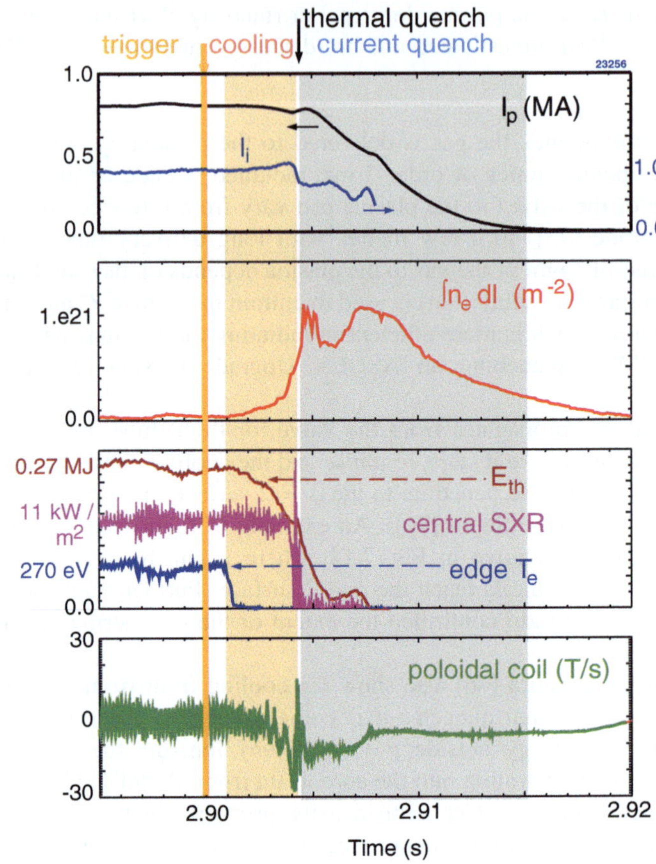

Fig. 7.13 ASDEX Upgrade pulse with MGI triggered at 2.9 s with a valve just 10 cm from the plasma. After a short flight time for injected gas the edge electron temperature ('edge Te') drops and then the outer region of the plasmas cools causing a drop in the plasma thermal energy (Eth), this is followed by a rapid loss of plasma energy (as shown by the central Soft X-ray) known as the thermal quench. Magnetic instabilities measured on poloidal coil outside the plasma increase near the thermal quench time. From [73] (© IOP Publishing. Reproduced by permission of IOP Publishing. All rights reserved)

The large increase in impurities during and immediately following the TQ phase leads to a strong increase in plasma resistivity, which (as in disruptions without MGI) causes a rapid quenching of the plasma current. With the MGI the rate of current quench tends to be larger initially than in non-mitigated pulses [74].

The three aims of MGI are to reduce disruption heat loads to surrounding components, to reduce disruption EM forces and mitigate runaways. Reduction of heat loads is achieved by the MGI increasing the radiated power fraction, which spreads the heat loads more uniformly. Increases in radiated fractions with MGI have been observed *inter alia* on DIII-D [80], JET [74], Alcator C-Mod [83], ASDEX Upgrade [73] and MAST [84]—in general the radiated fraction rises with

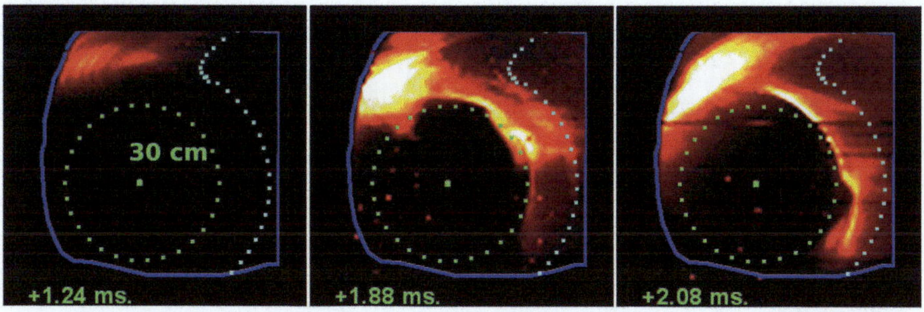

Fig. 7.14 Emission from deuterium and argon MGI before the thermal quench in TEXTOR. The camera was equipped with the Ar II interference filter ($\lambda \approx 611$ nm). The camera viewing window (*solid blue line*), TEXTOR inner wall (*dotted blue line*), the $q = 2$ flux surface (*dotted green circle*) and the plasma centre are marked in the figure. The time is relative to the MGI valve trigger [75].

Fig. 7.15 Data from MAST showing **a** build-up of electron density near $q = 2$ before the thermal quench—the data is assembled from a series of repeated MGI discharges. The related electron temperature is shown in (**b**). The broken lines indicate the error-bars on the $q = 3$ and 2 locations. From [78]

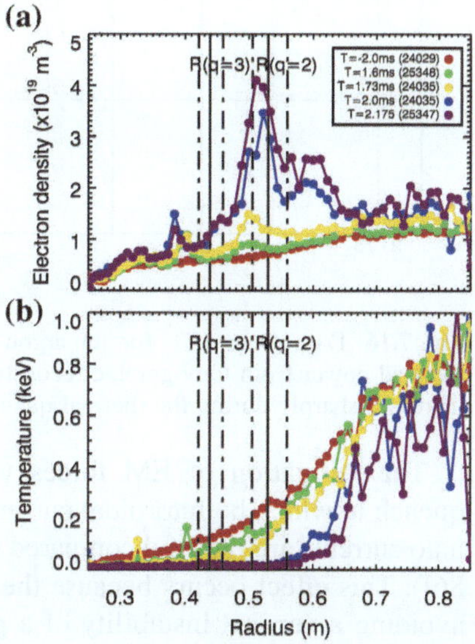

the atomic mass of the impurity and with the quantity of injected impurity. Direct observations on MAST using infra-red imaging confirm the reduction in heat loads in the divertor due to MGI [84]. An issue is that injection from a single valve introduces significant asymmetries into the radiated power—particularly up to and including part of the thermal quench phase [74, 80, 85, 86]. Experiments using 2 almost toroidally opposite gas valves on Alcator C-Mod show a significant reduction in asymmetry is possible during the pre-thermal quench phase [87], though modelling suggests the macroscopic instabilities that facilitate the penetration of the injected gas may lead to an underlying asymmetry effect [88].

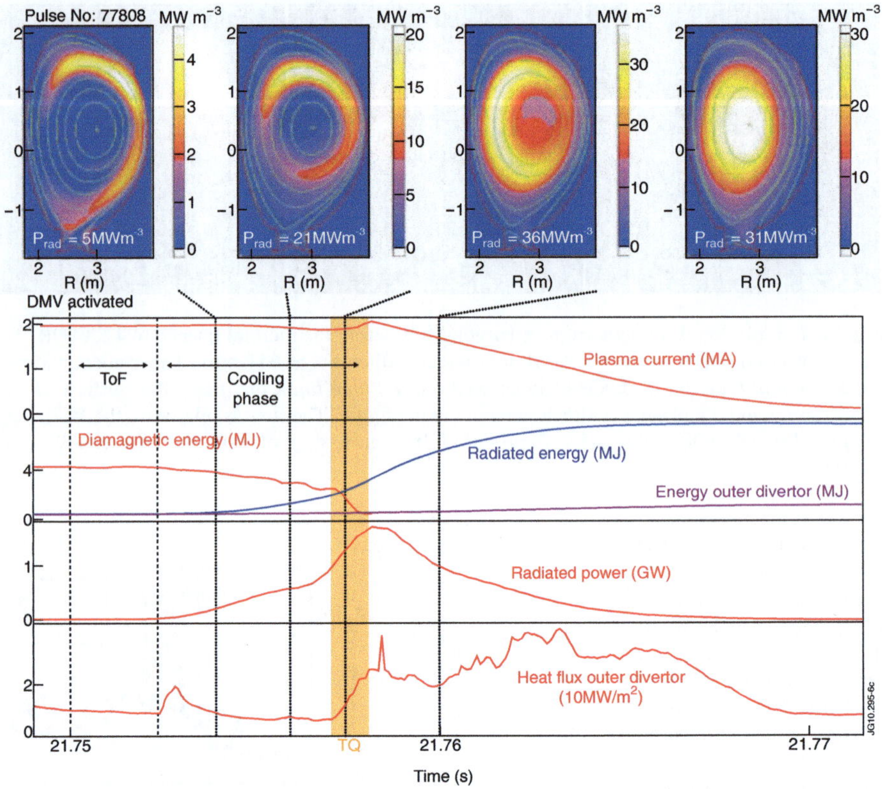

Fig. 7.16 Data from JET for an argon-deuterium MGI mixture, showing contour plots of radiated power from tomographic reconstructions, which illustrate how the core radiated power increases sharply during the thermal quench (TQ) phase. From [74]

The mitigation of EM forces with MGI is achieved by inducing a current quench in which both toroidal currents in the stabilising structures (vessel etc.) and halo currents are reduced compared to a natural disruption (see for example [74,83, 86]). This effect occurs because the MGI causes the current to decay earlier, thus avoiding a vertical instability of a plasma carrying a slow decaying toroidal current. Figure 7.17 is an example from ASDEX Upgrade showing the effect of injected neon gas on the reducing the halo current [89]—it can be seen that a factor of almost 2 reduction of the halo current can be achieved by MGI.

The data on the suppression of runaway electrons with MGI is less satisfactory at present. For secondary runaways (see Sect. 7.3.3) a critical density (n_c) exists above which complete suppression occurs [90]—for typical parameters $n_c(10^{20}\ \mathrm{m}^{-3}) \approx 11\ E\ (\mathrm{Vm}^{-1})$ [2] and this requires the plasma density is increased by around 2 orders of magnitude relative to the pre-disruption value. As yet no tokamak has approached $n/n_c \sim 1$ using MGI. In ASDEX Upgrade values of up to 24 % of n_c have been achieved [73], values of up to 15 % are reported for DIII-D in [91] and in JET less 1 % of n_c has been achieved [74].

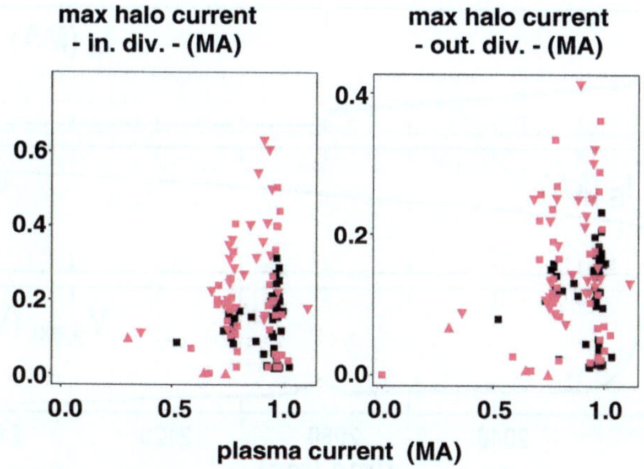

Fig. 7.17 Effect of the injection of neon gas on the reduction of the maximum halo current through the inner and outer divertor plates as a function of the plasma current. *Black* and *pink symbols* represent mitigated and unmitigated disruptions, respectively. *Triangles* represent *upwards* or *downwards* VDEs. *Squares* represent disruptions followed by vertical instability. From [89]

Overall MGI is a successful method for mitigation of disruption heat loads and EM forces, but its effectiveness for runaway electron mitigation has yet to be demonstrated.

b. Killer pellets

An alternate scheme for disruption mitigation, that pre-dates MGI, is by the injection of frozen gas *killer* pellets. Experiments were conducted on several tokamaks, including Alcator C-Mod [92], ASDEX Upgrade [93], DIII-D [94] and JT-60U [95] using Neon, Argon or Deuterium pellets. As with MGI the killer pellets were successful in mitigating heat loads and halo currents, but there was a tendency to produce runaway electrons, with the clearest example being from JT-60U where in pulses with low macroscopic magnetic fluctuation levels, discharges with the current carried by runaway electrons resulted after killer pellet injection [95]. The tendency to produce runways coupled with the technological challenges of having frozen gas pellets on stand-by to fire into discharges with high reliability, led to the more intense consideration of MGI. However, MGI suffers from a low fuelling efficiency (ratio injected to assimilated gas) and so there have been further experiments on DIII-D using shattered pellets [96, 97]. Here 2 plates are used to shatter the relatively large pellet just before it enters the plasma—this avoids any risk of the pellet passing through the plasma and impacting the tokamak wall and also increases the surface area allowing more efficient ablation of the pellet.

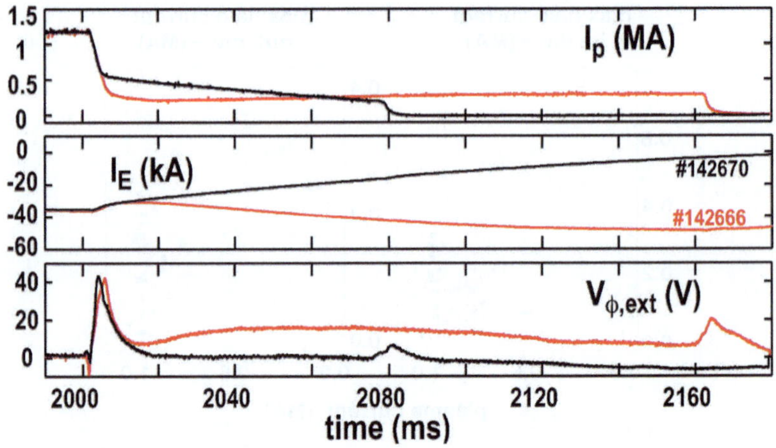

Fig. 7.18 Post-disruption discharges with the current (*top trace*) carried by runaway electrons. The discharges differ primarily in whether the solenoid current (*middle trace*) and the induced voltage (*bottom trace*) are positive or negative. From [99]

c. Other methods for runaway electron control

Two other methods have been applied to control runaways:

- Feedback control of the runaway column position to allow time for an applied negative voltage to wind-down the plasma current (that is being by the runaway electrons)
- Application of non-symmetric magnetic fields that cause the runaway electrons to be lost from the plasma.

The technique of position control of the runaway electron column and a slow reduction of the runaway current by applying an opposing electric field has been demonstrated on JT-60U [98] and on DIII-D [99]. Figure 7.18 shows a comparison of 2 pulses from DIII-D and illustrates the ability to achieve a controlled ramp-down of a plasma current that is carried by runaway electrons.

The key issue with this technique is whether the post-disruption plasma, in which the plasma current is carried by runaway electrons can be position controlled—since this is unlikely to be guaranteed with sufficient reliability a back-up technique (such MGI or shattered pellets) is likely to be needed.

An alternate technique is to apply low toroidal mode number (n) fields—these fields can effect the confinement of the runaway electrons. It was observed in JT-60U that there was a correlation between $n = 1$ instability fluctuation amplitudes and the disruption current quench rate needed to form runaways; at higher $n = 1$ amplitude a higher dI_p/dt is needed [100]. With fields applied using coils external to the plasma a reduction in runaway electrons has been observed in TEXTOR [101] ($n = 1$ and $n = 2$ applied fields) and in DIII-D [91] ($n = 3$ field).

7.7 Conclusions

Disruptions are a primary design constraint in large tokamaks. Significant progress has been made in quantifying the effects of disruptions (heat loads, halo currents and runaway electrons) allowing extrapolations to be made in designing future large tokamaks. Much of this quantification is empirical and would benefit from further development of underlying theory models to underpin the extrapolations.

A lot of the effort on disruptions concerns mitigating their consequences. Though much can be done to reduce the likelihood of a disruption, those arising from causes like flakes of material falling off the wall, or a power supply failure, can never be completely eliminated. Massive gas injection and killer pellets have both proved effective in giving the required level of mitigation of the EM-forces arising from halo currents. These techniques are also effective in reducing heat loads, though issues of toroidal asymmetries, arising from the local nature of the MGI or pellet injection, need to be fully understood. More problematic is the mitigation of runaway electrons; the surest method is raise the plasma density to such high level that the collisions inhibit secondary runaway growth—such densities have not yet been achieved. Further research is needed to understand if such high densities need to be achieved to mitigate runaways, when all the mechanisms damping runaway growth are considered. Mitigation requires reliable prediction of an impending disruption. The most developed technique to predict disruptions is the neural network, but they suffer from needing a training set of disruptions and do not work well outside the domain of their training. Ways to overcome these issues are being studied, but further developments are needed to achieve the required degree of reliability in predicting disruptions.

Overall, with continued research in this area the necessary understanding and mitigation methods to make disruptions acceptable, from the viewpoints of their frequency and consequences, for reactor scale tokamaks, looks very likely to be achieved.

References

1. G.F. Matthews et al., in *Proceedings of the 19th International Conference on Fusion Energy, Lyon, 2002* (IAEA, Vienna, 2002). Paper EX/D1-1 http://www-pub.iaea.org/mtcd/publications/pdf/csp_019c/pdf/exd1_1.pdf
2. T.C. Hender et al., Chapter 3: MHD stability, operational limits and disruptions. Nucl. Fusion **47**, S128 (2007)
3. M. Sugihara et al., Disruption scenarios, their mitigation and operation window in ITER. Nucl. Fusion **47**, 337 (2007)
4. S. Konovalov et al., Characterization of runaway electrons in ITER, in *Proceedings of the 23rd International Conference on Fusion Energy, Daejeon, 2010* (IAEA, Vienna, 2010). Paper ITR/P1-32
5. M. Murakami, J.D. Callen, L.A. Berry, Nucl. Fusion **16**, 347 (1976)

6. P.E. Stott et al., Controlled Fusion and Plasma Physics, in *Proceedings of the 8th European Conference, Prague*, vol. 1 (European Physical Society, 1979), p. 151
7. M. Greenwald, Density limits in toroidal plasmas. Plasma Phys. Control. Fusion **44**, R27–R80 (2002)
8. F.C. Schuller, Disruptions in tokamaks. Plasma Phys. Control. Fusion **37**, A135–A162 (1995)
9. D.A. Gates, L. Delgado-Aparicio, Origin of tokamak density limit scalings. Phys. Rev. Lett. **108**, 165004 (2012)
10. P.C. de Vries et al., Statistical analysis of disruptions in JET. Nucl. Fusion **49**, 055011 (2009)
11. M.F.F. Nave, J.A. Wesson, Mode locking in tokamaks. Nucl. Fusion **30**, 2575 (1990)
12. J.W. Connor et al., A review of internal transport barrier physics for steady-state operation of tokamaks. Nucl. Fusion **44**, R1 (2004)
13. G.T.A. Huysmans et al., MHD stability of optimized shear discharges in JET. Nucl. Fusion **39**, 1489 (1999)
14. R.S. Granetz et al., Disruptions and halo currents in Alcator C-Mod. Nucl. Fusion **36**, 545 (1996)
15. V. Riccardo, A. Loarte, and the JET EFDA Contributors, Timescale and magnitude of plasma thermal energy loss before and during disruptions in JET. Nucl. Fusion **45**, 1427 (2005)
16. P.C. de Vries et al., The impact of the ITER-like wall at JET on disruptions. Plasma Phys. Control. Fusion **54**, 124032 (2012)
17. J.I. Paley, Energy low during disruptions. Ph.D. thesis, University of London (Imperial College), 2006
18. Y.-K.M. Peng, D.J. Strickler, Features of spherical torus plasmas. Nucl. Fusion **26**, 769 (1986)
19. S.P. Gerhardt, Dynamics of the disruption halo current toroidal asymmetry in NSTX. Nucl. Fusion **53**, 023005 (2013)
20. V. Riccardo et al., Progress in understanding halo current at JET. Nucl. Fusion **49**, 055012 (2009)
21. G. Pautasso et al., The halo current in ASDEX Upgrade. Nucl. Fusion **51**, 043010 (2011)
22. D.A. Humphreys, A.G. Kellman, Analytic modeling of axisymmetric disruption halo currents. Phys. Plasmas **6**, 2742 (1999)
23. M. Sugihara et al., Analysis of disruption scenarios and their possible mitigation in ITER. in *Proceedings of the 20th International Conference on Fusion Energy, Vilamoura 2004* (IAEA, Vienna, 2004). CD-ROM IT/P3-29
24. L.E. Zakharov et al., The theory of the kink mode during the vertical plasma disruption events in tokamaks. Phys. Plasmas **15**, 062507 (2008)
25. H.R. Strauss, R. Paccagnella, J. Breslau, Wall forces produced during ITER disruptions. Phys. Plasmas **17**, 082505 (2010)
26. H. Dreicer, Electron and ion runaway in a fully ionized gas. Phys. Rev. **117**, 329–342 (1960)
27. H.M. Smith, E. Verwichte, Hot tail runaway electron generation in tokamak disruptions. Phys. Plasmas **15**, 072502 (2008)
28. R.D. Gill et al., Direct observations of runaway electrons during disruptions in the JET tokamak. Nucl. Fus **40**, 163 (2000)
29. M.N. Rosenbluth, S.V. Putvinski, Theory for avalanche of runaway electrons in tokamaks. Nucl. Fusion **37**, 1355 (1997)
30. J.J. Cordier, Tore Supra experience on actively cooled high heat flux components. Fusion Eng Design **61–62**, 71–80 (2002)
31. G. Martin et al., Disruption mitigation on Tore Supra, in *Proceedings of the 20th International Conference on Fusion Energy, Vilamoura, 2004* (IAEA, Vienna, 2004). CD-ROM file EX/10-6Rc
32. A. Loarte et al., Magnetic energy flows during the current quench and termination of disruptions with runaway current plateau formation in JET and implications for ITER. Nucl. Fusion **51**, 073004 (2011)

33. M. Jouve et al., Real-Time Protection of the "ITER-Like Wall at JET" *EFDA JET preprint* EFDA–JET–CP(11)06/01 (2011)
34. C.M. Bishop, *Neural Networks for Pattern Recognition* (Clarendon, Oxford, 1995)
35. J.V. Hernandez et al., Neural network prediction of some classes of tokamak disruption. Nucl. Fusion **36**, 1009–1017 (1996)
36. A. Vannucci et al., Forecast of TEXT plasma disruptions using soft X rays as input signal in neural network. Nucl. Fusion **39**, 255–262 (1999)
37. A. Sengupta, P. Ranjan, Forecasting disruptions in the ADITYA tokamak using neural networks. Nucl. Fusion **40**, 1993–2008 (2000)
38. D. Wroblewski et al., Tokamak disruption alarm based on neural network model of high-beta limit. Nucl. Fusion **37**, 725–741 (1997)
39. R. Yoshino, Neural-net disruption predictor in JT-60U. Nucl. Fusion **43**, 1771–1786 (2003)
40. R. Yoshino, Neural-net predictor for beta limit disruptions in JT-60U. Nucl. Fusion **45**, 1232–1246 (2005)
41. F. Milani, Disruption Prediction at JET. Ph.D. thesis, University of Aston, Birmingham UK, 1998
42. B. Cannas et al., Disruptions forecasting at JET using neural networks. Nucl. Fusion **44**, 68–76 (2004)
43. B. Cannas et al., Neural approaches to disruption prediction at JET, in *31st EPS Conference on Plasma Physics* (London, UK), vol. 28 G1.167 (2004)
44. T. Kohonen, Self Organ. Map Proc. IEEE **78**, 1464–1480 (1990)
45. B. Cannas et al., A prediction tool for real-time application in the disruption prediction system at JET. Nucl. Fusion **47**, 1559–1569 (2007)
46. G. Pautasso, O. Gruber, Study of disruptions in ASDEX Upgrade. Fusion Sci. Tech. **44**, 716 (2003)
47. B. Cannas et al., An adaptive real-time disruption predictor for ASDEX Upgrade. Nucl. Fusion **50**, 075004 (2010)
48. C.G. Windsor et al., A cross-tokamak neural network disruption predictor for the JET and ASDEX Upgrade tokamaks. Nucl. Fusion **45**, 337–350 (2005)
49. G. Vaglisindi et al., A disruption predictor based on fuzzy logic applied to JET database. IEEE Trans. Plas. Sci. **36**, 253–262 (2008)
50. A. Murari et al., Latest developments in data analysis tools for disruption prediction and for the exploration of multimachine operational spaces, in *Proceedings of the 24th International Conference on Fusion Energy, San Diego, 2012* (IAEA, Vienna, 2013). CD-ROM EX/P8-04
51. Y. Zhang et al., Prediction of disruptions on ASDEX Upgrade using discriminant analysis. Nucl. Fusion **51**, 063039 (2011)
52. S.P. Gerhardt et al., Detection of disruptions in the high-β spherical torus NSTX. Nucl. Fusion **53**, 063021 (2013)
53. J. Vega et al, Results of the JET real-time disruption predictor in the ITER-like wall campaigns. Fus. Eng. Des. **88**, 1228 (2013)
54. G.A. Rattá et al., Feature extraction for improved disruption prediction analysis at JET. Rev. Sci Instrum. **79**, 10F328 (2008)
55. JM. Lopez et al., Implementation of the disruption predictor APODIS in JET's real-time network using the MARTe Framework, EFDA JET report EFDA-JET-CP(12)03/04 (2012)
56. S. Dormido-Canto et al., Development of an efficient real-time disruption predictor from scratch on JET and implications for ITER. Nucl. Fusion **53**, 113001 (2013)
57. A. Murari et al., Clustering based on the geodesic distance on Gaussian manifolds for the automatic classification of disruptions. Nucl. Fusion **53**, 033006 (2013)
58. A.H. Boozer, Theory of tokamak disruptions. Phys. Plasmas **19**, 058101 (2012)
59. J.P. Goedbloed et al., MHD spectroscopy: free boundary modes (ELMs) and external excitation of TAE modes. Plasma Phys. Control. Fusion **35**, B277 (1993)
60. A. Fasoli et al., MHD spectroscopy. Plasma Phys. Control. Fusion **44**, B159 (2002)

61. H. Reimerdes et al., Measurement of the resistive-wall-mode stability in a rotating plasma using active MHD spectroscopy. Phys. Rev. Lett. **93**, 135002 (2004)
62. D. Testa et al., Real-time measurements of damping rates and instability limits for MHD modes on the JET Tokamak, in *Proceedings of the 27th EPS Conference on Controlled Fusion and Plasma Physics*, vol. 24B (Budapest Hungary, 2000), p. 1429
63. A. Kraemer-Flecken et al., Nucl. Fusion **43**, 1437 (2003)
64. V. Mertens et al., Fusion Sci. Technol. **44**, 593 (2003)
65. A. Kallenbach et al., Optimized tokamak power exhaust with double radiative feedback in ASDEX Upgrade. Nucl. Fusion **52**, 122003 (2012)
66. T. Oikawa et al., Development of plasma stored energy feedback control and its application to high performance discharges on JT-60U. Fusion Eng. Des. **70**, 175 (2004)
67. T.C. Luce et al., Long pulse high performance discharges in the DIII-D tokamak. Nucl. Fusion **41**, 1585 (2001)
68. A.W. Morris, Feedback stabilization of disruption precursors in a tokamak. Phys. Rev. Lett. **64**, 1254–1257 (1990)
69. B. Esposito et al., Disruption avoidance in the Frascati Tokamak Upgrade by means of magnetohydrodynamic mode stabilisation using electron-cyclotron resonance heating. Phys. Rev. Lett. **100**, 045006 (2008)
70. B. Esposito et al., Avoidance of disruptions at high β_N in ASDEX Upgrade with off-axis ECRH. Nucl. Fusion **51**, 083051 (2011)
71. K. Hoshino et al., Avoidance of $q_a = 3$ disruption by electron-cyclotron heating in the JFT-2 M tokamak. Phys. Rev. Lett. **69**, 2208 (1992)
72. M. Sugihara et al., Disruption impacts and their mitigation target values for ITER operation and machine protection, in *Proceedings of the 24th International Conference on Fusion Energy, San Diego, 2012* (IAEA, Vienna, 2012). Paper ITR/P1-14
73. G. Pautasso et al., Disruption studies in ASDEX Upgrade in view of ITER. Plasma Phys. Control. Fusion **51**, 124056 (2009)
74. M. Lehnen et al., Disruption mitigation by massive gas injection in JET. Nucl. Fusion **51**, 123010 (2011)
75. S.A. Bozhenkov et al., Generation and suppression of runaway electrons in disruption mitigation experiments in TEXTOR. Plasma Phys. Control. Fusion **50**, 105007 (2008)
76. C. Reux et al., Experimental study of disruption mitigation using massive injection of noble gases on Tore Supra. Nucl. Fusion **50**, 095006 (2010)
77. S.A. Bozhenkov et al., Runaway electrons after massive gas injections in TEXTOR: importance of the gas mixing and of the resonant magnetic perturbations, in *35th EPS Conference on Plasma Physics Hersonissos, 9–13 June 2008 ECA*, vol. 32D, (European Physical Society, 2008), p. P-1.079
78. A.J. Thornton et al., Plasma profile evolution during disruption mitigation via massive gas injection on MAST. Nucl. Fusion **52**, 063018 (2012)
79. E.M. Hollmann et al., Observation of q-profile dependence in noble gas injection radiative shutdown times in DIII-D. Phys. Plasmas **14**, 012502 (2007)
80. E.M. Hollmann et al., Measurements of injected impurity assimilation during massive gas injection experiments in DIII-D. Nucl. Fusion **48**, 115007 (2008)
81. C.R. Sovinec et al., Nonlinear magnetohydrodynamics simulation using high-order finite elements. J. Comput. Phys. **195**, 355 (2004)
82. V.A. Izzo et al., Magnetohydrodynamic simulations of massive gas injection into Alcator C-Mod and DIII-D plasmas. Phys. Plasmas **15**, 056109 (2008)
83. R. Granetz et al., Gas jet disruption mitigation studies on Alcator C-Mod. Nucl. Fusion **46**, 1001–1008 (2006)
84. A.J. Thornton et al., Characterization of disruption mitigation via massive gas injection on MAST. Plasma. Phys. Control. Fusion **54**, 125007 (2012)
85. M.L. Reinke et al., Toroidally resolved radiation dynamics during a gas jet mitigated disruption on Alcator C-Mod. Nucl. Fusion **48**, 125004 (2008)

86. G. Pautasso et al., Contribution of ASDEX Upgrade to disruption studies for ITER. Nucl. Fusion **51**, 103009 (2011)
87. R. Granetz et al., Disruption mitigation experiments with two gas jets on Alcator C-Mod, in *Proceedings of the 24th International Conference on Fusion Energy, San Diego, 2012* (IAEA, Vienna, 2012). Paper EX/P8-09
88. V.A. Izzo et al., Impurity mixing and radiation asymmetry in massive gas injection simulations of DIII-D. Phys. Plasmas **20**, 056107 (2013)
89. G. Pautasso et al., Plasma shut-down with fast impurity puff on ASDEX Upgrade. Nucl. Fusion **47**, 900–913 (2007)
90. M.N. Rosenbluth, S.V. Putvisnki, Theory for avalanche of runaway electrons in tokamaks. Nucl. Fusion **37**, 1355 (1997)
91. E.M. Hollmann et al., Experiments in DIII-D toward achieving rapid shutdown with runaway electron suppression. Phys. Plasmas **17**, 056117 (2010)
92. R.S. Granetz et al., Disruptions, halo currents and killer pellets in Alcator C-Mod, in *Proceedings of the 16th International Conference on Fusion Energy, Montreal, 1996*, vol. 1 (IAEA, Vienna, 1997), pp. 757–762
93. G. Pautasso et al., Use of impurity pellets to control energy dissipation during disruption. Nucl. Fusion **36**, 1291 (1996)
94. P.L. Taylor et al., Disruption mitigation studies in DIII-D. Phys. Plasmas **6**, 1872 (1999)
95. R. Yoshino et al., Fast plasma shutdown by killer pellet injection in JT-60U with reduced heat flux on the divertor plate and avoiding runaway generation. Plasma Phys. Control. Fusion **39**, 313 (1997)
96. N. Commaux et al., Demonstration of rapid shutdown using large shattered deuterium pellet injection in DIII-D. Nucl. Fusion **50**, 112001 (2010)
97. N. Commaux et al., Novel rapid shutdown strategies for runaway electron suppression in DIII-D. Nucl. Fusion **51**, 103001 (2011)
98. R. Yoshino et al., Generation and termination of runaway electrons at major disruptions in JT-60U. Nucl. Fusion **39**, 151 (1999)
99. E.M. Hollmann et al., Effect of applied toroidal electric field on the growth/decay of plateau-phase runaway electron currents in DIII-D. Nucl. Fusion **51**, 103026 (2011)
100. R. Yoshino et al., Runaway electrons in magnetic turbulence and runaway current termination in tokamak discharges. Nucl. Fusion **40**, 1293 (2000)
101. M. Lehnen et al., Suppression of runaway electrons by resonant magnetic perturbations in TEXTOR disruptions. Phys. Rev. Lett. **100**, 255003 (2008)

Chapter 8
Neoclassical Tearing Mode (NTM)

Marc Maraschek

Abstract Neoclassically driven tearing modes (NTM) are a major problem for tokamaks operating in a conventional ELMy H-mode scenario. Depending on the mode numbers these pressure driven perturbations cause a mild to strong reduction in the plasma pressure, thus limiting the maximum achievable normalized plasma pressure$\beta_N = \beta_t/(\frac{I_p}{aB_t})$, or can even lead to disruptions at low edge safety factor, q_{95} . A control of such modes in highβ_N plasmas is therefore of vital interest for tokamaks. The control consists of two major approaches, namely the avoidance of the excitation of these modes and the removal, or at least mitigation of the impact of these modes, if the avoidance did not succeed. For both routes examples will be given and the applicability of these approaches to ITER will be discussed.

8.1 Introduction

A Neoclassical Tearing Mode (NTM) is a pressure driven MHD instability, which typically occurs at high performance plasma discharges with a high pressure inside the plasma. The plasma pressure is typically described by the ratio between the volume averaged kinetic plasma pressure $\langle p\rangle_V$ and the volume averaged magnetic pressure $\langle p_{mag}\rangle_V := \langle B^2\rangle_V/(2\mu_0)$. Here $\langle\ldots\rangle_V$ denotes a volume averaging over the entire plasma volume. The ratio $\beta = \langle p\rangle_V/\langle p_{mag}\rangle_V$ describes the ability of a magnetic configuration to confine a certain amount of kinetic plasma pressure. For the normalized plasma pressure with respect to the toroidal field follows $\beta_t = \langle p\rangle_V/(\langle B_t^2\rangle_V/2\mu_0) \approx \langle p\rangle_V/(B_t^2/2\mu_0)$. β_t is usually normalized by the definition of $\beta_N := \beta_t/(I_p[MA]/(a[m]B_t[T]))$ [1, 2]. Here I_p denotes the plasma current in MA, B_t the vacuum toroidal field in T, a the minor radius of the plasma in m (see exact definition later in this chapter) and $\langle p\rangle_V$ the volume averaged total

M. Maraschek (✉)
Max Planck Institute for Plasma Physics, Boltzmannstr. 2, 85748 Garching, Germany
e-mail: Maraschek@ipp.mpg.de

V. Igochine (ed.), *Active Control of Magneto-hydrodynamic Instabilities in Hot Plasmas*, Springer Series on Atomic, Optical, and Plasma Physics 83,
DOI 10.1007/978-3-662-44222-7_8

pressure in Pa. The second normalization describes the proximity to the theoretically achievable plasma pressure and makes different experiments directly comparable. β_N itself has to be maximized, as the producible fusion power, P_{fusion}, is proportional to β_N^2 ($P_{fusion} \sim \langle p \rangle_V^2$).

NTMs are an issue, because they limit the maximum achievable plasma pressure. The onset of an NTM leads to a significant degradation in confinement, but does not have a hard limit in β_N, as one would expect for an ideal MHD limit. (3,2) NTMs with poloidal and toroidal mode number of $m = 3$ and $n = 2$ lead to a confinement loss of up to 20 %. Especially at low q_{95}, (2,1) NTMs lead to an even greater loss in confinement and can also cause disruptions. A control of NTMs in high β_N plasmas is therefore of vital interest. The control consists of two major approaches, namely to avoid the excitation of NTMs and the removal, or at least mitigation, once an excitation could not be avoided. For both routes examples will be given in this chapter and the applicability of these approaches to ITER will be discussed.

The most commonly applied theory describing the behaviour of the NTMs is based on a generalized Rutherford equation including the additional neoclassical term, which drives the island. This main driving force originates from the local bootstrap current at the resonant surface $\vec{j}_{bs}(r_{res})$. It is governed by the local pressure gradient $\vec{\nabla} p$, which is mainly proportional to the local poloidal normalized pressure $\beta_p(r_{res})$ at the resonant surface of the mode. Based on this theory, a stabilization of an existing NTM is possible by local current drive and heating at the resonant surface. This has been shown experimentally on various fusion experiments and will be a major part of this chapter.

A brief description of the underlying physics of current driven tearing modes in general, and its generalization towards NTMs is given in Sects. 8.2 and 8.3, respectively. Based on this, the physics of NTM removal and avoidance itself is also discussed in Sect. 8.3. In Sect. 8.4 the available tools for achieving this target are briefly addressed. An overview over the most important experimental steps for the stabilization of NTMs and their overall avoidance will be discussed in Sects. 8.5 and 8.6, respectively. If the modes can be neither avoided nor suppressed, the question, if one could mitigate or live with existing NTMs while maintaining good plasma performance is addressed in Sect. 8.7. In Sect. 8.8 ideas are provided on how in the future the shown tools can be combined. Finally, a discussion of the transferability and applicability for ITER is given in Sect. 8.9, which is followed by the summary Sect. 8.10 highlighting open questions in this area.

8.2 Description of the Tearing Mode

The classical tearing mode is a current driven instability, which is accompanied by magnetic reconnection. It can only occur when the local resistivity is sufficiently high, so that energy can be dissipated during the reconnection process changing the magnetic field topology. This modified field topology with nested flux surfaces along the original field lines of the resonant surface is also called a magnetic island. The drive for this instability is the radial gradient of the toroidal current density $dj_{0,\phi}/dr$. The equilibrium current $\vec{j}_0$, as well as the equilibrium magnetic field $\vec{B}_0$ completely lie within its corresponding flux surface. They can be represented as $\vec{j}_0(\vec{r}) = j_{0,\phi}(\vec{r}) \cdot \vec{e}_\phi + j_{0,\theta}(\vec{r}) \cdot \vec{e}_\theta$, and $\vec{B}_0(\vec{r}) = B_{0,\phi}(\vec{r}) \cdot \vec{e}_\phi + B_{0,\theta}(\vec{r}) \cdot \vec{e}_\theta$. This representation is even valid, when the base vectors $\vec{e}_\phi$ and $\vec{e}_\theta$ are the base vectors of so-called flux coordinates, which are aligned to the flux surfaces. In the following discussion all quantities of the tearing mode are considered as small perturbations to the equilibrium quantities. The equilibrium and the mode perturbation quantities are all indexed with 0 and 1, respectively. The following description of the tearing mode stability is derived for a circular plasma cross section in the large aspect ratio approximation, i.e. $R_0/a \gg 1$, with a denoting the minor radius of the plasma separatrix or last closed flux surface and R_0 the major radius of the plasma.

8.2.1 Current Driven Instabilities

The perturbation field $\vec{B}_1$ of a magnetic island only has components perpendicular to the main toroidal field $B_{0,\phi}$: $\vec{B}_1 = B_{1,r} \cdot \vec{e}_r + B_{1,\theta} \cdot \vec{e}_\theta$. $\vec{B}_1$ can therefore be expressed using a helical scalar flux function $\psi_1 = \psi_1(\vec{r})$ via

$$
\begin{aligned}
\vec{B}_1 &= \vec{\nabla}\psi_1 \times \vec{e}_\phi, \\
\text{or by components:} \quad B_{1,r} &= -\frac{1}{r}\frac{\partial \psi_1}{\partial \theta}, \\
B_{1,\theta} &= \frac{\partial \psi_1}{\partial r}.
\end{aligned}
\tag{8.1}
$$

This flux function is described instead of a (r, θ, ϕ) coordinate system in a helical coordinate system $(r, \chi := \theta - \frac{n}{m}\phi, \phi)$.

8.2.2 Tearing Mode Equation

The tearing mode equation connects the unperturbed plasma equilibrium with the perturbation flux ψ_1 of an arbitrary mode with the mode numbers (m,n) at the $q = m/n$ surface. The local force balance $\vec{\nabla} p = \vec{j} \times \vec{B}$ is solved for the perturbation current $\vec{j}_1$ and the perturbation field $\vec{B}_1$. For all perturbation quantities the ansatz

$$
\vec{A}_1(\vec{r}) = \vec{A}_1(r) \cdot e^{i(m\theta + n\phi - \varphi_0)} \tag{8.2}
$$

is used. Note the different functions $\vec{A}_1$ depending on their arguments. The spatial dependency is decomposed in its toroidal and poloidal Fourier components. Each poloidal and toroidal harmonic m and n can now be treated independently. However, the poloidal decomposition in m is only possible in the cylindrical approximation, i.e. in the large aspect ratio approximation ($R_0/a \gg 1$). Inserting the ansatz (8.2) in the local force balance equation and linearizing the result around the resonant $q = m/n$ surface with the minor radius r_{res}, a differential equation for the helical perturbation flux ψ_1 can be derived

$$
\begin{aligned}
\Delta\psi_1 + \frac{dj_{0,\phi}(r)/dr}{B_{0,\theta}(r)\cdot\left(1-\frac{nq(r)}{m}\right)}\psi_1 = 0, \text{ or:} \\
\left(\frac{\partial^2}{\partial r^2} + \frac{1}{r}\frac{\partial}{\partial r} - \frac{m^2}{r^2}\right)\psi_1 + \frac{dj_{0,\phi}(r)/dr}{B_{0,\theta}(r)\cdot\left(1-\frac{nq(r)}{m}\right)}\psi_1 = 0.
\end{aligned}
\tag{8.3}
$$

Equation (8.3) for the helical perturbation flux ψ_1 is valid only in cylindrical coordinates as applied for the ansatz of (8.2). The second form explicitly describes the representation of the Laplace operator Δ in these coordinates. The calculation is done Fourier component wise in a helical coordinate system, which is aligned along the field lines of the resonant surface. It is connected to the equilibrium via the $q(r)$ profile, the poloidal magnetic field $B_{0,\theta}(r)$ and, most important, via the radial gradient of the toroidal current profile, $dj_{0,\phi}(r)/dr$. Note again, that ψ_1 is defined only in this system.

Equation (8.3) describes the solution outside the singular layer, where the plasma can be treated ideally with negligible resistivity. Because of a singularity at the surface $q = m/n$ it can only be solved piecewise in an inner and outer region, i.e. for $[0, r_{res} - \delta]$ and $[r_{res} + \delta, \infty] = [r_{res} + \delta, a]$ independently. The two independent solutions of ψ_1 outside the so-called singular layer around the resonant surface r_{res} have to be continuous. The treatment for the area $[r_{res} - \delta, r_{res} + \delta]$ has to be done differently for the linear and non-linear case, as discussed in the following.

8.2.3 Stability of the Tearing Mode—Rutherford Equation

8.2.3.1 Linear Solution for Small Perturbations

For small perturbations with an infinitesimally small resistive layer ($\delta \to 0$) around the resonant surface, the stability of a tearing mode can be treated linearly. The perturbation flux ψ_1 does not vary over this resistive layer and the inner and outer solution ψ_1 for $\lim_{\delta\to 0}[0, r_{res} - \delta]$ and $\lim_{\delta\to 0}[r_{res} + \delta, a]$ can be joined with the boundary condition $\lim_{\delta\to 0}\psi_1(r_{res}-\delta) = \lim_{\delta\to 0}\psi_1(r_{res}+\delta)$. The island growth is governed in this linear regime by the linear stability parameter

$$\Delta_0' = \lim_{\delta\to 0}\left(\left(\frac{\partial}{\partial r}\ln(\psi_1)\right)\Bigg|_{r_{res}-\delta}^{r_{res}+\delta}\right) = \lim_{\delta\to 0}\left(\left(\frac{\psi_1'}{\psi_1}\right)\Bigg|_{r_{res}-\delta}^{r_{res}+\delta}\right). \tag{8.4}$$

With W being the full island width in radial direction at the O-point of the island, its linear island growth rate $\gamma_0 = \frac{dW}{dt}$ is connected to this linear Δ_0' parameter. For the linear regime the growth rate γ_0 finally results to [3]

$$\gamma_0 = \frac{dW}{dt} = \frac{0.55}{\tau_R^{3/5}\tau_A^{2/5}}\left(\frac{a}{R_0}\frac{nadq/dr}{q}\right)^{2/5}(a\Delta_0')^{4/5} \sim \frac{(a\Delta_0')^{4/5}}{\tau_R^{3/5}\tau_A^{2/5}}. \tag{8.5}$$

For $\Delta_0' > 0$, generated by the yet unperturbed equilibrium, a tearing mode starts to grow (see Fig. 8.1b). For $\Delta_0' < 0$ no mode is supported by the equilibrium and a possibly existing tearing mode shrinks until the mode disappears (see Fig. 8.1a). The linear solution in this constant ψ_1 approximation is only valid for small islands (less than 1 cm in typical fusion plasmas).

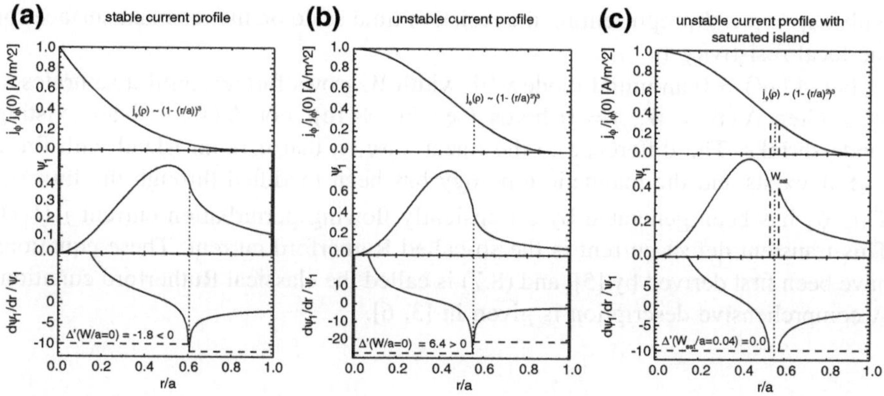

Fig. 8.1 From top to bottom the equilibrium current profile density, the calculated perturbation flux, $\psi_1(\rho)$ and the quantity $\frac{d\psi_1}{dr}/\psi_1 = \psi_1'/\psi_1$ for a $(m,n) = (2,1)$ tearing mode in a circular plasma with $q_{95} = 5$ are shown. In case (**a**) a stable profile $j_{0,\phi} \sim (1-(r/a))^3$ with $\Delta' < 0$, in case (**b**) an unstable profile $j_{0,\phi} \sim (1-(r/a)^2)^3$ with $\Delta' > 0$ and in (**c**) this unstable profile with a saturated island W_{sat} is presented. During the growth of an island with finite size, $\Delta'(W)$ starts to shrink from $\Delta'(W=0) = \Delta_0' > 0$ (case b) until $\Delta'(W \to W_{sat}) = 0$ is reached (case c)

8.2.3.2 Non-linear Solution for Large Islands

The subsequent evolution towards larger islands, which is typically observed with diagnostics, has to be calculated non-linearly. The solution for perturbation flux ψ_1 varies over the region of the island width W and has to be self consistently included. Within this region the perturbation flux ψ_1 is approximated as constant. Therefore this assumption is called the constant ψ_1 approximation, and has been derived in [4].

The transition $\lim_{\delta\to 0}$, i.e. the singular resistive layer, is replaced by the fixed boundaries of the macroscopic resistive region $r_{res} - W/2$ and $r_{res} + W/2$. A third continuous solution in the region $[r_{res} - W/2, r_{res} + W/2]$ has to be additionally solved. This can, also in cylindrical approximation, only be solved numerically. The Δ'_0 term is finally replaced by a $\Delta'(W)$ term for the finite resistive radial region $[r_{res} - W/2, r_{res} + W/2]$

$$\Delta'(W) = \left(\frac{\partial}{\partial r}\ln(\psi_1)\right)\Bigg|_{r_{res}-W/2}^{r_{res}+W/2} = \left(\frac{\psi_1'}{\psi_1}\right)\Bigg|_{r_{res}-W/2}^{r_{res}+W/2}, \tag{8.6}$$

and explicitly depends on the island size W itself. This $\Delta'(W)$ is connected to the island growth rate via the following equation

$$\frac{\tau_{res}}{r_{res}}\frac{dW}{dt} = r_{res}\Delta'(W), \tag{8.7}$$

with $\tau_{res} = \mu_0 a^2/\eta$ representing the resistive time scale on the resonant surface for the local resistivity η.

For $\Delta'(W) > 0$ an initial mode with width W grows further, until it saturates at W_{sat} when $\Delta'(W = W_{sat}) = 0$ holds (see Fig. 8.1c). For $\Delta'(W) < 0$ an existing mode shrinks. The difference to the linear case is, that here an island with finite size W exists and the magnetic topology has been modified through the flux ψ_1. This ψ_1 has been generated by a transiently flowing perturbation current $\vec{j}_1(\vec{r}, t)$. This transient defect current is the so-called Rutherford current. These equations have been first derived by [5], and (8.7) is called the classical Rutherford equation. A comprehensive description is given in [3, 6].

8.3 Description of the NTM Physics and Their Suppression or Avoidance

8.3.1 Generalized Rutherford Equation Describing NTMs

The generalized Rutherford equation describes the stability of an NTM in contrast to the above described current driven classical tearing mode. It extends the original Rutherford equation (8.7) with its classical stability parameter $\Delta'(W)$ [5] by additional stabilizing and destabilizing terms [7, 8] (see the footnote[1] and references in the subsequent paragraphs for the additional definitions)

$$\begin{aligned}\frac{\tau_{res}}{r_{res}}\frac{dW}{dt} &= r_{res}\Delta'(W) \\ &+ r_{res}\beta_p\left(a_2\sqrt{\varepsilon}\frac{L_q}{L_p}\frac{W}{W^2+W_0^2} - a_3\frac{r_{res}}{R_0^2}\frac{L_q^2}{L_p}\frac{1}{\sqrt{W^2+0.2W_{d,e}^2}}\right. \\ &\left. - a_4 g(\varepsilon,\nu_{ii},\omega_e^*,m)\left(\rho_{pi}\frac{L_q}{L_p}\right)^2\frac{1}{W^3}\right).\end{aligned} \tag{8.8}$$

The gradient length for all quantities α is calculated according to $L_\alpha = \alpha/\nabla\alpha$, with $\nabla = \vec{\nabla}_r = \frac{d}{dr}$ representing the radial component of the gradient $\vec{\nabla}$, with r the cylindrical minor radius in m. Hence the gradients are described in 1/m and for the gradient length $[L_\alpha] = m$ holds. L_p and L_q denote the pressure gradient and the q-gradient scale length, respectively.

The local normalized poloidal ion gyro radius, ρ_{pi}^*, at the resonant surface r_{res} of the mode is defined as

$$\rho_{pi}^*(r_{res}) = \rho_{pi}(r_{res})/a, \quad \text{with } \rho_{pi}(r_{res}) = \frac{\sqrt{2m_i kT_i(r_{res})}}{e\langle B_\theta(r_{res})\rangle_S}. \tag{8.9}$$

It has to be pointed out here, that in literature also the definition $\rho_{pi}^*(r_{res}) = \rho_{pi}(r_{res})/r_{res}$ is found. The local flux surface averaged normalized poloidal plasma pressure at the resonant surface r_{res} is defined as

$$\beta_p(r_{res}) = \frac{\langle p(r_{res})\rangle_S}{\langle B_\theta(r_{res})^2\rangle_S/2\mu_0}. \tag{8.10}$$

The required poloidal magnetic field $\langle B_\theta(r_{res})\rangle_S$ and its square $\langle B_\theta(r_{res})^2\rangle_S$ are defined as the flux surface average $\langle\ldots\rangle_S$ of $B_\theta(r_{res},\theta)$ and $B_\theta(r_{res},\theta)^2$ on the resonant surface.

8.3.1.1 Destabilizing Neoclassical Bootstrap Drive

The term with the coefficient a_2 describes the neoclassical bootstrap drive. For $W_0 = 0$ (neglecting transport effects) this term is proportional to the inverse island width ($\sim 1/W$). Due to the flattening of the pressure, temperature and density within the island the resulting lack of bootstrap current, j_{BS}, further drives the

[1] a_i: numerical constants of the order of unity; $\varepsilon = r_{res}/R_0$: inverse aspect ratio of the resonant surface; R_0: major radius of the geometric axis of the resonant surface.

island (a_2-term > 0). For very small islands ($W \rightarrow 0$) this becomes even more effective and the mode should be always unstable. (For $\frac{\tau_{res}}{r_{res}}\frac{dW}{dt} \approx r_{res}\beta_p a_2 \sqrt{\varepsilon}\frac{L_p}{L_q}\frac{1}{W}$ follows: $W\frac{dW}{dt} = const$ and hence $\frac{d}{dt}(W^2) = 2 \cdot const$. With $W \sim \sqrt{B_1}$ it follows $\frac{dB_1}{dt} = const$, i.e. the perturbation field B_1 should grow linearly with time.) This is not observed and hence some further stabilizing effects are needed to explain the observed mode behaviour correctly.

It is important to note, that the density gradient, ∇n, and the temperature gradient, ∇T, have a differently weighted impact on the variation of the local bootstrap current, $j_{bs}(r_{res})$, as applied in [9]. The corrected pressure gradient length according to [10] results in

$$\frac{1}{L_p^{corr}} \approx \frac{1}{1+\alpha} \cdot \left(\frac{1}{L_T} + \alpha \cdot \frac{1}{L_n}\right), \text{ with: } \alpha = 2.5. \tag{8.11}$$

The weighting factor α can be calculated to be around 2.5 and has been experimentally fitted to be of that order [9]. The $T \cdot \nabla n$ term thus contributes stronger to the bootstrap current than the $n \cdot \nabla T$ term. As shown in [11], the onset and marginal β scalings and the time history of an NTM could only be reproduced by including this distinction between the influence of the temperature and the density gradient.

All terms discussed in the following become dominant only for small islands, whereas for large islands the bootstrap effect dominates the mode behaviour. Therefore they are summarized as small island terms.

8.3.1.2 Stabilizing Finite Parallel Heat Conductivity ($\chi_\perp/\chi_\parallel$-Correction)

Usually the particle as well as the heat conductivity $\chi_\parallel$ along a field line within a flux surface is much larger, but still finite, compared to the perpendicular heat conductivity $\chi_\perp$ in the radial direction. This fast parallel transport flattens the profile within the island region, i.e. inside the island's separatrix, as it connects regions inside ($r < r_{res} - W/2$) and outside ($r > r_{res} + W/2$) of the island and equilibrates the temperature $T(r, \theta)$. However, as shown in [12], close to the island's separatrix the effective parallel conductivity $\chi_\parallel$ is reduced, due to the formation of an X-point. The stagnation of the field lines radially close to the X-point increases the so-called connection length along the field lines within flux surfaces in the island. This suppression of the parallel transport $\chi_\parallel$ increases the ratio $\frac{\chi_\parallel(\nabla_\parallel T)}{\chi_\perp(\nabla_\perp T)}$. For small islands a complete flattening is no longer reached, i.e. a parallel temperature gradient is created close to the separatrix. The critical island size W_0, below which this effect becomes dominant, has been calculated in cylindrical approximation as $W_0 \approx 5.1(\chi_\perp/\chi_\parallel)^{1/4}\sqrt{(R_0 q L_q/m)}$ [12]. For incomplete flattening for such small islands ($W < W_0$) the neoclassical drive is reduced. This effect is included in (8.8) in the W_0 correction term of the bootstrap drive term (coefficient a_2). Similar arguments, but with different numerical values, hold for the parallel and perpendicular particle diffusivity, $D_{\parallel}$ and $D_{\perp}$, and the density distribution $n(r, \theta)$ within the island.

Further refinements of such transport effects are discussed by various authors in literature. For example, in (8.8) a small ion banana width w_b is assumed, i.e. $w_b \ll W$. This is however true only for the saturated island at its maximum size, but questionable for a small initial island size at the mode onset and also for small islands in low β_p phases. Therefore the theory has been extended to account for this finite banana orbit width of the thermal ions. As in some sense trapped ions average over regions inside and outside the island, the results concerning the ionic part of the bootstrap current, as well as the polarization current (see below) are strongly modified for $W \approx w_b$. In [13] it is shown that a significant amount of the ion bootstrap current still survives inside the island even if the temperature and density profile are completely flattened over the island. This corresponds to an additional stabilizing effect for small islands, not included in (8.8). The strong increase of this current for small islands ($\sim 1/W^3$) is weakened [14].

8.3.1.3 Stabilizing Glasser Effect

The term with the coefficient a_3 represents the stabilizing Glasser effect. It describes the helical component of Pfirsch-Schlüter currents induced by toroidicity and shaping of the poloidal cross-section [15–18]. A detailed discussion of this effect goes beyond the scope of this book and is left for studying the literature.

8.3.1.4 Stabilizing Polarization Currents

The term with the coefficient a_4 describes the stabilizing polarization currents induced by the island rotation within the background ion fluid [19–22]. As the island rotates through the fluid of heavy ions, which cannot follow the quickly varying local electrical fields from the electrons, a polarization current in the island fluid is induced. This current is phase shifted and leads to an additional stabilizing helical current.

The polarization current term, as given in (8.8), diverges for $W \rightarrow 0$. A full kinetic treatment (Monte Carlo δf approach) [23–25] shows that finite-orbit effects weaken the impact of the polarization current on the island stability, although a complete description of the related physics is still lacking. As done in (8.1) in [26], the divergence of the polarization current contribution can be removed by replacing $\sim 1/W^3$ with $\sim W/(W^4 + W_{d,pol}^4)$.

Additional terms, originating, for example from non fully ionized and therefore radiating impurities accumulating inside the O-point of the island, can be added to (8.8). Finally it should be noted, that the theory of NTM excitation, which has to take into account all small island effects, is still under discussion. In particular the impact of rotation is presently discussed.

8.3.2 Natural Development of an NTM

In Fig. 8.2 the characteristic values of an NTM are indicated (a) in the stability diagram and (b) the time traces of an experiment measuring the marginal $\beta_{p,marg}$ of the mode. The stability diagram represents the solution of (8.8) for different values of the plasma pressure represented by the local $\beta_p(r_{res})$ at the resonant surface r_{res} of the mode, as defined in (8.10). Note that the driving force is not proportional to β_N, but rather to the local $\beta_p(r_{res})$ [27, 28], as it is governed by the local perturbed bootstrap current density j_{bs}.

At the highest $\beta_p = \beta_{p,onset} > \beta_{p,marg}$ two roots for W of (8.8) with $dW/dt = 0$ exist. The larger root represents the saturated island size W_{sat} and the smaller root the minimal, so-called seed island size W_{seed}. For an NTM to get excited, this W_{seed} has to be provided on the resonant surface by an external perturbation, such as a sawtooth. Once such a trigger generating an initial island W with $W \geq W_{seed}$ has occurred (point 1 in Fig. 8.2a, b), the further growth ($dW/dt > 0$) of the NTM is initiated (line between point 1 → 2 in Fig. 8.2). The flattening of the pressure profile over this initial island reduces the bootstrap current locally within this seed island. For conventional monotonic q-profiles, this defect current acts as the major drive of the NTM.

The island itself degrades the confinement. Therefore, β_p is reduced during the subsequent mode growth, until a new equilibrium is reached at $\beta_{p,sat} < \beta_{p,onset}$ while the island size $W_{sat}(\beta_p)$ is reached (point 2 in Fig. 8.2).

When the external heating power is continuously reduced, the island size $W_{sat}(\beta_p)$ follows mainly the local $(L_q/L_p^{corr}) \cdot \beta_p$ values [11, 26] and continuously shrinks (points 2 → 3 → 4 in Fig. 8.2a, b). This can be approximated with

$$W_{sat}(\beta_p) \sim \frac{1}{L_p^{corr}} \beta_p \sim \beta_p. \tag{8.12}$$

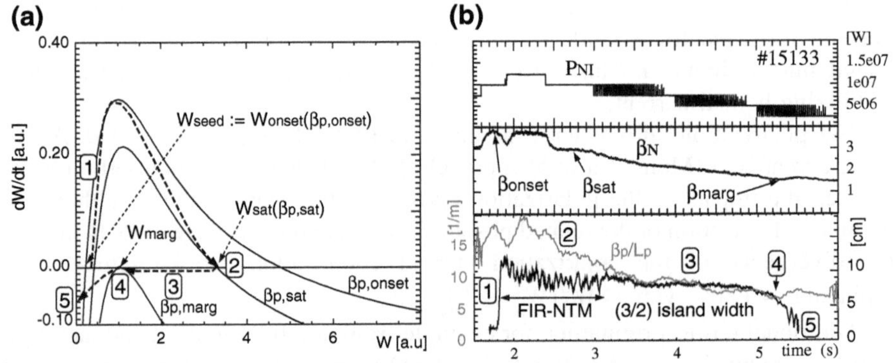

Fig. 8.2 Stability diagram for a naturally occurring NTM in terms of the island growth rate dW/dt as function of the island width W for 3 different β_p values, $\beta_{p,onset}$, $\beta_{p,sat}$ and $\beta_{p,marg}$. Characteristic points in terms of β_p and W are marked. **b** Corresponding time traces of a natural NTM with the characteristic time points marked (modified Fig. 1 from [29])

The critical value of β_p, for which $W_{sat}(\beta_p)$ and $W_{seed}(\beta_p)$ become one identical root, is defined as the marginal β_p, i.e. $\beta_{p,marg}$. This single root is defined as the marginal island width $W_{marg} := W_{sat}(\beta_{p,marg}) \equiv W_{seed}(\beta_{p,marg})$ (point 4 in Fig. 8.2). When β_p is reduced below $\beta_{p,marg}$, the island decays away ($dW/dt < 0$ for all W and β_p, when $\beta_p < \beta_{p,marg}$, point 4 → 5 in Fig. 8.2). The dependence $W(\beta_p) \sim \beta_p$ no longer holds for $\beta_p < \beta_{p,marg}$.

When the NTM has disappeared, β_p can rise again due to the now recovered confinement. It can even exceed again $\beta_{p,marg}$ such that a large external trigger may excite a new NTM. In Fig. 8.2b, β_p even rises while the heating power is still ramped down. Exact formulas for these characteristic quantities can be found in [26, 30–32]. In [32] the quantities are used in a way that they can be applied directly to predictions towards ITER and the NTM stabilization there.

8.3.2.1 Frequently Interrupted Regime (FIR)

For higher local β_p values, just after the onset of the (3,2) NTM, a phase with frequent interruptions of the (3,2) NTM growth can be observed, which is caused by nonlinear coupling to a (4,3) and a (1,1)-mode. This behaviour can not be described by (8.8), but requires additional non-linear calculations [33]. This coupling reduces the (3,2) NTM amplitude on a very short time scale. Thus for a repetition time of these rapid amplitude drops smaller than the NTM growth time ($\tau_{FIR\text{-}drop} < \tau_{growth}$), the average island width W gets reduced and does no longer follow the local β_p values, as shown in [11, 26]. This behaviour is described as **F**requently **I**nterupted **R**egime, FIR-NTM [33–35]. In Fig. 8.2b this happens in the time between the mode's onset and roughly 3.1 s. At lower β_p values, the (4,3) mode remains stable, i.e. the FIR phase disappears and the (3,2) mode follows the local β_p values. The threshold for this FIR behaviour is described by an empirically determined critical β_N at the mode's onset, $\beta_{N,onset} \geq \beta_{N,FIR} = 2.3$. Similarly the amplitude of a (4,3)-NTM can be transiently reduced by the non-linear interaction with a (1,1) and an ideal (5,4)-mode, i.e. also (4,3) FIR-NTMs are observed [34, 35]. A corresponding (2,1) FIR-NTM with interaction with a (1,1) and a (3,2)-mode has not been observed.

8.3.3 Distinction Between Current and Neoclassically Driven Tearing Modes

NTMs can be experimentally clearly distinguished from classically current driven tearing modes (Sect. 8.2.2).

1. Typically NTMs appear only in high β_N scenarios, where a sufficiently large pressure gradient at the resonant surface is present and can dominate the drive of the mode. These discharges turn out to be $\Delta'(W)$ stable ($\Delta'(W) < 0$ when the

data are available and $\Delta'(W)$ can be calculated). There is no current gradient drive of the instability available and no tearing mode should be present.

2. A clear trigger is observed exciting the mode to a finite initial island size $W > W_{seed}$. It has to be noted, that also trigger less classical tearing modes (current driven) can get excited and only get later additionally driven by an increased pressure and hence become an NTM at higher β_p.
3. An NTM grows typically linearly with time, namely $B_{pert}^{NTM} \sim t$. For a small linearly growing current driven tearing mode (TM) $W \sim e^{\gamma t}$, and hence $(W \sim \sqrt{B_1})$ $B_1^{TM,lin} \sim e^{\gamma t 2}$ holds. In the non-linear regime $W \sim (1 - e^{-\gamma t}) \approx \gamma t$ and hence $B_1^{TM,non-lin} \sim t^2$ holds.
4. During an applied variation of β_N with a saturated NTM, the island size W_{sat} follows the applied β_p linearly, i.e. $W_{sat} \sim \beta_p / L_p^{corr} \sim \beta_p$ holds.
5. The marginal point $\beta_{p,marg}$, at which the NTM disappears when β_p is reduced shows a clear hysteresis compared to the onset $\beta_{p,onset}$, i.e. $\beta_{p,marg} \ll \beta_{p,onset}$.

8.3.4 Stabilization of Excited NTMs

The dominant driving term for the NTM is the lack of bootstrap current within the island, as the pressure profile is flattened and therefore the bootstrap current is reduced. For the unperturbed bootstrap current $j_{bs} \sim \nabla p / \langle B_\theta \rangle$ holds. This opens the possibility to directly remove an NTM by locally replacing the missing current by external means, such as local Electron Cyclotron Current Drive (ECCD, see Sect. 8.4) [31, 36]. The generalized Rutherford equation (8.8) has to be extended by another term describing this externally driven current [37, 38]

$$\begin{aligned} \frac{\tau_{res}}{r_{res}} \frac{dW}{dt} = & r_{res} \Delta'(W) \\ & + r_{res} \beta_p \left(a_2 \sqrt{\varepsilon} \frac{L_q}{L_p} \frac{W}{W^2 + W_0^2} - a_3 \frac{r_{res}}{R_0^2} \frac{L_q^2}{L_p} \frac{1}{\sqrt{W^2 + 0.2 W_{d,e}^2}} \right. \\ & \left. - a_4 g(\varepsilon, \nu_{ii}, \omega_e^*, m) \left(\rho_{pi} \frac{L_q}{L_p} \right)^2 \frac{1}{W^3} \right) \\ & - c_{stab} 16 \sqrt{\pi} \mu_0 r_{res} L_q \frac{\eta_{ECCD} \cdot j_{ECCD}}{\langle B_\theta (r_{res}) \rangle} \frac{d_{dep}}{W^2}. \end{aligned} \tag{8.13}$$

Here c_{stab} is an empirical coefficient of the order of unity, d_{dep} the deposition width at $1/e$ of the maximum of a Gaussian radial deposition profile of the externally driven ECCD current j_{ECCD} and η_{ECCD} the helically flowing fraction of j_{ECCD} along the helicity on the resonant surface. The current drive efficiency for NTM stabilization is defined by the ratio of the helically driven current normalized to the lost equilibrium bootstrap current $\eta_{NTM}\!: = j_{cd,hel}/j_{bs} = (\eta_{ECCD} \cdot j_{ECCD})/j_{bs}$.

In the stability diagram and the time traces this process is illustrated in Fig. 8.3a, b. The external local current drive is turned on, while the plasma is in a stationary equilibrium at $\beta_{p,sat}$ with an island size of $W_{sat}(\beta_{p,sat})$ (point 2 in Fig. 8.3a, b). The additional ECCD term (c_{stab}-term) in the generalized Rutherford equation (8.13) now shifts the stability curve, while keeping the high $\beta_{p,sat}$, to a lower value in the dW/dt-direction. Within the c_{stab}-term, d_{dep} denotes the deposition width at $1/e$ of the maximum of the Gaussian radial profile of the externally driven ECCD current $j_{ECCD}(r) = j_{ECCD,0} \cdot e^{-((r-r_{dep})/(2d_{dep}))^2}$. Throughout this chapter always this definition is used. The larger root of (8.13) W_{sat} is reduced and the island shrinks with increasing size of the c_{stab}-term (points 2 → 3 → 4 in Fig. 8.3). Having reached $W_{sat}(\beta_{p,sat}, ECCD) = W_{marg}$, the island decays away, as in the previous case when β_p had been reduced (point 4 → 5 in Fig. 8.2).

In the example shown in Fig. 8.3b the magnetic field is slowly varied until the resonant surface is hit correctly by the ECCD. The deposition location in R depends linearly on B_t. This variation increases the c_{stab}-term, when the ECCD comes closer to the resonant surface. During the size reduction and the final removal of the NTM, β rises again, as desired.

Equation (8.13) can be simplified by summarizing the ECCD independent factors a_i into only one experimentally fitted parameter c_{sat} [38], so that only c_{sat} and the ECCD parameter c_{stab} are kept as free parameters

$$\begin{aligned}\frac{\tau_{res}}{r_{res}}\frac{dW}{dt} =& r_{res}\Delta'(W) \\ &+ c_{sat}\left(r_{res}3.17\mu_0 L_q \frac{j_{bs}}{B_\theta}\left(\frac{W}{W^2+W_d^2}+\frac{W}{W^2+28W_b^2}\right) - \frac{r_{res}6,35\mu_0 D_r}{\sqrt{W^2+0.65W_d^2}}\right) \\ &- c_{stab}16\sqrt{\pi}\mu_0 r_{res} L_q \frac{\eta_{ECCD}\cdot j_{ECCD}}{\langle B_\theta(r_{res})\rangle}\frac{d_{dep}}{W^2}.\end{aligned} \tag{8.14}$$

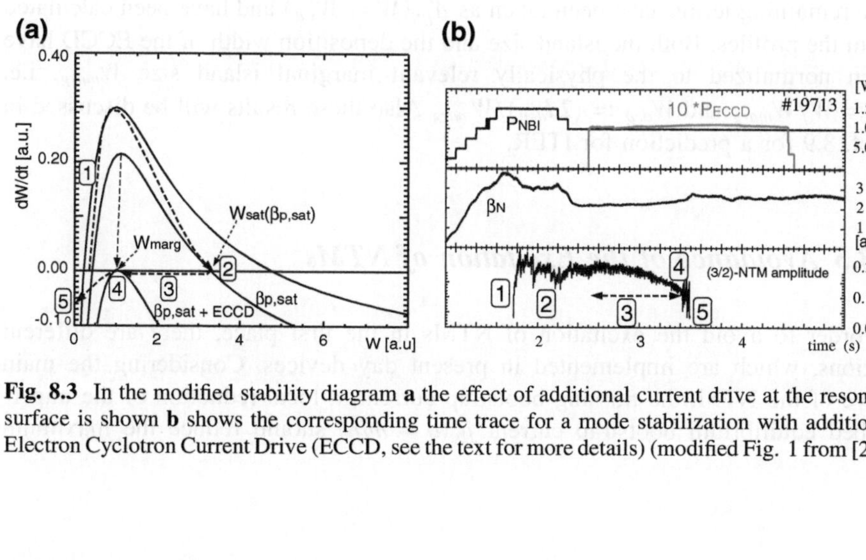

Fig. 8.3 In the modified stability diagram **a** the effect of additional current drive at the resonant surface is shown **b** shows the corresponding time trace for a mode stabilization with additional Electron Cyclotron Current Drive (ECCD, see the text for more details) (modified Fig. 1 from [29])

With this simplification of the generalized Rutherford equation (8.14), ASDEX Upgrade and JT-60U data of stabilization experiments have been used to calculate the fit parameters to $c_{sat} = 0.81 \pm 0.13$ and $c_{stab} = 0.68 \pm 0.22$. The resulting predictions for ITER will be discussed in Sect. 8.9.

A crucial quantity for NTM stabilization, as defined above, is the ratio between the maximum externally driven helical current density, $j_{cd,hel}$, at the start of the stabilization process and the defect bootstrap current density j_{bs} at the saturated phase, $\eta_{NTM} := j_{cd,hel}/j_{bs} = (\eta_{ECCD} \cdot j_{ECCD})/j_{bs}$. The requirement for ITER has been estimated based on experimental data both for the (3,2) and the (2,1) NTM to $\eta_{NTM} \geq 1.2$ for large β_p at the onset, i.e. $\beta_{p,onset} \gg \beta_{p,marg}$ [39]. This results in a large island size at the onset ($W_{onset} \gg W_{seed}$). From JT-60U and DIII-D, different requirements on η_{NTM} for ITER have been reported, based on experimental data [40–42]. The reduced requirement on η_{NTM} there might be due to the consideration of an island size just above the marginal island size, i.e. only $W_{onset} > W_{marg}$. Only a β moderately higher than β_{marg} has been assumed at the modes onset. Also depending on the full deposition width $2d_{dep}$ compared to the marginal island width W_{marg}, a variation of the requirements might be needed [32, 43, 44]

In an alternative approach an analytical expression has been derived for η_{NTM} without the need for empirically fitted coefficients in the generalized Rutherford equation [45]. This calculation takes advantage of the fact, that most of the terms included in η_{NTM} originate from the same generic term for a helical current perturbation. The two cases of a dominating transport term ($\chi_\perp/\chi_\parallel$-term) or for a dominating polarization current term are included by distinct analytical formulas for η_{NTM}^{tra} and η_{NTM}^{pol}. The formulas for η_{NTM}^{tra} and η_{NTM}^{pol} include the calculation for $j_{bs} = j_{bs}^{tra}$ and $j_{bs} = j_{bs}^{pol}$ in the corresponding limit. The remaining experimentally free parameters are the saturated island size W_{sat} and the marginal island size W_{marg}. The classical tearing mode stability parameter $\Delta'(W)$ has been approximated in terms of the fully saturated island size W_{sat} without any additional ECRH or ECCD applied (c_{stab}-term $= 0, j_{ECCD} = 0$) as $\Delta'(W) := -\Delta'_{BS}(W = W_{sat})$. For this determination also $dW/dt = 0$ has been used (stationary, i.e. saturated island). The remaining terms have been taken as $\Delta'_{BS}(W = W_{sat})$ and have been calculated from the profiles. Both the island size and the deposition width of the ECCD have been normalized to the physically relevant marginal island size W_{marg}, i.e. $\bar{W} := W/W_{marg}$ and $\bar{W}_{dep} := (2d_{dep})/W_{marg}$ Also these results will be discussed in Sect. 8.9 for a prediction for ITER.

8.3.5 Avoidance of the Excitation of NTMs

In order to avoid the excitation of NTMs in the first place, there are different options, which are implemented in present day devices. Considering the main drive of the NTM, the missing bootstrap current, a local reduction of the unperturbed equilibrium bootstrap current $j_{bs}(q = m/n)$ should reduce the maximum reachable island size and in particular the excitability of the mode. Considering the different influence of the density gradient and the temperature gradient on the bootstrap current, a reduction of the local density gradient $\nabla n_e(q = m/n)$ at the resonant surface reduces the probability for an NTM excitation. This can be done without any local current drive, as it relies on the profiles, which can be achieved within the considered discharge scenario.

The second approach for avoiding NTMs, lies in the avoidance of other MHD instabilities, which provide a trigger for NTMs, i.e. avoid the seed islands. This concentrates mainly on the suppression of sawteeth and fishbones at the $q = 1$ surface. In practice a destabilization of sawteeth is done, in order to achieve high frequent small sawteeth, which are not able to generate a large enough perturbation to trigger the NTM. A variation of the stability of the ($m = 1$, $n = 1$) activity at the q = 1 surface is the key for this task [46, 47].

An alternative way is the tailoring of the global shape of the q-profile with, typically broader, external current drive. Here, the resonant surface with $q = m/n$ is completely avoided in the plasma and hence the corresponding (m,n) NTM can not be excited. This idea can not only be applied for the avoidance of the NTM, but also for the avoidance of the triggering MHD, namely the avoidance of sawteeth and fishbones by avoiding the $q = 1$ surface.

8.4 Relevant Systems for Controlling and Detecting NTMs

Both for the stabilization of existing NTMs, as well as for their avoidance, a local modification of MHD stability is required. Replacing the missing bootstrap current at a specific resonant surface is one possibility. The most widely used tool at present experiments is the Electron Cyclotron Resonance Heating (ECRH) and in particular the Electron Cyclotron Current Drive (ECCD).

8.4.1 Electron Cyclotron Resonance Heating and Current Drive (ECRH/ECCD)

The key advantages of this technique are narrow localization in the order of centimetres in the radial direction, which is smaller than the typical saturated island width W_{sat} of a (3,2) NTM and a (2,1) NTM. In the vertical direction the deposition width is defined by the width of the propagating beam. Moreover, the deposition radius of ECRH and ECCD in terms of the minor radius can be controlled with a steerable mirror system. Via a modulation of the gyrotron voltage the emitted power can be modulated in the range of the rotation frequency of an NTM, i.e. in the 10–20 kHz range in present experiments. For ITER, this frequency is expected to be in the order of 2–3 kHz, due to the lower expected plasma rotation.

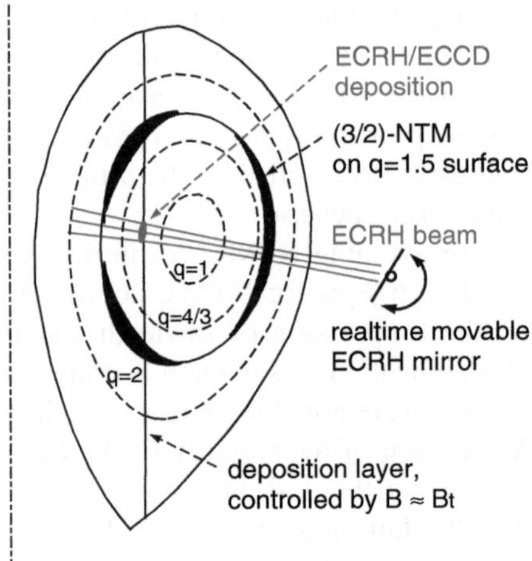

Fig. 8.4 Scheme for controlling the ECRH/ECCD deposition for a stabilization of a (3,2) NTM. The localization of the ECCD is primarily governed by the total magnetic field $|B|$, which can be approximated by the main toroidal field B_t. The exact localization is determined by the poloidal angle of the ECRH/ECCD launching mirror (Fig. 2 from [29])

The modulation becomes essential to fully stabilize the mode, when the full ECCD deposition width $2d_{dep}$ is larger than the marginal island width W_{marg} [48] (see below).

8.4.1.1 Control of the ECCD Deposition on a Resonant Surface

The deposition location of the ECRH wave propagating through the plasma (see Fig. 8.4) is mainly governed by the total magnetic field $B = \sqrt{B_t^2 + B_r^2 + B_\theta^2}$, including the response of the plasma itself. As this field is dominated by the toroidal field component B_t, the deposition mainly occurs at a fixed major radius R, as $B_t(R, z) \approx B_0 \cdot (R_0/R)$ holds.[2] When the beam propagation has a component perpendicular to the density gradient, its propagation is modified and typically bent away from the region of high density. Therefore, in general the deposition has to be calculated by a beam/ray-tracing code, such as TORBEAM [49–51], mainly used at ASDEX Upgrade, or TORAY-GA [52], mainly used at DIII-D. The deposition on a specific resonant surface, such as the $q = 3/2$, 2/1, 4/3 surface, can be achieved with a variation of the poloidal angle of the launching mirror. The toroidal angle (perpendicular to the drawing plane in Fig. 8.4) to lowest order controls the amount of driven current and its width in the radial direction.

[2] R_0 = major plasma radius.

Before dynamically steerable mirrors were available, the global plasma position (e.g. at DIII-D) or the magnetic field B_t has been varied (e.g. at ASDEX Upgrade and JT-60U), in order to ensure the ECCD deposition hits the relevant resonant surface. For the first time, a direct control of the ECRH mirror [53] and a feedback control [54] has been used at JT-60U.

8.4.1.2 Control of the ECCD Phase in O-Point of the NTM

The above consideration assumed a rotating island with an island width W larger than the deposition width $2d_{dep}$ of the ECCD ($2d_{dep} < W$). The relevant term for the stabilization in (8.13) and (8.14) is the helical part η_{ECCD} of the driven ECCD current j_{ECCD} inside the island, i.e. $\eta_{ECCD} \cdot j_{ECCD}$ (first sketch in Fig. 8.5). For

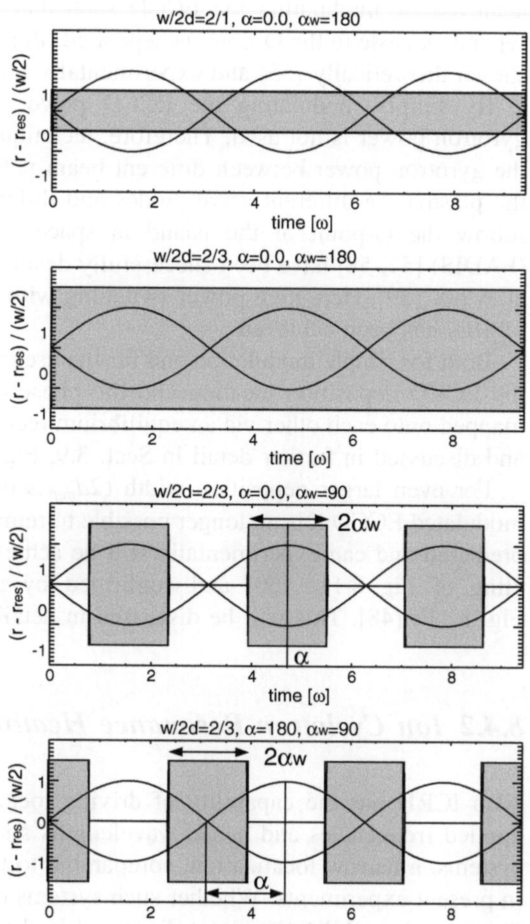

Fig. 8.5 Sketch of a (3,2) island propagating in front of one gyrotron launcher for a full rotation of the mode. The deposition phase is indicated by the *shaded area*. From top to bottom: Narrow ECCD deposition compared to the island width ($2d_{dep} < W$), broad ECCD deposition ($2d_{dep} > W$), broad ECCD deposition with favourable modulation around the O-point and broad ECCD deposition with unfavourable modulation around the X-point. Note that only the fraction η_{ECCD} of the driven current j_{ECCD} flowing helically inside the island contributes to the stabilization (Fig. 3 from [29])

small deposition most of the current is deposited inside the island ($\eta_{ECCD} \rightarrow 1$) and its size is further reduced. The fraction $1 - \eta_{ECCD}$ driven outside the island in the vicinity of the X-point can be neglected. When the island size can be reduced below the marginal size ($2d_{dep} \leq W_{marg}$ and $W \leq W_{marg}$), an unmodulated ECCD is sufficient for a complete removal of an NTM. Once $W < W_{marg}$ is achieved the island decays away independently of the ECCD.

For a wider deposition the island might become smaller than the deposition width ($2d_{dep} \geq W$) during the stabilization process. In this case an increasing amount of current is driven outside the island's separatrix and the efficiency of the stabilization process is significantly reduced ($\eta_{ECCD} < 1 \ldots \eta_{ECCD} \ll 1$, second sketch of Fig. 8.5). Hence the island size can hardly be further reduced. If the marginal island size W_{marg} is also smaller than the deposition width ($2d_{dep} > W_{marg}$), the mode can no longer be completely removed [48, 55]. For a further stabilization the fraction $1 - \eta_{ECCD}$ current driven outside the island has to be minimized or in other words η_{ECCD} has to be increased again. This can be achieved by modulating the ECCD such that mainly power inside the island's separatrix close to the O-point is deposited (third sketch in Fig. 8.5), as it has been shown theoretically [55] and experimentally [48, 56].

By simply modulating the ECCD power, typically 50 % of the available gyrotron power is not used. Therefore, techniques have been developed to switch the gyrotron power between different beam paths. The different paths now reach the plasma via different wave guides and different launching mirrors in order to follow the O-point of the island in space. These **FA**st **DI**rectional **S**witches (FADIS) [57, 58] have been successfully developed and tested at the ECRH setup at W7-X [59]. Here high power switching with up to 20 kHz for a pulse duration of 10 s has been achieved.

Both for simple modulation and for the use of multiple beam paths the phase of the ECCD deposition location and the phase of the island have to be carefully mapped onto each other via an equilibrium reconstruction, as indicated in Fig. 8.6 and discussed in further detail in Sect. 3.9, Fig. 3.22.

For even larger deposition width ($2d_{dep} \gg W_{marg}$ and $2d_{dep} > W_{sat}$) even with modulated ECCD it is no longer possible to remove the mode. A small reduction is predicted and can experimentally still be achieved. This is indicated in the modeling of Fig. 8.11a [60] and confirmed by the experimental observations in Fig. 8.11b [48]. This will be discussed in detail in Sect. 8.5.2.

8.4.2 Ion Cyclotron Resonance Heating (ICRH)

Also ICRH has the capability of driving local current in a plasma. Due to the applied frequencies and hence wavelengths and the size of the antennas of such systems, a narrow localization, comparable to the ECRH systems, is not possible in present experiments. Whether such systems can be used in ITER or DEMO for direct mode stabilization must discussed in the future.

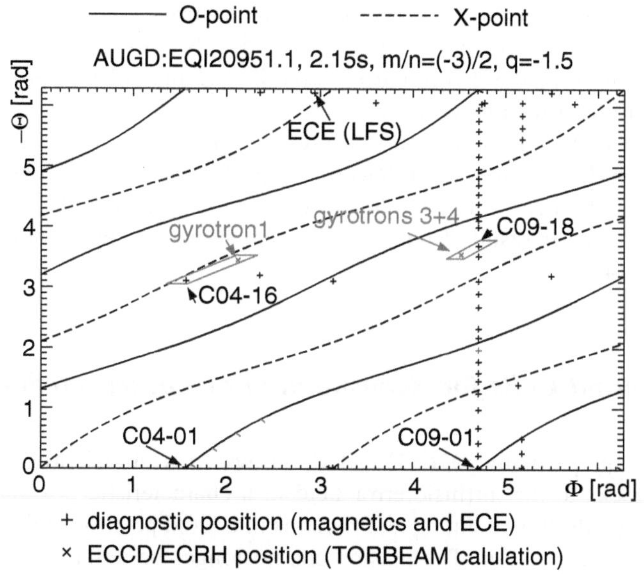

Fig. 8.6 Localization of the ECCD deposition positions (*tilted crosses*) with respect to the magnetic field lines of the X (*dashed lines*) and the O-point (*solid lines*) of the (3,2) mode on the resonant surface according to an equilibrium reconstruction. Individual magnetic pickup coils (*crosses*) in this example are located close to the deposition location in 3d space (indicated by the *grey rhombs*), which has been calculated with TORBEAM. The ECE diagnostic measurement position is also marked as a reference point for the mode detection (adapted Fig. 5 from [48])

The more important role of the ICRH lies in tailoring the sawtooth stability, which is an important trigger of NTMs, via a modification of the fast particle distribution (see Sect. 8.6.4). The interaction between such fast ion populations and the ideal internal kink mode at the $q = 1$ surface is considered as an important explanation for a modification of the sawtooth behaviour [61] and is supported by JET data [62]. In ITER and in particular in DEMO the plasma pressure generated by the fast particles from the fusion products, β_{fast}^{fusion}, might dominate and the applicability of such schemes needs to be carefully discussed ($\beta_{fast}^{fusion} \gg \beta_{fast}^{ICRH}$). The global current drive capabilities of ICRH might be an alternative tool for controlling the current profile for NTM avoidance scenarios (see Sect. 8.6.2).

8.4.3 Lower Hybrid Current Drive (LHCD)

At COMPASS-D successful stabilization of NTMs with Lower Hybrid Current Drive (LHCD) could be achieved [63, 64]. In these experiments it is reported, that the main effect for the stabilization of a (2,1) NTM comes from the variation of the current gradient at the resonant surface, i.e. a reduction of the $r_{res}\Delta'(W)$-term with roughly 10 % additional LHCD power compared to the background heating.

At JET a range of experiments has been performed with LHCD. In these experiments no successful removal of an existing NTM could be achieved. The radial localization of the driven current could not a priori be controlled and was typically too broad for this purpose.

The effect of LHCD on the current profile can be exploited in a more global way as it has been done at JT-60U for q-profile tailoring, as described later in detail. The absence of a $q = m/n$ surface obviously avoids the excitation of a (m, n) NTM. Newer results from JT-60U even show the suppression of existing (2,1) NTMs [40].

8.4.4 External Coils for Repositioning of Locked Modes

If an NTM, in particular a (2,1) NTM, grows to a sufficiently large size, it eventually locks to the intrinsic error field in a characteristic position. In such a situation the mode stops rotating and remains in a fixed position and can even stop the entire plasma rotation. The plasma can no longer be described as a 2d equilibrium, but has to be treated as a full 3d equilibrium containing the (2,1) locked mode as a perturbation. Especially for low q_{95} this situation often leads to a disruption (see Fig. 8.12a).

Applying ECCD on the resonant surface with a locked mode, is not sufficient to stabilize the mode. The ECCD deposition spots have to be aligned with the O-point of the island. This can be analyzed via a mapping with the help of the plasma equilibrium, as indicated in Fig. 8.6. Selecting either the upper or lower intersection between the resonant surface r_{res} and the ECCD resonance layer mainly determined by the magnetic field B (see Fig. 8.4) via different mirror paths allows for some adjustment, but does not in general assure current drive centred around the O-point.

A possible way to resolve this problem lies in the modification of the locking position of the mode itself. This can be achieved by controlling the total error field of the experiment. At DIII-D the total error field can be modified by additional internal coils [65]. This system can generate a static or a rotating error field with a toroidal mode number of $n \leq 3$. JET has $n \leq 2$ error field correction. ASDEX Upgrade has started operating an set of internal coils in 2011 with $n \leq 2$, which has been extended in the 2012/2013 campaign to $n \leq 4$ [66–69]. With such artificial internal perturbation at DIII-D the locking position of a mode has been controlled [56].

8.4.5 Relevant Measurements for Detecting NTMs

All of the described tools are actuators for an integrated NTM controller. Such a controller, however, must get information from the plasma about the existence, the poloidal and toroidal mode number m and n and in particular the radial location of the mode with respect to the minor radius. For modulated injection in the O-point of the island, also the phase of the mode is required. Obviously all these signals have to be provided in realtime and communicated to the control system, on which the NTM controller is running on.

For the detection of the existence of NTMs, most importantly the (3,2) and the (2,1) NTM, magnetic pick-up coils are being most widely used. An integrated and weighted sum of Mirnov coil measurements (dB_θ/dt or B_θ) provides the magnetic perturbation amplitude for a specific (m,n) mode (spatial Fourier filtering). Applying a set of thresholds on such (m,n) filtered signals with an appropriate hysteresis, ($B^{detection}_{\theta,(m,n)} \gg B^{removal}_{\theta,(m,n)}$), gives the information when a certain (m,n) NTM gets excited and when it has been successfully removed.

Additionally these magnetic signals provide the phase information of a rotating mode, as described in Sect. 3.9.2 in Fig. 3.21. Such information is needed for fast modulation of the ECCD power in order to centre the driven current only inside the island separatrix, i.e. around the island's O-point (see Sect. 8.5.2). For locked modes large saddle coils measuring B_r are used, as shown in Sect. 3.9.1 in Fig. 3.20.

The radial localization of the mode with respect to the flux surface is most easily and directly possible with a local temperature measurement $T_e(\rho)$ with the ECE diagnostic [70, 71]. The FFT amplitude and phase profile for the dominant mode frequency extracted from the filtered Mirnov signals reveal the mode localization (phase jump of π at the resonant surface in $T_e(\rho)$) and the island width directly (see Fig. 3.19 in Sect. 3.6.5.3). At various experiments also a correlation analysis within the ECE data alone and in connection with the magnetic signal has been implemented ([72–75] and references therein). The ECE diagnostic also provides a direct measurement of the ECRH deposition radial localization for slowly modulated power ($f_{modulation} \ll f_{(m,n)}$), thus possibly removing the requirement of a realtime raytracing code. As this modulation only needs a reduction of the power, a stabilizing effect can be maintained throughout and could be even combined with the fast modulation with the mode frequency.

For a proper mapping of the ECE measurements a reliable and well defined equilibrium is needed (for example in [76]). Ideally this would be based on a direct current density measurement. The Motional Stark Effect (MSE) measurement provides directly pitch angles of the local magnetic field (for example in [77]), which helps to constrain the equilibrium reconstruction.

For calculating the expected deposition of the ECRH/ECCD system a raytracing code has to be operated in realtime. For this also electron density and temperature profiles have to be available. Such systems are presently indeed available at multiple experiments. As soon as the ECCD is really switched on, the ECE measurement provides a complementary information, which might also serve as a correction for the calculated values of the deposition.

A possible way around the mapping procedures is the in-line (or oblique or line-of-sight) ECE. In this case an additional ECE diagnostic is embedded in the mirror system measuring the emitted microwave radiation from the plasma in the vicinity of the deposition location of the ECRH system directly. Any mapping between the different diagnostic systems via the plasma equilibrium becomes unnecessary. Such a system has been implemented and successfully used for mode stabilization at TEXTOR [78, 79] and DIII-D [56].

Also a soft X-ray (SXR) diagnostic may provide additional information on the radial mode localization. However, this is more complicated, as the SXR diagnostic provides integrated line of sight measurements from multiple cameras. Even without a full tomography, a correlation analysis with the magnetic measurements might provide complementary information to an NTM controller.

8.5 Stabilization of Excited NTMs

Over the last years on many experiments, which have an ECRH/ECCD system available, experiments have been performed, which have shown that both (3,2) and (2,1) NTMs in high β_N discharges can be reliably removed. In this section an overview of these experiments will be given.

8.5.1 Removal of Rotating (3,2) and (2,1) NTMs

First experiments for a removal of rotating (3,2) NTMs were performed with a modulated injection, targeting for the O-point [80] (Fig. 8.7). A variation of the relative phases between O-point and X-point showed, that only with O-point phasing a reduction of the island size by 40 % is possible with 4–8 % additional ECCD power compared to the background NBI power. The radial location of the ECCD deposition has been varied on a shot-to-shot basis by varying the main toroidal field B_t. About 5 % of β_N could be recovered for the first time by such a scheme. In these initial attempts it turned out, that unmodulated co-ECCD was almost as efficient as modulated ECCD with deposition in the O-point. Most of the following experiments have been performed only with unmodulated ECCD. This could be done, as for most of the experiments a deposition width $2d_{dep}$ smaller than the marginal island width W_{marg} could be easily achieved. The different behaviour for O and X-point phasing could be well understood with the help of a non-linear MHD stability code including self-consistently the local bootstrap current [81–83].

Later experiments with an increased amount of unmodulated co-ECCD power showed for the first time the possibility of a complete removal of a (3,2) NTM [84]. With typically 10 % additional co-ECCD power it is possible to recover the discharge during a pre-programmed magnetic field ramp, finally reaching a β_N, which is identical to a discharge without any NTMs (Fig. 8.8). The global β_N in these experiments slowly decreases due to the applied magnetic field ramp. For (3,2) NTMs the decrease of performance was the driving issue, why a stabilization is required.

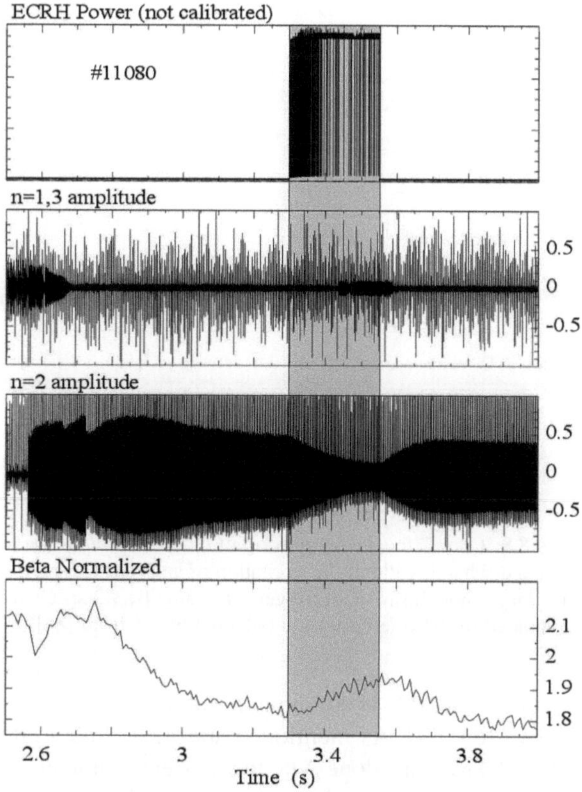

Fig. 8.7 First clear reduction of a (3,2) NTM with modulated ECCD (from 3.3 s till ≈3.4 s), which becomes unmodulated at small island sizes, due to a failure of the trigger system for the modulation (from ≈3.4 s on, Fig. 2 from [80])

Consequently the next step is the stabilization of the more dangerous (2,1) NTM, as this mode often causes disruptions. The complete stabilization of a (2,1) NTM has been achieved for the first time at DIII-D [41]. In these experiments both feedforward programmed ramps of the toroidal field B_t (Fig. 8.9), as well as the SEARCH-AND-SUPPRESS algorithm [72] have been used with the toroidal magnetic field as actuator to hit the resonant surface. As sensor for the size of the NTM the amplitude of an integrated Mirnov coil ($\int \frac{dB_\theta}{dt}\, dt$) has been used.

At ASDEX Upgrade, the (2,1) NTM has been stabilized without feedback control, again with a feedforward programmed magnetic field ramp [85]. The ECCD was in this case applied to a locked (2,1) NTM, which for $q_{95} \geq 3.3$ typically does not lead to disruptions [86]. The phasing could not be actively controlled with respect to the locking position of the mode, as ASDEX Upgrade at that time was not equipped with internal coils for generating a dedicated perturbation field. The phase of the locked mode could only be deduced after the discharge via the poloidal field distribution of the integrated Mirnov coils (B_θ). The gyrotron system and the mirror position was located in a position, where the typical locking position allowed to hit the O-point at least approximately.

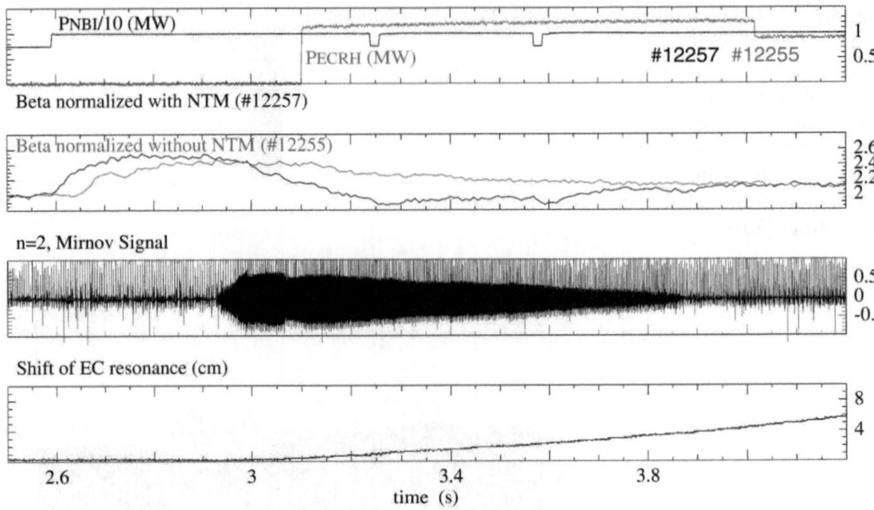

Fig. 8.8 Complete stabilization of a (3,2) NTM (*black traces* in box 2, #12257) in comparison with a discharge without the excitation of an NTM (*grey traces* in box 2, #12255) in the absence of a large enough sawtooth trigger. The same β_N values could be reached after the NTM removal compared to the reference case (adapted Fig. 1 from [84])

In COMPASS-D a complete stabilization of a (2,1) NTM with LHCD has been shown [63, 64]. As mentioned above, the main effect for the stabilization of the (2,1) NTM was done via the current gradient at the resonant surface, i.e. a reduction of the $r_{res}\Delta'(W)$-term. Roughly 10 % additional LHCD power was needed to achieve this removal. Experiments at JT-60U were performed with mechanically steerable launcher mirrors [53] and a feedback control to keep the deposition at the resonant surface [54].

8.5.2 Experiments Steering the ECCD Phase in the Islands O-Point

It has been shown experimentally at ASDEX Upgrade, that when the island becomes narrower than the ECCD deposition width, unmodulated ECCD is not able to reduce the island below its marginal width W_{marg} and hence can not completely remove the island for the same ECCD power applied (see Fig. 8.10a) [48, 80]. The ECCD deposition has been deliberately broadened by increasing the toroidal launching angle. The ECCD has been modulated by varying the emitted gyrotron power by using an $n = 2$ filtered Mirnov coil ($dB_\theta/dt_{n=2}$) signal for the modulation control. With an additional voltage comparator, which has been implemented in the hardware, a binary signal has been generated for that purpose. With such a setup and the favourable phasing with the ECCD in the O-point, the

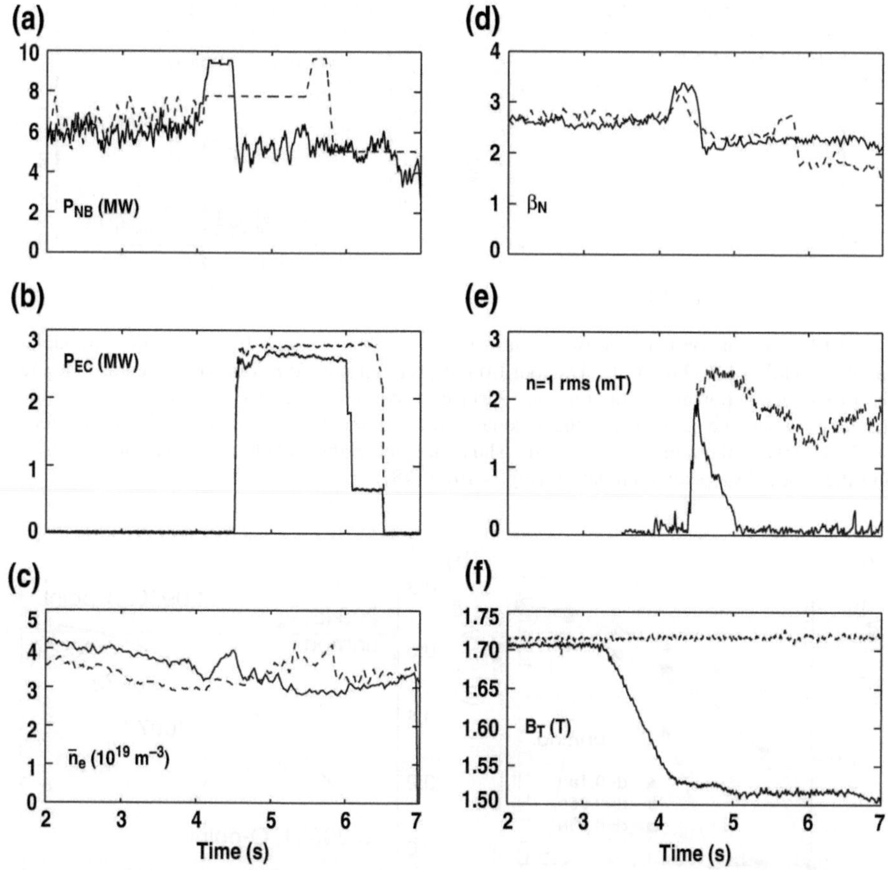

Fig. 8.9 Complete stabilization of a (2,1) NTM at DIII-D. The boxes show the applied NBI power P_{NB} (**a**), the applied ECCD power P_{EC} (**b**), the line averaged density $\bar{n}_e$ (**c**), the achieved β_N (**d**), the $n = 1$ magnetic field perturbation from the NTM (**e**) and the magnetic field B_t (**f**). The *dashed traces* represent a discharge with constant magnetic field B_t, where the ECCD did not hit the $q = 2$ surface and the NTM could not be removed. The *solid traces* represent a discharge, where the magnetic field B_t has been used as an actuator by the control system for depositing the ECCD on the $q = 2$ surface. In this case the NTM could be completely removed in less than one second. In both cases the applied NBI heating power has been feedback controlled to keep β_N constant (Fig. 9 from [41])

mode could be completely removed in spite of the broadened deposition width (see Fig. 8.10b). A dedicated phase scan of the deposition has shown, that the ECCD deposition around the X-point is only slightly worse compared to the unmodulated case (see Fig. 8.11b).

At DIII-D at this point another trigger scheme has been used [56]. An oblique ECE diagnostic embedded in the mirror system of the ECRH heating system provides the trigger signal for ECCD modulation. This has the advantage, that no

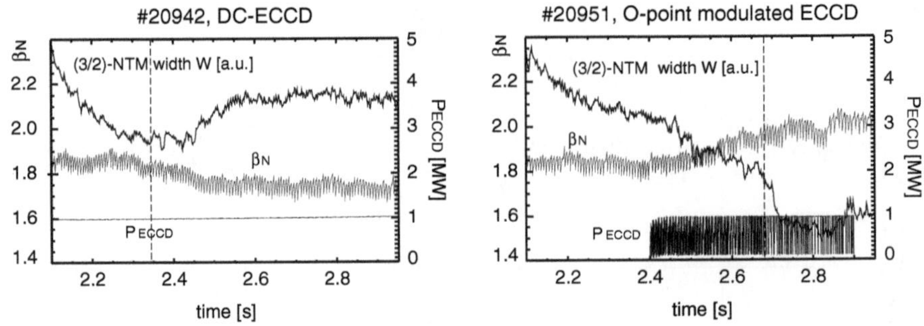

Fig. 8.10 Comparison between two identical discharges at ASDEX Upgrade with unmodulated (**a**) and modulated (**b**) broad ECCD deposition. A preprogrammed B_t-ramp has been used to move the ECCD deposition across the resonant surface of the mode. The *vertical dashed lines* indicate the times when the ECCD hits the resonant surface, and the maximal stabilization effect is reached. Only in the modulated case the island size gets reduced below the marginal island size and the mode disappears completely (Fig. 4 from [48])

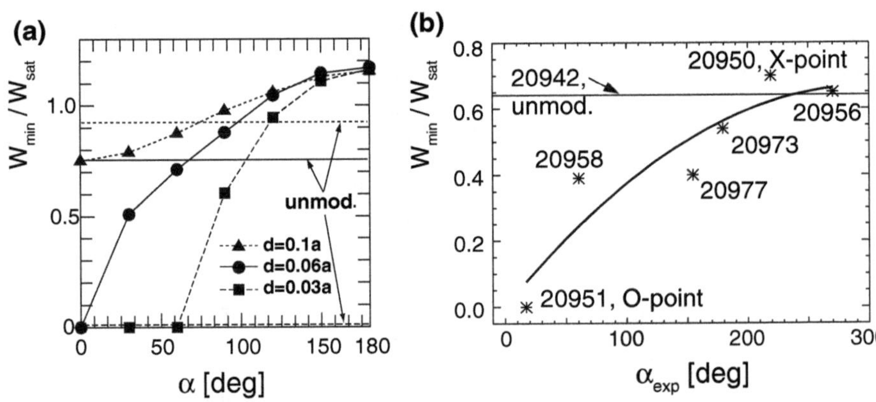

Fig. 8.11 Figure (**a**) shows the predicted achievable island size reduction W_{min}/W_{sat} for different ratios of the deposition width $2d_{dep}$ in units of the minor plasma radius a as function of the relative phase of the ECCD against the O-point of the island (see also Fig. 8.5). The corresponding predicted W_{min}/W_{sat} values for unmodulated ECCD are indicated by a *horizontal line*. In figure (**b**) the experimentally achievable island size reduction as function of the phase is shown. Only for O-point deposition $W_{min} \rightarrow 0$ could be achieved. The predicted increase $W_{min}/W_{sat} > 1$ could not be observed, due the additional heating effect and imperfection in the helical alignment. For the X-point a higher accuracy is required to observe the predicted effect (Figs. 2 and 7 from [48])

phase mapping via a realtime equilibrium, as indicated in Fig. 8.6, is needed. The increased efficiency has been quantified by a reduction of 10 % of the peak ECCD power and a reduction of 30 % of the time averaged ECCD power to completely remove the island.

Subsequent experiments at JT-60U performed a modulated stabilization of the more dangerous (2,1) NTM [40]. The triggering signal for the modulation in these experiments was also a magnetic pickup coil. It could be shown, that the island decay rate of the (2,1) NTM is about 1/3 faster compared to the unmodulated case. A dedicated scan of the relative phase clearly shows the advantageous effect of deposition in the O-point. In addition, a detrimental effect of modulated deposition in the X-point ($\eta_{ECCD} \ll 1$) could be shown.

8.5.3 Locked (2,1) NTMs Before Disruptions—Disruption Avoidance

Typically at low q_{95} and at low collisionality $\bar{\nu}_{ii}$ [88] (2,1) NTMs get excited. However, the role of the collisionality is rather complex. The polarization current term in (8.8) is one possible candidate. The interplay between n_e and T_e-profile and its impact on the bootstrap current [see (8.11)] and hence the NTM drive is another possible explanation. The particle transport, which is obviously influenced by the collisionality, might create a modified overall pressure and q-profile at low collisionality. Under such plasma conditions, a (2,1) NTM can lead to mode locking, which can also stop the entire plasma rotation. The perturbation field of the mode interacts with the vessel wall and with intrinsic error fields and brakes the plasma rotation. Finally this situation can lead to a disruption, as shown in the example in Fig. 8.12a.

Experiments on ASDEX Upgrade have shown, that also local heating alone in the vicinity of the resonant $q = 2$ surface is able to unlock the (2,1) NTM and avoid the disruption (Fig. 8.12b) [87, 89]. DIII-D experiments on the other hand have shown a general clear beneficial effect of ECCD over pure ECRH [56, 90], i.e. only for ECCD the island could be completely removed and β_N recovers. In terms of possible extension of the pulse length, for DIII-D and ASDEX Upgrade, the pure heating seems to be sufficient to salvage the discharge from a disruption, even though the mode is still present.

It should be noted, that mode stabilization close to disruptions using pure ECRH seems to be less sensitive to radial misalignment with respect to the resonant surface compared to ECCD. A co-current drive inside the resonant surface ($\rho_{dep} < \rho_{res}$) further increases the current gradient $dj_{0,\phi}(r)/dr$ and hence increases the classical stability parameter $\Delta'(W)$ (see (8.3)) and can lead to an even faster growth of the mode, as seen on the ASDEX Upgrade experiments.

In the DIII-D experiments additionally a variation of the phasing between the applied ECCD and the phase of the island has been performed. In this part of the experiment a clear beneficial effect of the deposition in the O-point has been observed. Only in this case the locked mode could be removed completely while remaining in the locked position. The X-point phasing is comparable to the case without applying any ECRH or ECCD. In both cases the discharge disrupts approximately at the same time. The locking position of the mode has been actively controlled by an externally applied static error field [90, 91].

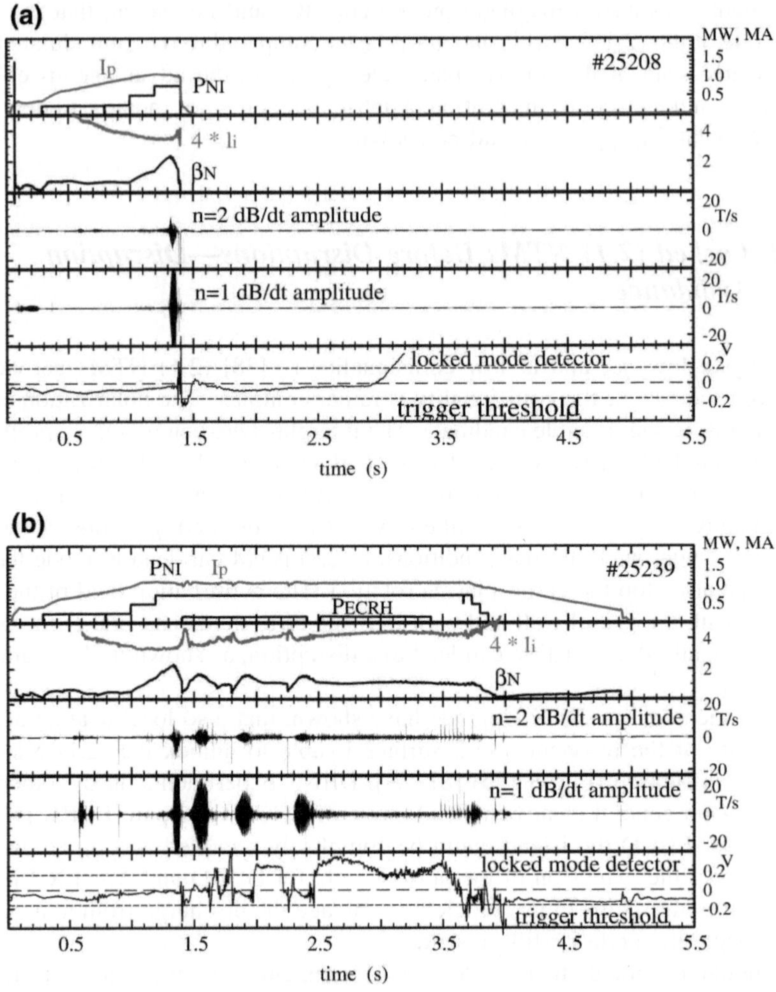

Fig. 8.12 **a** High β_N discharge at ASDEX Upgrade with low $q_{95} = 3.8$ resulting in a disruption due to the locking of a (2,1) NTM. From top to bottom the plasma current I_p and the injected beam power P_{NBI}, the achieved β_N and the ideal limit $4l_i$, the dB_θ/dt $(n = 2)$ signal of a set of Mirnov coils, the $dB_\theta/dt(n = 1)$ signal and the locked mode signal with its trigger threshold is shown. The locked mode signal is clearly above the preprogrammed trigger threshold for ECRH application, which has deliberately not been active in this case. **b** In an identical discharge the trigger has been used to switch on the ECRH twice for a preprogrammed phase of 1 s. The (2,1) NTM unlocks several times due to its reduction in size and does no longer lead to a disruption (adapted Fig. 1 from [87])

The possibility of a complete disruption avoidance with local ECCD/ECRH, in particular in high β_N-discharges, is very important for safe tokamak operation. Within a global control scheme the ECCD/ECRH based disruption avoidance will play an important role, as it is not only restricted to high β_N scenarios, but also, for example, for discharges with increased impurity content and hence radiation, as well as for high density cases with the formation of a MARFE. For most disruption types a locking (2,1) mode occurs, which is often a classically driven tearing mode. A deeper understanding of the mechanisms of the disruption avoidance in these transient states shortly before the disruption is still lacking and further theoretical and experimental work is needed. For other disruptive paths alternative avoidance or at least mitigation schemes will have to be included in a control system. This discussion however goes beyond the scope of this book.

8.6 Avoidance of the Excitation of NTMs

8.6.1 Preemptive ECCD at Resonant Surface(S)

At JT-60U local ECCD has been applied at the resonant surface before the onset of a (3,2) NTM [92, 93]. In these experiments the onset of the (3,2) NTM could be significantly delayed. For the same amount of applied ECCD power before the excitation of the mode, the saturated island size remained smaller compared to the case when the ECCD is applied after the mode has reached its naturally saturated size (Fig. 8.13 and left part of Fig. 8.14). The accuracy requirements on the exact radial localization of the ECCD with respect to the resonant surface r_{res} are identical as in the usual application of the ECCD after the NTM onset (right part of Fig. 8.14). The (3,2) NTM could be reliably avoided, if the radial misalignment of the deposition radius Δr_{dep} was in the order of d_{dep} ($\Delta r_{dep} \approx d_{dep}$). The radial deposition has been controlled during the NTM lifetime via a realtime controlled steering of the launching mirror system, while the starting value was based on experience from previous discharges.

At DIII-D similar experiments could completely avoid the excitation of a (3,2) NTM by preemptive ECCD at the resonant surface. As shown in Fig. 8.15, the excitation could be completely avoided while β_N is ramped up to $4l_i$, the assumed ideal limit. Sawteeth and fishbones occur, which would normally trigger an NTM [94]. In these experiments the ECCD deposition has been controlled also via the launching mirror system using a full realtime equilibrium reconstruction for estimating the resonant surface before the mode excitation.

For the (2,1) NTM at DIII-D a complete avoidance of the mode could also be achieved in the presence of an already excited (3,2) NTM [95], as shown in Fig. 8.16. In this experiment instead of moving the mirrors the toroidal magnetic field B_t has been used as the actuator for moving the ECCD deposition onto the resonant surface.

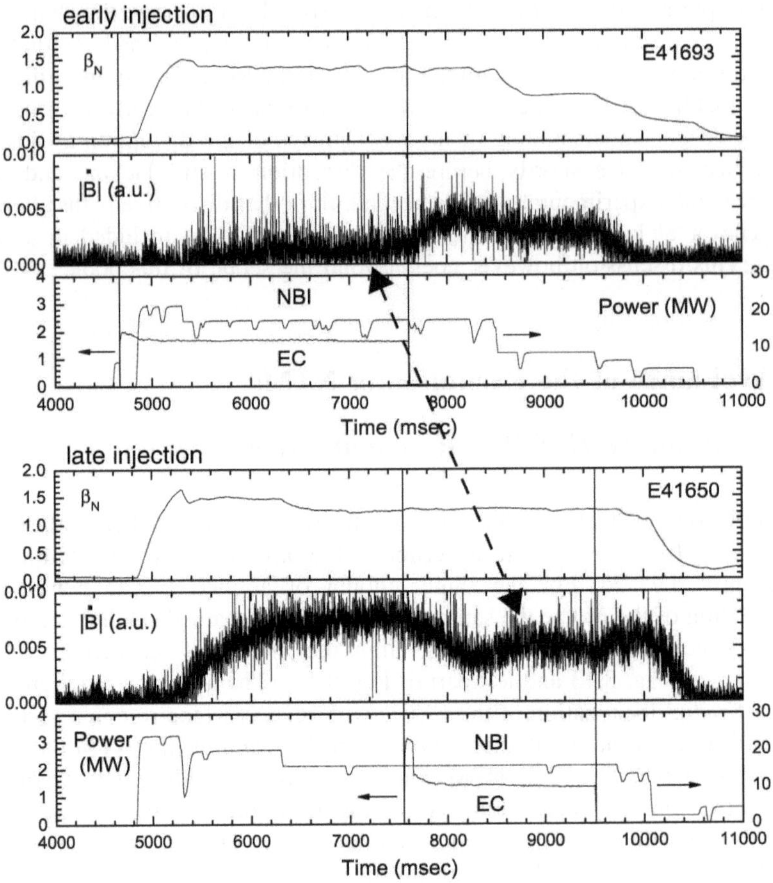

Fig. 8.13 Time traces for a comparison between early and late ECCD injection from JT-60U. With early ECCD (before the onset of the mode) the saturated island size never becomes as large as in the late ECCD case. In this case the ECCD deposition is kept fixed at the pre-calculated $q = 3/2$ surface. In the case of late ECCD the deposition is controlled via a steerable mirror on the resonant surface, where the mode is located (taken from Fig. 1 from [92])

8.6.2 Profile Tailoring with Wave Heating

As indicated above the main drive for an NTM is the pressure gradient at the resonant surface. In particular, the dominant part of the pressure gradient is the local density gradient [10, 96]. A local reduction of the density gradient dn_e/dr should therefore be beneficial for a reduced saturated island size W_{sat}. The excitation of an NTM by a large enough trigger should be less likely.

With a centrally flattened n_e profile and hence reduced bootstrap drive the excitation of NTMs could be avoided entirely in experiments at ASDEX Upgrade [9, 97]. In Fig. 8.17 an example is shown, where in the presence of central ICRH

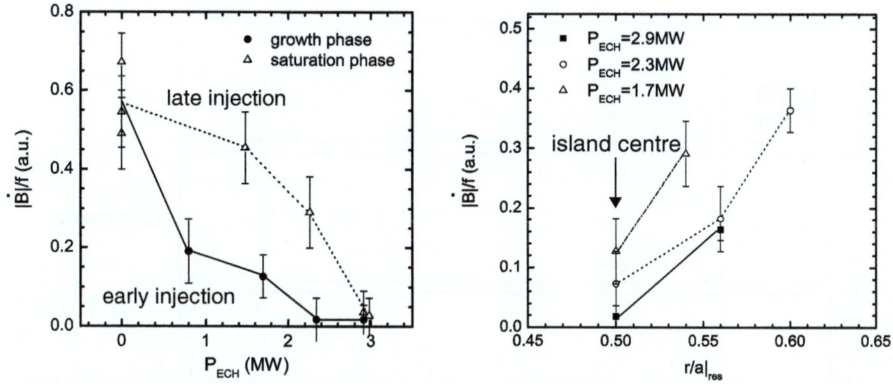

Fig. 8.14 *Left part* For early ECCD the maximum island size remains more than a factor of 2 smaller. *Right part* The radial localization remains a critical issue, both for early and late ECCD application (taken from Figs. 2a, 3a from [92])

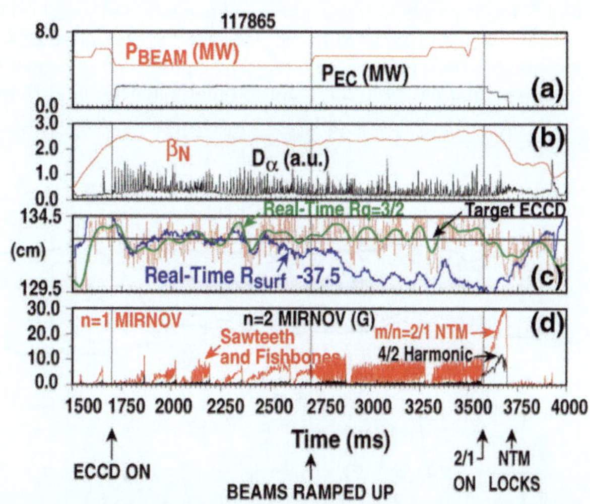

Fig. 8.15 Complete prevention of the excitation of a (3,2) NTM at DIII-D in the presence of preemptive ECCD at the $q = 3/2$ surface. During the preemptive ECCD phase only sawteeth and fishbones are observed, which would typically trigger a (3,2) NTM in such a scenario (taken from Fig. 2 from [94])

the excitation of a (3,2) NTM appears at a significantly higher β_N. With central electron heating, in this case via ICRH, an increased outward particle transport can be observed depending on the value of the collisionality. This can be understood in terms of an interplay between Trapped Electron Modes (TEM) and Ion Temperature Gradient driven modes (ITG), which is theoretically described [98] and experimentally observed [99].

However, as already indicated, such approaches are only possible in the appropriate collisionality range (see [98] for details). The central electron heating in a fusion reactor will be supplied by the α-particle heating. Whether the appropriate collisionality range is reached for such an effect, is an open question and has to be shown experimentally in the future. The available electron heating by ECRH might be insufficient to overcome the dominant α-particle self heating

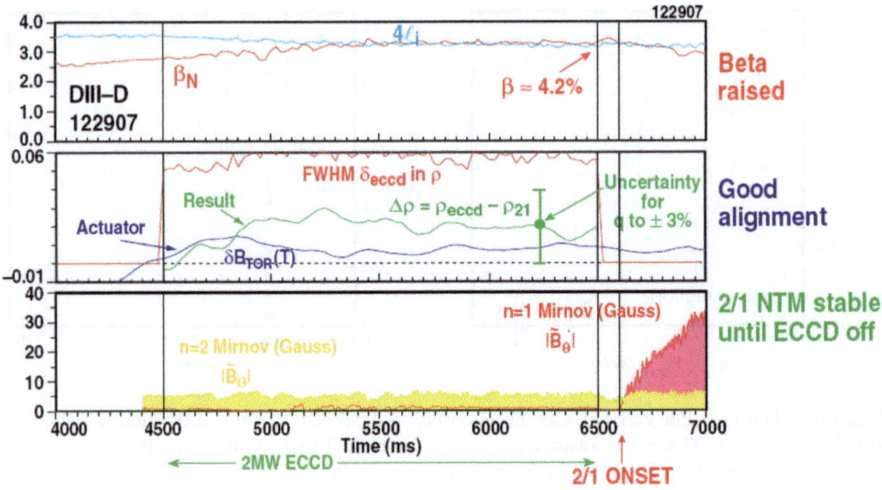

Fig. 8.16 Prevention of the additional excitation of a (2,1) NTM at DIII-D in the presence of an already excited (3,2) NTM in the presence of preemptive ECCD at the $q = 2$ surface. The mode location has been tracked via a realtime equilibrium information. The actuator for controlling the radial position has been the main toroidal magnetic field B_t (adapted Fig. 5 from [95])

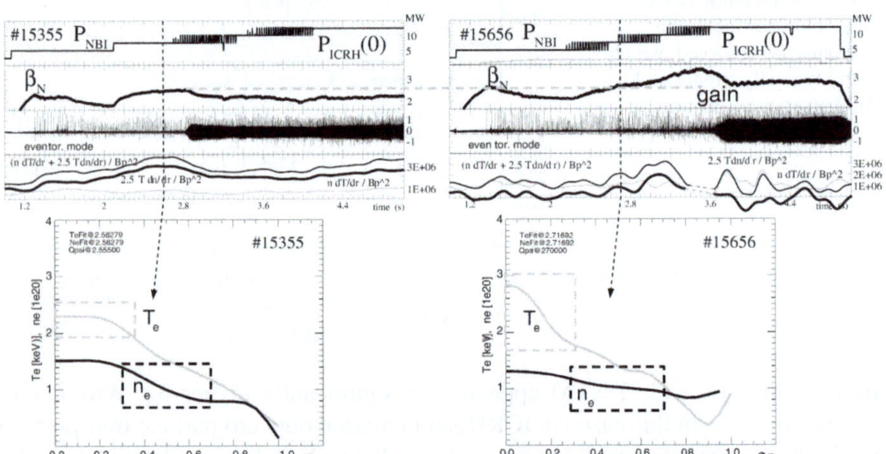

Fig. 8.17 Two otherwise identical discharges at ASDEX Upgrade without (*left*) and with central (*right*) electron heating via ICRH are compared. From top to bottom the applied heating power, the achieved β_N, the even magnetic amplitude $dB_\theta(n = 2)/dt$, the total corrected pressure gradient and the contributions from $T_e \nabla n_e$ and $n_e \nabla T_e$ are shown. At the bottom the n_e and T_e profiles at the indicated time points are shown for the two cases (adapted Fig. 14 from [29])

provided by the plasma. In this case ECRH might not be possible control tool for density peaking. A dedicated reduction of the peaking of the central density might also be detrimental for fusion power generation in a later fusion device.

Another approach lies in the combined tailoring of the q-profile and the overall pressure profile, which has been performed at JT-60U [100]. The current profile has been modified by LHCD in combination with off axis NBI heating, in such a way that both the $q = 3/2$ and the $q = 2/1$ are radially located in a region with an reduced pressure gradient. At $q_{95} = 4.5$ a stationary high $\beta_N \approx 2.4$ could be achieved for a duration of 5.8 s ($\approx 2.5\tau_R$). An extension of this profile tailoring leads to a scenario with extremely low $q_{95} \approx 2.2$ at $\beta_N \approx 3$ stationary for $\approx 6\ s$ (see Fig. 8.18). The extremely low q_{95} was achieved by drastically reducing the plasma cross section while still maintaining the full plasma current I_p and toroidal field B_t. Both the $q = 3/2$ and $q = 2$ surface have been shifted into a region with $\rho_{pol} \geq 0.7$. Due to the off-axis NBI heating a broad pressure profile has been established, such that the relevant resonant surfaces are in a region with reduced pressure. The central q-profile becomes flat and no sawteeth or fishbones are observed, i.e. no triggering MHD is available. All these ingredients are considered to contribute to this NTM free discharge.

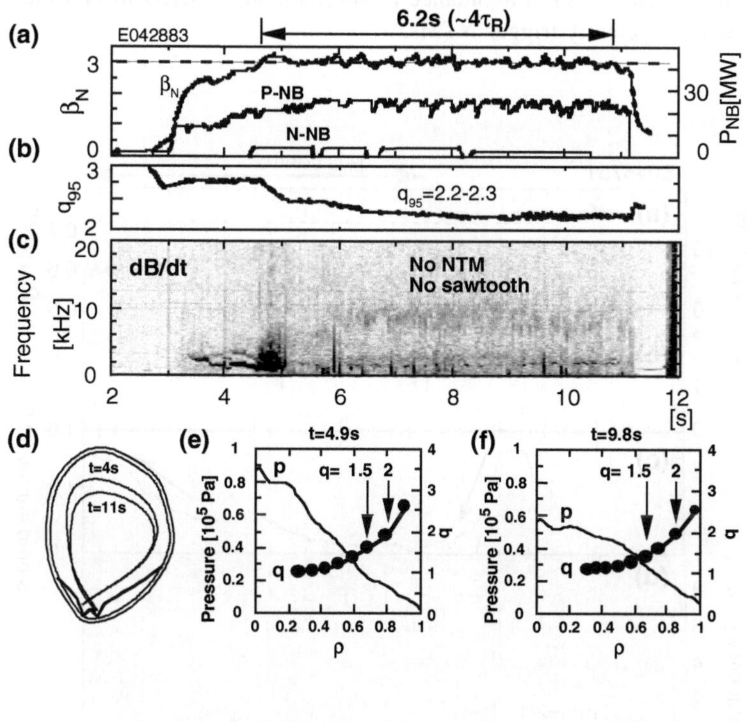

Fig. 8.18 Time traces of the achieved stationary β_N and the applied heating powers **a** for a high β_N, low q_{95} discharge at JT-60U. The q_{95} (**b**) is lowered by a shrinking of the plasma cross section (**d**). During the low $q_{95} \approx 2.2$ phase no MHD is observed (**c**). The pressure and q-profiles are adjusted in a way that the $q = 3/2$ and $q = 2$ surface are located radially in a region ($\rho_{pol} \geq 0.7$) with reduced pressure and pressure gradient (**e, f**) (Reprint with permission from Fig. 5 from [100], Copyright 2005, American Institute of Physics.)

8.6.3 Current Profile Control with LHCD with Excited NTM

At JT-60U experiments with LHCD were performed with an entirely different approach. The LHCD was used to establish a current profile, which does not support the excitation of an NTM at all [101]. Obviously the absence of a certain resonant surface with $q = m/n$, removes the possibility for a (m,n) NTM to occur. In the reported experiments the minimal q value, q_{min}, has been controlled via off-axis LHCD in feedback operation after the NTM onset. As soon as q_{min} rises above $q = 2/1$, the existing (2,1) NTM disappears and the now removed confinement degradation allows for a recovery of β_N (see Fig. 8.19).

8.6.4 Avoidance of NTM Triggering MHD

Another approach for NTM avoidance is the removal of MHD instabilities in the plasma, which typically trigger NTMs.

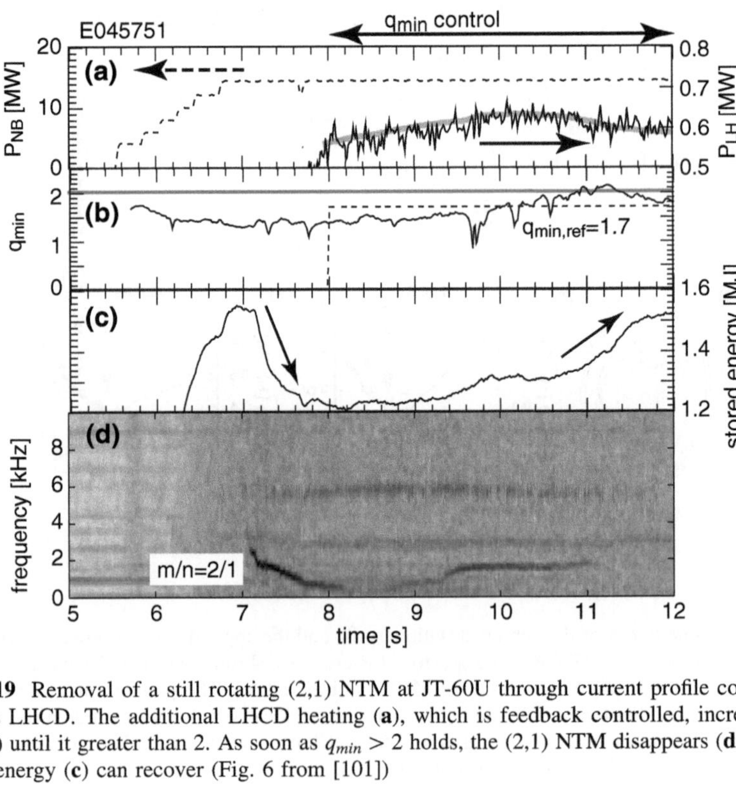

Fig. 8.19 Removal of a still rotating (2,1) NTM at JT-60U through current profile control via off-axis LHCD. The additional LHCD heating (**a**), which is feedback controlled, increases the q_{min} (**b**) until it greater than 2. As soon as $q_{min} > 2$ holds, the (2,1) NTM disappears (**d**) and the stored energy (**c**) can recover (Fig. 6 from [101])

At JET experiments with ICRH and ICCD around the $q = 1$ surface have been performed [46, 102]. It is possible to stabilize sawteeth, i.e. create large and less frequent sawteeth (large τ_{st} between two subsequent sawteeth), as well as to destabilize them and create smaller and more frequent sawteeth (small τ_{st}). With large τ_{st} and large sawtooth crashes NTMs have been triggered at very low β_N values, whereas for small τ_{st} and small sawteeth NTMs have been avoided almost reaching the ideal limit. Only a further increase of the applied NBI heating power could eventually trigger an NTM. These experiments have been interpreted later in more detail in terms of interaction between fast ion population and the ideal internal kink mode at the $q = 1$ surface [61] and are supported by experimental data from JET [62].

At TCV experiments with local ECRH and co and counter-ECCD have been performed in order to modify the sawtooth period and hence the size of the crashes. Both experimentally and in particular theoretically the effect of heating and current drive could be disentangled. The importance of the magnetic shear around the $q = 1$ surface has been shown [103]. These considerations have been performed in terms of the linear resistive stability threshold of the internal kink.

Based on a complete radial scan of the deposition of pure ECRH heating, co-ECCD and counter-ECCD on a shot-by-shot basis, a clear characterization of the sawtooth behaviour as function of the deposition radius has been achieved at ASDEX Upgrade [104, 105]. The local current drive around the $q = 1$ surface leads to a variation of the q-profile and hence the local gradient dq/dr at the resonant surface. Co-ECCD with its additional heating effect and hence its reduced resistivity has been most effective in modifying the sawtooth behaviour. Pure ECRH heating showed similar results as co-ECCD, but had a smaller effect on the size and frequency of the sawteeth. Counter-ECCD case was less effective in changing the sawteeth characteristics.

For current drive with ECCD the heating effect is always present, which has a similar effect as co-ECCD due to the increased inductively driven current when the plasma becomes locally hotter. The variation of the sawtooth stability could be explained in terms of a destabilization of sawteeth by an increase/decrease of dq/dr at the $q = 1$ surface. Both a decrease of the sawtooth size with decreasing τ_{st}, as well as an increase of the sawtooth size with increasing τ_{st}, even up to a complete avoidance of sawteeth could be achieved by a variation of dq/dr. This scheme has been used to avoid sawtooth triggered NTMs in a high β_N discharge during the full pulse length of co-ECCD [106]. Only after the ECCD around the $q = 1$ surface was switched off, the sawteeth revert back to their normal size and a (3,2) NTM is promptly triggered. Sawtooth tailoring needs control of the ECCD deposition with respect to the $q = 1$ surface, but with a reduced demand on the radial accuracy compared to NTM removal.

TCV [107] experiments and modeling results [108] have shown two new approaches, namely sawtooth pacing and locking respectively. A combination of sawtooth detection via central Soft X-ray measurements and realtime control of the feedback controlled ECCD power in the vicinity of the $q = 1$ surface allows a control of the occurrence of the next sawtooth crash, i.e. apply sawtooth pacing [107]. Such a scheme significantly reduces the required ECCD power, as the power is no longer required continuously. New simulation results propose a new scheme, sawtooth locking, where the sawtooth cycle could be phase locked to an externally applied modulated ECCD deposition [108]. In this case no feedback control is required. This approach has been experimentally confirmed and demonstrated in TCV [109]. For both approaches models have been discussed explaining the observed behaviour of the sawteeth. Additionally to the sawtooth tailoring current has been driven at the resonant surface of the mode for a pre-emptive avoidance of the NTM excitation.

At ASDEX Upgrade a comparison between different NBI sources with different injection angles and deposition profiles has been done. A significant variation of the individual sawtooth size and the sawtooth period has been observed [105]. For the most tangential and off-axis deposition large sawteeth with large τ_{st} (stabilization) has been observed, whereas for more radial and central injection small sawteeth with small τ_{st} (destabilization) has been observed. Later experiments have used a variation of the deposition radius of an individual NBI source for producing large sawteeth in combination with ICRH heating [110]. The detailed distribution of NBI generated fast ions around the $q = 1$ surface plays a crucial role there. The application of ECCD is able to destabilize these artificially stabilized large sawteeth, which potentially trigger NTMs [111]. A comprehensive overview over the control of sawteeth is given in [112] and in Chap. 4.

It must be noted however, that in the absence of sawteeth and fishbones NTMs can be triggered at higher β_N values also by other MHD events, such as ELMs at the plasma edge. At even higher values of β_N, NTMs can also be initiated without any obvious trigger, and grow “out of the noise”. For these consideration the following inequality holds $\beta_{onset}(\text{sawtooth}) < \beta_{onset}(\text{fishbone}) < \beta_{onset}(\text{ELM}) < \beta_{onset}(\text{trigger} - \text{less})$ [113, 114]. For cases with $\Delta' > 0$, trigger-less NTMs can also occur at lower β_N-values.

8.7 Effect and Mitigation of Unavoidable NTMs

There might be cases when neither avoidance nor stabilization of excited NTMs is possible. This might be due to a lack of ECRH power for avoidance or complete removal. The question arises, whether it is still possible to partially recover the confinement loss due to the NTM or whether it is possible to design scenarios where the NTM induced confinement loss is tolerable.

8.7.1 Triggering the FIR Regime

As described in Sect. 8.3.1, for high enough $\beta_{N,onset} \geq 2.3$ the FIR regime with a nonlinear coupling between a (3,2) NTM, an ideal (4,3) infernal mode and a (1,1) mode can be reached. In this FIR regime the confinement loss due to a (3,2) NTM

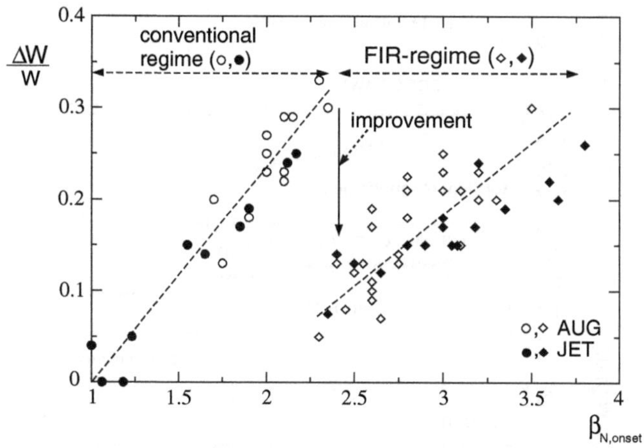

Fig. 8.20 Confinement loss $\Delta W/W_{onset}$ due to an existing (3,2) NTM as function of the β_N value at the NTM onset for a combined data set from ASDEX Upgrade (*open symbols*) and JET (*full symbols*). The *circles* below the critical threshold $\beta_N \approx 2.3$ represent the conventional occurrence of NTMs, whereas the *diamonds* above $\beta_N \approx 2.3$ represent the FIR regime, with a reduced loss $\Delta W/W_{onset}$ (adapted Fig. 4 from [33])

is reduced from the usually expected 20 % down to a level of ≈ 10 % (see Fig. 8.20) [33–35].

With the application of co-ECCD or counter-ECCD, for this purpose at the $q = 4/3$ surface, it is possible to destabilize or stabilize the (4,3)-mode externally (see Fig. 8.21). With the stabilization, the entry into this beneficial regime can be suppressed, whereas with destabilization of the (4,3)-mode, access can be gained already at lower β_N values (see Figs. 8 and 9 in [33] and Fig. 8.21). Although higher confinement values are accessible with such a modification of the NTM behaviour, the applicability for a reactor is at present not developed to a point as direct NTM stabilization as discussed above. The complete removal of a (3,2) NTM and in particular of a (2,1) NTM is still the most attractive approach, as it promises the largest gain in confinement and β_N. However, the access to the FIR-regime should be considered as a possible mitigation choice.

8.7.2 Beneficial Effect of NTMs in Improved H-Mode/Hybrid Scenario

In the conventional ELMy H-mode with a monotonic q-profile and $q_0 < 1$, (3,2) NTMs generate the confinement loss discussed above. However, in the so-called improved H-mode (ASDEX Upgrade) or hybrid scenario (DIII-D) the presence of a (4,3) NTM or a (3,2) NTM only has little impact on the confinement [115–117]. In this scenario one typically reaches a flat central q-profile with $q_0 \approx 1$. This current profile is clamped by the presence of these modes. Alternatively, stationary

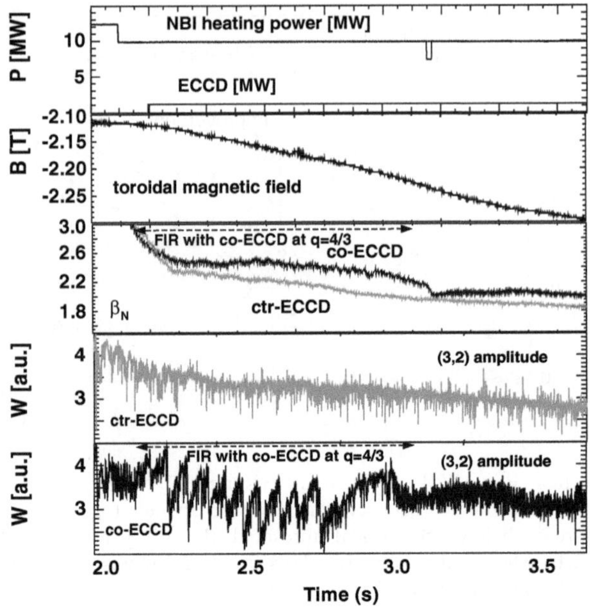

Fig. 8.21 Artificial suppression or triggering of the FIR-regime by local counter (*grey traces*) or co-ECCD (*black traces*) at the $q = 4/3$ surface in two otherwise identical discharges at ASDEX Upgrade. From top to bottom the time traces show the identically applied NBI and ECRH power, the applied magnetic field ramp, the varying β_N traces, the island width of the (3,2) NTM in the non-FIR and in the FIR case respectively. The applied B_t-ramp is used to scan for the resonance with the (4,3)-surface. In the presence of the (4,3)-mode the FIR-regime, with its characteristic amplitude drops, is accessible in the case of co-ECCD (*black traces*). An increase of almost 20 % in β_N can be achieved in comparison to the non FIR-regime with counter-ECCD (*grey traces*, indicated phase $t \approx 2.3$–2.8 s in 3rd and the 5th box) (adapted Fig. 8 from [33])

(1,1)-fishbone activity can occur, which also maintains a rather flat central current profile. For the hybrid scenario at DIII-D [118–120] and JET [121] findings are fully consistent with this picture. Similar MHD behaviour and impact on the confinement is reported. All these scenarios have in common, that the much more detrimental and dangerous (2,1) NTM does not occur. Complementary it has been shown, that also in such scenarios a complete stabilization of (3,2) NTMs is possible with a moderate improvement in confinement [122]. The formation of such scenarios is left to references, as this goes beyond the scope of this book.

8.8 Combining the Sensors and Actuators into a Control Scheme

All the above described detection and intervention concepts have to be combined into an integrated scheme for controlling the occurrence, the mitigation or ultimately the removal of NTMs.

These approaches contain tailoring of the q-profile, the j_{bs}-profile and hence mainly the n_e-profile, in order to remove the relevant resonant surface or the drive for NTMs, respectively. For both avoidance schemes, avoidance of the resonant surface or the drive for the mode, central or slightly off-axis current drive is the appropriate method. However, this might be impossible under the constraints of an energy producing high performance plasma, as a high central pressure is needed for a high fusion power output. Alternatively the avoidance of the triggering MHD is a possibility, which mainly consists of the control of (1,1) activity at the $q = 1$ surface.

For removal of unavoidable NTMs the local current drive with ECCD seems to be an appropriate choice. For small islands the current has to be modulated, in order to deposit power only in the O-point region. In this approach at least two independent current drive tools, such as two gyrotrons and two mirrors, for the (3,2) and the (2,1) NTM respectively, have to be ready to operate and aim at the resonant surfaces. As soon as an NTM appears they have to fire until the NTM gets stabilized, while tracking and correcting for the radial location of the mode.

The analysis and the decision process for this has to be implemented as a real-time algorithm dealing with fast realtime signals and has to control actions of the external heating systems. Presently such kind of algorithms are being implemented or are already in parts in operation on multiple experiments. However, most of these approaches presently do not contain the complete set of actions in a fully automatic way. Such a complete integration is still an important task for the future, in order to gain operation experience for ITER with a much larger τ_R, i.e. much smaller growth times of the modes.

At DIII-D a sophisticated system is implemented, which uses the so-called SEARCH-AND-SUPPRESS algorithm [72]. This algorithm originally used the magnetic field and the radial plasma position as the main actuator. Later this scheme has been adapted to also control the poloidal mirror position of the ECRH launchers. One core ingredient is the MSE supported equilibrium reconstruction providing an estimation for the localization of the resonant surfaces. At ASDEX Upgrade a very generic integrated control and realtime diagnostic system has been implemented [75, 123]. This system includes any new measurement as a potential realtime diagnostic, which can announce new signals to the system. This freely programmable scheme allows for an flexible application of the system for NTM control and eventually for disruption avoidance.

8.9 Implication and Outlook for ITER

In a sequence of publications an attempt has been made to quantify the impact of NTMs and their stabilization schemes on the efficiency of an energy producing tokamak, in particular for ITER [32, 124, 125]. The efficiency of an energy releasing system is defined through its gain $Q := P_{fusion}/P_{input}$. As output $P_{fusion} = 5 \cdot P_\alpha$ is taken, as the total fusion power is distributed on the α-particles

($P_\alpha = 0.2 \cdot P_{fusion}$) and the neutrons ($P_n = 0.8 \cdot P_{fusion}$) according to their mass ratios. For the calculations $Q = 10$, $P_\alpha = 80$ MW, $\beta_N = 1.8$ in the ITER scenario 2 [126] has been assumed. For the energy confinement time τ_E without an NTM $\tau_E = H_H \cdot 3.7$ s and $H_H = 1$ has been used. $H_H = \tau_{exp}/\tau_{scaling}$ denotes the normalized energy confinement time with respect to the scaling law for a sawtoothing ELMy H-mode. In Fig. 8.22 the resulting operation curves for constant H_H-factors in the range from 0.75 to 1.25 are shown as function of the additionally applied ECCD power and the Q-factor.

At the onset the discharge sits on the curve with $H_H = 1$ without any additional ECCD power applied. After the NTM-onset Q and H_H drop significantly down to $Q(3,2) = 6.9$ and $Q(2,1) = 4.7$ and to $H_H(3,2) = 0.85$ and $H_H(2,1) = 0.75$ for a (3,2) and (2,1) NTM respectively (points A and B in Fig. 8.22). If the (2,1) NTM locks the situation becomes worse and $H_H(2,1)$ becomes even lower. Applying now additional 20 MW ECCD power for the removal of the corresponding modes, one gets back to full energy confinement time τ_E, i.e. $H_H = 1$, but now at lower Q values. The additional ECCD power at reduced heating efficiency due to its off-axis localization has been included in the energy balance. For 20 MW one arrives at $Q \approx 7$, for the optimistic case with only 10 MW needed, one arrives at $Q \approx 8.5$. Once the NTM has been stabilized, one can arrive along the $H_H = 1$ curve again towards $Q = 10$, by switching off the gyrotrons. When the next NTM gets triggered, this loop is followed again and the gyrotrons are applied.

An incomplete stabilization of the NTM with only a partial confinement recovery, i.e. only a reduction in the island size, will result in a working point in the range between the broken curves. Triggering the FIR-regime would also be in this area of the diagram. A continuously applied preemptive ECCD will reduce the achievable Q value, correspondingly.

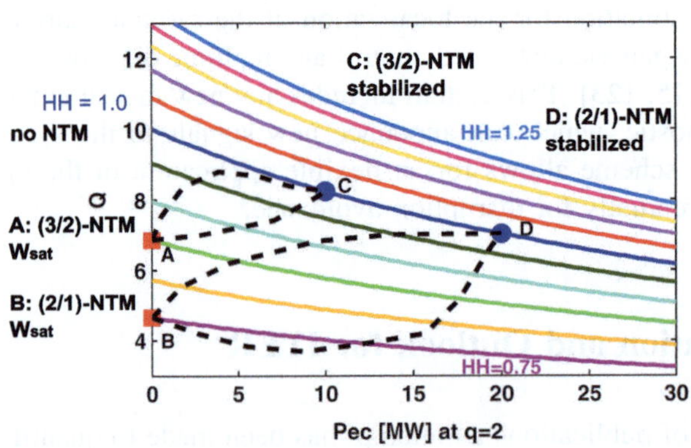

Fig. 8.22 Energy gain Q as function of the additionally required ECCD power for a complete or partial removal of an existing NTM. The *curves* indicate the operational points for a fixed H_H-factor in the range $H_H = 0.75\ldots1.25$ in steps of 0.05 (Fig. 1 taken from [32], Copyright 2010, IOP Publishing, Reproduced with permission. All rights reserved.)

In ITER such a control scheme will have to be an integral part of the control system. This part will mainly be responsible for the steering of the mirrors and the deposited power from the connected gyrotrons. The low field side equatorial launcher is optimized for central heating and current drive in a radial range of $\rho_{pol} \approx 0\ldots0.5$, whereas the upper launcher is optimized for $\rho_{pol} \approx 0.3\ldots0.9$. The upper launcher is primarily designed for the purpose of (3,2) and (2,1) NTM control and sawtooth tailoring. It consists of two different designs for a set of upper steering mirrors and a set of lower steering mirrors. A comprehensive overview over the present design is given in [127]. The system has been optimized for a narrow deposition (target: $2d_{dep} < W_{marg}$) with a maximization of $\eta_{NTM} = j_{cd}/j_{bs}$, in order to avoid the need for a modulation of the gyrotrons. The gyrotrons provide a total power of 24 MW at 170 MHz, from which 20 MW are assumed to be deposited in the plasma. These optimizations are an ongoing process driven by new experimental and theoretical input until the design for the ITER launchers are fixed.

From the combined ASDEX Upgrade and JT-60U data the following predictions for the required ECCD power and alignment have been made [38]. For the ITER scenario 2 [126] and a deposition width of 3–4 cm with perfect alignment onto the resonant surface, an estimation for the required ECCD power for an unmodulated and a modulated case has been given. For the (3,2) NTM 10 and 7 MW, for the (2,1) NTM 10 and 9 MW are predicted for a complete stabilization. A possible misalignment is expressed in terms of the deposition width $x_{mis}/(2d_{dep})$. With the presently planned 20 MW of ECCD power, for the (3,2) NTM a misalignment of $x_{mis}(3,2), unmod.)/(2d_{dep}) = 0.4$ for the unmodulated case, and $x_{mis}(3, 2), mod.)/d_{dep} = 0.6$ for the modulated case is possible while still stabilizing the mode. For the (2,1) NTM $x_{mis}(2, 1), unmod.)/(2d_{dep}) = 0.5$ and $x_{mis}(2,1,mod.)/(2d_{dep}) = 0.8$ has been estimated. From these considerations, the ECCD system should be able to fulfill its purpose.

In an analytical approach for η_{NTM} also the ITER scenario 2 has been used as reference [45]. A different criterion for the ECCD system has been derived

$$2d_{dep} \leq 5 \text{ cm}, \quad \text{and} \quad (2d_{dep}) \cdot \eta_{NTM} \geq 5 \text{ cm}. \tag{8.15}$$

Calculating with these constraints with the TORBEAM raytracing code the resulting needs on the mirror system, a modification of the steering capability towards a larger toroidal launching angle is suggested. These modifications should result in a reduction of the needed ECCD power by 25 % for the lower steering mirror and 10 % for the upper steering mirror.

The discussion of details of the ECCD system, in particular the upper launcher is still a field were new input might require some modifications. A design change however, should be based only on the most reliably understood theoretical considerations and experimentally verified information.

8.10 Summary

In this chapter an attempt has been made to summarize the present status of experiments on NTM control. This consists of two approaches, namely the avoidance of NTM excitation and the stabilization or mitigation of excited NTMs. For the avoidance different schemes of profile tailoring (n_e, p, q-profile) and avoidance of MHD, which can trigger NTMs, have been discussed. In the improved H-mode or hybrid scenario apart from the (2,1)-NTM, higher (m,n)-NTMs cause only a reduced problem and the usefulness of an NTM removal there needs to be further investigated. A mitigation by attempting a transition into the FIR-regime has been indicated. For all these avoidance, stabilization or mitigation scenarios, the application of ECCD at different resonant surfaces ($q \leq 1, q \approx 1$ for sawtooth tailoring, $q = 4/3$ for initiating the FIR-regime, $q = 3/2$ and $q = 2$ for NTM suppression and preemptive NTM avoidance) is an appropriate tool.

As the ITER design of the ECCD and their launcher system has been driven by the ongoing work on NTM control in present devices and current theoretical work, ITER seems to be well equipped for controlling NTMs. The available power of 20 MW at 170 GHz should be sufficient. Fine details on the upper launcher should be taken care of and possibly optimized for reduced demands on the system.

References

1. F. Troyon, R. Gruber, H. Saurenman, S. Semenzato, S. Succi, Plasma Phys. Control. Fusion **26**, 209 (1984)
2. F. Troyon, A. Roy, W.A. Cooper, F. Yassen, A. Turnbull, Plasma Phys. Control. Fusion **30**, 1597 (1988)
3. J. Wesson, *Tokamas*, 3rd edition (Clarendon Press, Oxford, 2004)
4. H.P. Furth, P.H. Rutherford, H. Selberg, Phys. Fluids **16**, 1054 (1973)
5. R.H. Rutherford, Phys. Fluids **16**, 1903 (1973)
6. G. Bateman, *MHD Instabilities* (MIT Press, Cambridge Mass, 1978)
7. H. Zohm et al., Nucl. Fusion **41**, 197 (2001)
8. H. Zohm et al., Phys. Plasmas **8**, 2009 (2001)
9. J. Stober et al., Plasma Phys. Control. Fusion **43**, 39 (2001)
10. O. Sauter, C. Angioni, Y.R. Lin-Liu, Phys. Plasmas **6**, 2834 (1999)
11. M. Maraschek, O. Sauter, S. Günter, H. Zohm, ASDEX Upgrade Team, Plasma Phys. Control. Fusion **45**, 1369 (2003)
12. R. Fitzpatrick, Phys. Plasmas **2**, 825 (1995)
13. E. Poli, A.G. Peeters, A. Bergmann, S. Günter, S.D. Pinches, Phys. Rev. Lett. **88**, 075001 (2002)
14. A. Bergmann, E. Poli, A.G. Peeters, Monte carlo δ f simulations of neoclassical phenomena in tokamak plasmas, in *Proceedings of the 19th IAEA Conference, Fusion Energy, Lyon, France,* October 2002, (CD-ROM), pages IAEA–CN–94/TH/P1–01. (IAEA, Vienna, 2002)
15. A.H. Glasser, J.M. Greene, J.L. Johnson, Phys. Fluids **18**, 875 (1975)
16. A.H. Glasser, J.M. Greene, J.L. Johnson, Phys. Fluids **19**, 567 (1976)
17. M. Kotschenreuther, R.D. Hazeltine, P.J. Morrison, Phys. Plasmas **28**, 294 (1985)
18. H. Lütjens, J.-F. Luciani, X. Garbet, Phys. Plasmas **8**, 4267 (2001)

19. H.R. Wilson, J.W. Connor, R.J. Hastie, C.C. Hegna, Phys. Plasmas **3**, 248 (1996)
20. H.R. Wilson et al., Plasma Phys. Control. Fusion **38**, 149 (1996)
21. F.L. Waelbroeck, R. Fitzpatrick, Phys. Rev. Lett. **78**, 1703 (1997)
22. J.W. Connor, F.L. Waelbroeck, H.R. Wilson, Phys. Plasmas **8**, 2835 (2001)
23. E. Poli, A. Bergmann, A.G. Peeters, Phys. Rev. Lett. **94**, 205001 (2005)
24. E. Poli, A. Bergmann, A.G. Peeters, L.C. Appel, S.D. Pinches, Nucl. Fusion **45**, 384 (2005)
25. M. Siccinio, E. Poli, F.J. Casson, W.A. Hornsby, A.G. Peeters, Phys. Plasmas **18**, 122506 (2011)
26. O. Sauter et al., Plasma Phys. Control. Fusion **44**, 1999 (2002)
27. R.J. Buttery et al., Nucl. Fusion **43**, 69 (2003)
28. M.E. Maraschek et al., Density dependence of the onset of neoclassical tearing modes in H-mode and pellet refuelled discharges on JET and ASDEX Upgrade, in *Europhysics Conference Abstracts (CD-ROM, Proceedings of the 28th EPS Conference on Controlled Fusion and Plasma Physics, Madeira 2001)*, ed. by C. Silva, C. Varandas, D. Campbell, vol. 25A. (EPS, Geneva, 2001), pp. 1801–1804
29. M. Maraschek, Nucl. Fusion **52**, 074007 (2012)
30. O. Sauter et al., Phys. Plasmas **4**, 1654 (1997)
31. H. Zohm, Phys. Plasmas **4**, 3433 (1997)
32. O. Sauter, M. Henderson, G. Ramponi, H. Zohm, C. Zucca, Plasma Phys. Control. Fusion **52**, 025002 (17 pp) (2010)
33. S. Günter et al., Nucl. Fusion **44**, 524 (2004)
34. A. Gude, M. Maraschek, H. Zohm, ASDEX Upgrade Team, Nucl. Fusion **42**, 833 (2002)
35. S. Günter et al., Phys. Rev. Lett. **87**, 275001 (2001)
36. A.W. Morris, Plasma Phys. Control. Fusion **34**, 1871 (1992)
37. G. Giruzzi et al., Nucl. Fusion **39**, 107 (1999)
38. L. Urso et al., Nucl. Fusion **50**, 025010 (12 pp) (2010)
39. H. Zohm et al., J. Phys: Conf. Ser. **25**, 234 (2005)
40. A. Isayama et al., Nucl. Fusion **49**, 055006 (9 pp) (2009)
41. C.C. Petty et al., Nucl. Fusion **44**, 243 (2004)
42. R.L. Haye et al., Nucl. Fusion **46**, 451 (2006)
43. H. Zohm et al., Plasma Phys. Control. Fusion **49**, B341 (2007)
44. A. Isayama, The JT60 Team, Nucl. Fusion **47**, 773 (2007)
45. N. Bertelli, D. De Lazzari, E. Westerhof, Nucl. Fusion **51**, 103007 (15 pp) (2011)
46. O. Sauter et al., Phys. Rev. Lett. **88**, 105001 (2002)
47. I.T. Chapman et al., Plasma Phys. Control. Fusion **55**, 065009 (2013)
48. M. Maraschek et al., Phys. Rev. Lett. **98**, 025005 (2007)
49. E. Poli, G.V. Pereverzev, A.G. Peeters, Phys. Plasmas **6**, 5 (1999)
50. E. Poli, A.G. Peeters, G.V. Pereverzev, Comput. Phys. Comm. **136**, 90 (2001)
51. E. Poli, G.V. Pereverzev, A.G. Peeters, M. Bornatici, Fusion Eng. Design **53**, 9 (2001)
52. K. Matsuda, IEEE Trans. Plasma Scien. **17**, 6 (1989)
53. A. Isayama et al., Plasma Phys. Control. Fusion **42**, 37 (2000)
54. A. Isayama et al., Nucl. Fusion **43**, 1272 (2003)
55. C.C. Hegna, J.D. Callen, Phys. Plasmas **4**, 2940 (1997)
56. F.A.G. Volpe et al., Phys. Plasmas **16**, 102502 (2009)
57. Y.I. Koshuriinov, V.G. Pavel'el, M.I. Petelin, I.V. Turchin, D.Y. Shchegol'kov, Tech. Phys. Lett. **31**, 709 (2005)
58. W.A. Bongers et al., ECE system on ASDEX-Upgrade placed inline at the high power waveguide based transmission system, in *2009 34th International Conference on Infrared, Millimeter, and Terahertz Waves.* (Busan, South Korea, 2009), pp. 590–591. http://www.tue.nl/en/publication/ep/p/d/233286/ or http://repository.tue.nl/664989 or http://ieeexplore.ieee.org/xpl/articleDetails.jsp?arnumber=5325585
59. V. Erckmann et al., Fusion Sci. Technol. **55**, 23 (2009)
60. Q. Yu, X.D. Zhang, S. Günter, Phys. Plasmas **11**, 1960 (2004)

61. J.P. Graves, I.T. Chapman, S. Coda, L.G. Erikson, T. Johnson, Phys. Rev. Lett. **102**, 065005 (2009)
62. J.P. Graves et al., Nucl. Fusion **50**, 052002 (6 pp) (2010)
63. C.D. Warrick et al., Phys. Rev. Lett. **85**, 574 (2001)
64. C.D. Warrick et al., Neo-classical tearing mode stabilisation and onset mechanisms in COMPASS-D, in *Europhysics Conference Abstracts (CD-ROM, Proceedings of the 28th EPS Conference on Controlled Fusion and Plasma Physics, Madeira 2001)*, ed. by C. Silva, C. Varandas, D. Campbell, vol. 25A. (EPS, Geneva, 2001), pp. 1817–1820
65. T.E. Evans et al., Phys. Rev. Lett. **92**, 235003 (2004)
66. W. Suttrop et al., Design of in-vessel saddle coils for MHD control in ASDEX Upgrade, in *Europhysics Conference Abstracts (CD-ROM, Proceedings of the 35th EPS Conference on Plasma Physics, Hersonissos, Crete, 2008)*, ed. by P. Lalousis, S. Moustaizis, vol. 32D. (EPS, Geneva, 2008), p. P–4.075
67. W. Suttrop et al., Fusion Eng. Des. **84**, 290 (2009)
68. W. Suttrop et al., Phys. Rev. Lett. **106**, 225004 (2011)
69. M. Maraschek et al., Measurement and impact of the n = 1 intrinsic error field at ASDEX Upgrade, in *Europhysics Conference Abstracts (CD-ROM, Proceedings of the 40th EPS Conference on Plasma Physics, Espoo, Finland, 2013)*, ed. by V. Naulin et al., vol. 37D of *ECA*. (European Physical Society, Geneva, 2013), p. P4.127
70. S. Cirrant et al., Mode coupling trigger of tearing modes in ECW heated discharges in FTU, in *Fusion Energy 2000 (Proceedings 18th International Conference, Sorrento, Italy, 2000)*, vol. IAEA-CSP-8/C, CD-ROM file EX/3-3. (IAEA, Vienna, 2001) http://www-pub.iaea.org/MTCD/publications/PDF/csp_008c/html/node1.htm
71. N. Hicks et al., Fusion Sci. Technol. **57**, 1 (2010)
72. D.A. Humphreys et al., Phys. Plasmas **13**, 056113 (9p) (2006)
73. M. Reich et al., Progress on ECCD-based NTM rt-control at ASDEX Upgrade, in *Europhysics Conference Abstracts (CD-ROM, Proceeding of the 38th EPS Conference on Plasma Physics, Strasbourg, France, 2011)*, ed. by C. McKenna, vol. 34A of *ECA*. (European Physical Society, Geneva, 2011), p. P5.102
74. M. Reich et al., ECCD-based NTM control using the ASDEX upgrade real-time system, in *Europhysics Conference Abstracts (CD-ROM, Proceedings of the 37th EPS Conference on Plasma Physics, Dublin, Ireland, 2010)*, ed. by C. McKenna, vol. 34A of *ECA*. (European Physical Society, Geneva, 2010), p. P2.189
75. M. Reich et al., Fusion Sci. Technol. **58**, 727 (2010)
76. L. Giannone et al., Fusion Eng. Des. **84**, 825 (2009)
77. M. Reich et al., Real-time current profile measurements for NTM control, in *Europhysics Conference Abstracts (CD-ROM, Proceedings of the 36th EPS Conference on Plasma Physics, Sofia, Bulgaria, 2009)*, ed. by M. Mateev, E. Benova, vol. 33E of *ECA*. (European Physical Society, Geneva, 2009), p. P–1.160
78. M.R. de Baar et al., Physics and real time control of tearing modes in TEXTOR, in *Fusion Energy 2008 (Proceedings 22nd International Conference, Geneva, Swizerland, 2008)*, vol. IAEA-CN-165, CD-ROM file EX/P9-12. (IAEA, Vienna, 2008) http://www-naweb.iaea.org/napc/physics/FEC/FEC2008/html/index.htm
79. B.A. Hennen et al., Nucl. Fusion **52**, 104006 (2010)
80. H. Zohm et al., Nucl. Fusion **39**, 577 (1999)
81. Q. Yu, S. Günter, Modelling of the neo-classical tearing mode and its stabilisation by ECCD/ECRH, in *Europhysics Conference Abstracts (CD-ROM, Proceedings of the 1998 ICPP & 25th EPS Conference on Controlled Fusion and Plasma Physics, Praha, 1998)*, ed. by P. Pavlo, vol. 22C. (EPS, Petit-Lancy, 1998), pp. 1899–1902
82. Q. Yu, S. Günter, Plasma Phys. Control. Fusion **40**, 1977 (1998)
83. Q. Yu, S. Günter, G. Giruzzi, K. Lackner, M. Zabiego, Phys. Plasmas **7**, 312 (2000)
84. G. Gantenbein et al., Phys. Rev. Lett. **85**, 1242 (2000)

85. G. Gantenbein et al., Investigations on $m = 2, n = 1$ tearing mode stabilisation with ECRH at ASDEX Upgrade, in *Europhysics Conference Abstracts (CD-ROM, Procedings of the 30th EPS Conference on Controlled Fusion and Plasma Physics, St. Petersburg, 2003)*, ed. by R. Koch, S. Lebedev, vol. 27A. (EPS, Geneva, 2003), p. P–1.187
86. H. Zohm et al., Plasma Phys. Control. Fusion **39**, B237 (1997)
87. B. Esposito et al., Nucl. Fusion **51**, 083051 (2011)
88. M. Maraschek, S. Günter, H. Zohm, ASDEX Upgrade Team, Plasma Phys. Control. Fusion **41**, 1 (1999)
89. B. Esposito et al., Nucl. Fusion **49**, 065014 (2009)
90. F.A.G. Volpe et al., Stabilization of disruptive locked modes at DIII-D by means of ECCD and magnetic perturbations, in *15th workshop on MHD stability control: US-Japan Workshop on 3D Magnetic Field Effects in MHD Control, Madison (WI), USA, Nov. 2010.* (2010) https://fusion.gat.com/conferences/mhd10/talks/Volpe_Wednesday.pdf or https://fusion.gat.com/conferences/mhd10/talks.php
91. F. Volpe, R.J. La Haye, Private communication, University of Madison (WI), USA and General Atomics, San Diego (CA), USA (2011)
92. K. Nagasaki et al., Nucl. Fusion **43**, 7 (2003)
93. K. Nagasaki et al., Nucl. Fusion **45**, 1608 (2005)
94. R.J. La Haye et al., Nucl. Fusion **45**, 37 (2005)
95. R. Prater et al., Nucl. Fusion **47**, 371 (2007)
96. M. Maraschek, O. Sauter, S. Günter, H. Zohm, ASDEX Upgrade Team, Scaling of the marginal β_p of neoclassical tearing modes during power ramp-down experiments in ASDEX Upgrade, in *Europhysics Conference Abstracts (CD-ROM, Proceedings of the 30th EPS Conference on Controlled Fusion and Plasma Physics, St. Petersburg, 2003)*, ed. by R. Koch, S. Lebedev, vol. 27A. (EPS, Geneva, 2003), p. P–1.120
97. J. Stober et al., Nucl. Fusion **41**, 1535 (2001)
98. C. Angioni et al., Phys. Plasmas **12**, 040701 (2005)
99. A. Mlynek et al., Nucl. Fusion **51**, 043002 (2011)
100. A. Isayama, The JT60 Team, Phys. Plasmas **12**, 056117 (10 pp) (2005)
101. T. Suzuki et al., Nucl. Fusion **48**, 045002 (9 pp) (2008)
102. E. Westerhof et al., Nucl. Fusion **42**, 1324 (2002)
103. C. Angioni, T.P. Goodman, M.A. Henderson, O. Sauter, Nucl. Fusion **43**, 455 (2003)
104. H. Zohm et al., Nucl. Fusion **43**, 1570 (2003)
105. A. Mück et al., Plasma Phys. Control. Fusion **47**, 1633 (2005)
106. H. Zohm et al., Plasma Phys. Control. Fusion **45**, A163 (2003)
107. T.P. Goodman, F. Felici, O. Sauter, J.P. Graves, The TCV Team, Phys. Rev. Lett. **106**, 245002 (2011)
108. G. Witvoet, M. Lauret, M.R. de Baar, E. Westerhof, M. Steinbuch, Nucl. Fusion **51**, 103043 (15 pp) (2011)
109. M. Lauret et al., Nucl. Fusion **52**, 062002 (2012)
110. I.T. Chapman et al., Phys. Plasmas **16**, 072506 (2009)
111. V. Igochine et al., Plasma Phys. Control. Fusion **53**, 022002 (2011)
112. I.T. Chapman, Nucl. Fusion **53**, 013001 (36 pp) (2011)
113. A. Gude, S. Günter, S. Sesnic, ASDEX Upgrade Team, Nucl. Fusion **39**, 127 (1999)
114. S. Fietz et al., Plasma Phys. Control. Fusion **55**, 085010 (2013)
115. O. Gruber et al., Plasma Phys. Control. Fusion **47**, 135 (2005)
116. A. Stäbler et al., Nucl. Fusion **45**, 617 (2005)
117. J. Stober et al., Nucl. Fusion **47**, 728 (2007)
118. T. Luce et al., Development of advanced inductive scenarios for ITER, in *Fusion Energy 2010 (Proceedings of 23rd International Conference, Daejon, Korea Rep., 2010)*, vol. IAEA-CN-180, CD-ROM file ITR/1-5. (IAEA, Vienna, 2010) http://www-naweb.iaea.org/napc/physics/FEC/FEC2010/html/index.htm or http://www-pub.iaea.org/iaeameetings/38091/23rd-IAEA-Fusion-Energy-Conference
119. M.R. Wade et al., Nucl. Fusion **45**, 407 (2005)

120. C.C. Petty et al., Phys. Rev. Lett. **102**, 045005 (2009)
121. E. Joffrin et al., Nucl. Fusion **45**, 626 (2005)
122. A.C.C. Sips et al., Nucl. Fusion **47**, 1485 (2007)
123. W. Treutterer et al., Fusion Eng. Des. **86**, 465 (2011)
124. O. Sauter, H. Zohm, Partial stabilisation of NTMs with ECCD for standard scenarios in ITER, in *Europhysics Conference Abstracts (CD-ROM, Proceedings of the 32nd EPS Conference on Plasma Physics, Tarragona, 2005)*, ed. by C. Hidalgo, B.P. van Milligen, vol. 29C. (EPS, Geneva, 2005), p. P–2.059
125. O. Sauter, M. Henderson, H. Zohm, C. Zucca, Partial stabilization and control of neoclassical tearing modes in burning plasmas, in *Fusion Energy 2006 (Proceedings of 21st International Conference, Chengdu, China, 2006)*, vol. IAEA-CN-149, CD-ROM file TH/P3-10. (IAEA, Vienna, 2006) http://www-naweb.iaea.org/napc/physics/FEC/FEC2006/html/index.htm
126. M. Shimada et al., Progress in the ITER physics basis chapter 1: Overview and summary, Nucl. Fusion **47**, S1 (2007)
127. M. Henderson et al., Nucl. Fusion **48**, 054013 (14 pp) (2008)

Chapter 9
Energetic Particle Driven Modes

Control of Energetic Particle Driven Modes in Fusion Plasmas

Simon D. Pinches and Sergei E. Sharapov

Abstract Twenty percent of the energy produced in a power station based around the self-sustaining thermonuclear fusion of deuterium and tritium ions will be released in the form of highly energetic (3.5 MeV) He^{2+} ions (alpha particles). In addition to this highly energetic population of ions that is expected to slow down on the rest of the bulk plasma, thereby heating it and sustaining the thermonuclear reaction, auxiliary heating systems that also produce energetic ions will be used to initiate and control the plasma burn. These different populations of fast particles have the potential to drive various instabilities in the plasma with possibly deleterious consequences for the reactor's performance and the integrity of the plant. The extent to which these energetic particle driven modes can be diagnosed and controlled is considered in this Chapter.

9.1 Introduction

Successfully producing energy from a system based on the magnetic confinement of hot plasmas will require many aspects of control. In this chapter the properties of the most unstable modes that can be driven by the free energy in the populations of fast ions present in the plasma is considered from the perspective of exploiting the underlying physics for control.

The fusion reactor concept is based upon the dominant self-heating of plasmas by alpha-particles born with an energy of 3.52 MeV in deuterium-tritium (D-T)

S.D. Pinches (✉)
ITER Organization, Route de Vinon-sur-Verdon, CS 90 046, 13067 St. Paul-lez-Durance Cedex, France
e-mail: Simon.Pinches@iter.org

S.E. Sharapov
CCFE, Culham Science Centre, Abingdon OX14 3DB, UK
e-mail: Sergei.Sharapov@ccfe.ac.uk

V. Igochine (ed.), *Active Control of Magneto-hydrodynamic Instabilities in Hot Plasmas*, Springer Series on Atomic, Optical, and Plasma Physics 83,
DOI 10.1007/978-3-662-44222-7_9

fusion reactions [1]. The transport properties of these highly energetic alpha-particles are fundamental to the success of any burning plasmas, since these properties will determine the plasma heating profiles, the plasma dilution due to the 'helium ash' accumulation and the power loading upon the first wall of the reactor. Any possible resonant interaction between the alpha-particle motion and Alfvén waves in the plasma may lead to alpha-particle behaviour that is different to that expected classically. An enhanced transport of alpha-particles caused by Alfvén instabilities could lead to less favourable heating profiles and to anomalous losses of alpha-particles, potentially limiting the lifetime of the first wall. To help identify possible strategies for controlling Alfvén instabilities, the relevant physics details are briefly recapped.

High-frequency ($f_{AE} \approx const \cdot V_A/qR_0 \sim$ 50–500 kHz) Alfvén Eigenmodes (AEs) can be excited by fusion-born alpha-particles provided that the following three main conditions are fulfilled:

(1) the drift frequency due to the radial gradient of the alpha-particle pressure P_a (which constitutes the free energy source for such instabilities) is larger than Alfvén frequency $\omega_{AE} = 2\pi f_{AE}$:

$$\omega_{*\alpha} \equiv -\frac{m}{r}\frac{T_\alpha}{e_\alpha B_0}\frac{d\ln p_\alpha}{dr} > \omega_{AE} \tag{9.1}$$

where m is the poloidal harmonic, r the minor radius, T_α the alpha particle temperature, e_α the alpha charge, B_0 the equilibrium magnetic field strength and p_α the alpha pressure

(2) the free energy of the alpha-particle radial gradient can be coupled to the wave energy of AEs via wave-particle resonances, e.g. via the Landau resonance

$$V_{\|a} = V_{A}, \tag{9.2}$$

(3) and the power transfer P_a from alpha-particles to the wave overcomes the wave damping by thermal plasma:

$$\gamma_a = \frac{P_a}{2W_{AE}} > |\gamma_{plasma}|. \tag{9.3}$$

Here, m is poloidal mode number, $T_a \approx 1$ MeV is the temperature of the alpha-particle population, e_a is the charge of the alpha-particle, r is the minor radius of the torus, B_0 is the value of equilibrium magnetic field, c is speed of light, W_{AE} is the wave energy, and γ_j is the growth/damping rate of the mode amplitude considered to be $A(t) \propto \exp(\gamma_j t)\exp(-i\omega_{AE}t)$.

Condition (1) implies that for a given alpha-particle pressure profile the particle-to-wave power transfer is positive if the wave has high poloidal mode number m and/or the wave frequency f_{AE} is low enough. Condition (2) is easily satisfied for alpha-particles in ITER since the phase velocity of Alfvén wave,

$V_A = B_0 / \sqrt{4\pi \sum_i m_i n_i} \approx 7 \times 10^6$ m/s, is just below the alpha-particle speed at birth, $V_A = 1.3 \times 10^7$ m/s (V_A is estimated for D:T = 50:50 mixture and for ITER parameters $n_e \approx n_i \approx 10^{20}$ m^{-3}, $B_0 \approx 5$ T). Finally, condition (3) selects Alfvén Eigenmodes with as small $|\gamma_{plasma}|$ as possible. This condition determines which particular types of Alfvén Eigenmodes are the easiest to excite [2]. In the baseline ITER scenarios with positive magnetic shear, the most weakly damped modes are the discrete Toroidicity-induced AEs (TAEs) [3].

The main consequence of Alfvén instabilities is radial re-distribution of the resonant alpha particles at nearly constant energy of these particles. The extent of the re-distribution depends on the saturation amplitude of the modes, the mode temporal evolution, and the overlap of the wave-particle resonances in the phase space [4]. The wave amplitude depends on the free energy source due to the radial gradient of the fast particle pressure and the type of mode excited, while the mode temporal evolution is determined by details of the mechanism replenishing the distribution function at the resonance, the width of the resonance, and nonlinear evolution of the mode damping. Several types of temporary evolutions of the mode amplitude and the mode frequency are observed experimentally on present-day machines:

(a) amplitude saturates at nearly constant frequency as determined by the Alfvén scaling $f \propto B_0/\sqrt{n_i}$;
(b) amplitude becomes modulated at nearly constant frequency giving rise to so-called "pitchfork splitting" and "chaotic" features in the mode Fourier spectrograms;
(c) amplitude exhibits bursts in time with the mode frequency sweeping (so-called "chirping" modes).

Depending on the type of the nonlinear scenario and on the maximum amplitude of the mode, the ratio between peak and time-averaged mode amplitudes and the fast ion redistribution and losses varies very significantly.

Due to the large size of the burning plasma in ITER, $\rho_a/a \leq 10^{-2}$, a single TAE cannot expel a resonant alpha-particle all the way across the plasma to the wall. The possible stochastic transport of alpha-particles caused by interaction with multiple unstable TAEs, which form a "corridor" from the plasma centre to the wall, represents a more likely scenario for alpha-losses to the wall. However, for the stochastic transport to occur, an overlap of the multiple wave-particle resonances is essential, which depends on the amplitude of the mode. The stochasticity threshold in the mode amplitude depends on the mode number and magnetic shear as explained in [5]:

$$\frac{\delta B_r}{B_0} \geq C \frac{r}{qR_0} \cdot \frac{1}{mS} \tag{9.4}$$

where $\delta B_r/B_0$ is the ratio of the radial component of the perturbed magnetic field to the unperturbed value at the magnetic axis, C is a constant ($\sim 1/64$), r is the minor radius where the mode is localised, q is the safety factor, R_0 is the major radius, m is the poloidal mode number, and S is the local magnetic shear ($S = rq'/q$).

For typical parameters in present-day experiments, the stochasticity threshold is about $\delta B_r/B_0 \cong 10^{-3}$ [6].

The problem of preventing and controlling the alpha-particle driven Alfvén instabilities and/or fast particle transport caused by these instabilities, is one of the most important problems for burning plasmas. Many questions arise when considering the problem of controlling energetic particle driven Alfvén instabilities, including:

i. What is the main aim of the control: Preventing fast ion loss to the first wall or suppressing Alfvén instabilities in the plasma core?
ii. Is it better to control the linear phase of the instabilities or to affect the enhanced alpha-particle transport in the nonlinear phase of the instability?
iii. Which region of energetic particle phase-space should be controlled?
iv. Is it possible to suppress Alfvén instabilities in a narrow radius and thereby create a "transport barrier" for alpha-particles?

This chapter presents some perspectives for the control of energetic particle driven Alfvén instabilities based on present understanding and views.

9.2 Fast Ions in Tokamaks

9.2.1 Sources of Fast Ions

In a fusion reactor there will be various sources of fast ions. They arise as a consequence of nuclear fusion, with the two most common reactions being D-D fusion (producing a 1 MeV triton and a 3 MeV proton in 50 % of the cases and an 820 keV He^3 ion and a 2.5 MeV neutron in the other 50 % of cases) and D-T fusion (which produces a 3.5 MeV alpha particle and a 14.1 MeV neutron). The latter reaction is the most attractive from the point of view of energy production because of its larger reaction cross-section [7].

To reach fusion relevant conditions, auxiliary heating systems are used to initially heat the plasma and to tailor the plasma profiles during the fusion burn. In present day devices, auxiliary heating systems are also used to simulate the behaviour of fusion produced alpha particles in future burning plasma experiments.

Energetic ions can be produced with neutral beam injection (NBI) and ICRH techniques capable of accelerating hydrogen isotopes H, D, T, and He^3 up to the MeV energy range [7–9]. A population of energetic He^4 ions will be produced by D-T fusion reactions with quite significant values of the fast ion density and energy contents [10, 11]. It is also possible to accelerate a population of He^4 ions up to the MeV energy range with NBI plus ICRH technique in helium (He^4) plasmas heated by helium NBI [12].

Among the various techniques of plasma heating, NBI heating powers at the level of 10 s of MW are frequently utilized in present devices [12]. The majority of high performance plasma scenarios in present tokamaks rely upon an NBI produced fast ion population as the principal source of plasma heating [13]. NBI has been used to inject hydrogen, deuterium, He^4, He^3 and tritium.

NBI-heating is ideally suited for heating the bulk ions if the injection energy is below the critical energy at which the pitch-angle scattering of the beam ions off other thermal ions dominates over electron drag. In addition to heating, NBI also fuels the plasma by acting as a source of (energetic) plasma ions. In particular, NBI injection of tritium at an energy of 160 keV was very effective in JET to penetrate into the plasma core in hot ion H modes thus providing tritium fueling close to the optimum D:T = 50:50 mixture in high fusion power D-T experiments [11]. In ITER, the dominant fuelling mechanism will be via pellet injection as the NBI are based on negative ion acceleration to MeV energies (in order to penetrate to the core plasma) and they do not provide significant fuelling to the plasma.

ICRH is also a very flexible heating scheme that can be used for heating both electrons and ions depending on the ratio between the critical energy and the tail temperature of ICRH-accelerated ions. ICRH experiments in JET have generated energetic ion tails in the MeV energy range, where these ions can mimic the fusion ions of burning plasma. Table 9.1 shows a comparison of some characteristic parameters of energetic ions in the MeV energy range obtained in high performance D-T fusion experiments and with different ICRH and ICRH plus NBI acceleration techniques in JET versus ITER parameters [14, 31].

9.2.2 Orbit Topology

In a toroidal magnetic confinement device, the toroidal field, B_t, which is usually larger than the poloidal field for most toroidal magnetic confinement schemes, varies in inverse proportion to the distance from the axis of symmetry of the torus:

$$B_t = \frac{\mu_0 I_{tf}}{2\pi R}. \tag{9.5}$$

where I_{tf} is the sum of the currents in the toroidal field coils. Including the poloidal field produced by the plasma current in the tokamak scheme, the magnetic field lines lie on nested surfaces of constant flux and slowly spiral around these surfaces. Without the smaller poloidal field, a vertical drift would occur in opposite directions for the ions and electrons. This would lead to charge separation and a subsequent vertical electric field giving rise to an $\mathbf{E} \times \mathbf{B}$ drift in the radial direction. This would then lead to an unacceptable loss of particles from the

Table 9.1 Characteristics of ICRH-accelerated ions and fusion-born alpha-particles in JET experiments: slowing down time, τ_S, heating power per unit volume at the magnetic axis, $P_f(0)$, ratio of the on-axis fast ion density to electron density, $n_f(0)/n_e(0)$, on-axis fast ion beta, $\beta_f(0)$, volume-averaged fast ion beta, $\langle\beta_f\rangle$ (%), and normalised radial gradient of fast ion beta, max$|R\nabla\beta_f|$ [31]

Reference	JET [7, 8]	JET [7, 8]	JET [12]	JET [10, 11]	ITER [15]
Type of fast ions	H	He^3	He^4	Alpha	Alpha
Source	ICRH tail	ICRH tail	ICRH tail	Fusion	Fusion
Mechanism	Minority	Minority	3rd harm of NBI	DT nuclear	DT nuclear
τ_S (s)	1.0	0.9	0.4	1.0	0.8
$P_f(0)$ (MW/m^3)	0.8	1.0	0.5	0.12	0.55
$n_f(0)/n_e(0)$ (%)	1.0	1.5	1.5	0.44	0.85
$\beta_f(0)$ (%)	2	2	3	0.7	1.2
$\langle\beta_f\rangle$ (%)	0.25	0.3	0.3	0.12	0.3
max$|R\nabla\beta_f|$ (%)	≈5	≈5	5	3.5	3.8

Predicted values of similar parameters are also given for alpha-particles in ITER

confinement system. By adding a small poloidal field the particles still primarily follow the field lines, but they now traverse the entire poloidal cross section before returning close to where they started. Each species still has a vertical drift associated with it but this now cancels in the upper and lower halves of the torus with the effect that there is no net drift.

The curvature of the field lines and the variation in the field strength leads to various distinct classes of particles which can be divided into two main classes; trapped and passing particles, as shown in Fig. 9.1.

The trapped particles have insufficient parallel kinetic energy compared with their perpendicular energy to penetrate into the high-field side of the torus and are consequently located in the outer low-field side of the tokamak. They bounce backwards and forwards between their mirror points experiencing a continual vertical drift due to the combined effects of field curvature and gradient. When projected into a poloidal plane the trajectories traced out by these particles earn them the appropriate name of banana orbits.

Passing particles are not reflected anywhere and spiral around the torus following the helical path of the field lines. Like the trapped particles they also experience a continual vertical ∇B and curvature drift independent of their direction of travel around the tokamak. This results in co-passing and counter-passing (particles travelling with and against the plasma current) appearing to drift radially inwards or outwards depending on the direction of the plasma current.

There are various natural frequencies associated with these particle orbits that are important for their resonant interaction with global instabilities as discussed in Sect. 9.3.1. In addition to the natural gyro-frequency of the ions, ω_{ci}, (the frequency at which they gyrate around their slowly drifting gyro, or guiding, centre) there is also the toroidal precession frequency, ω_φ, the frequency at which ions circumnavigate the tokamak toroidally, and the poloidal transit frequency, ω_θ, the frequency at which they complete a closed trajectory in the poloidal plane.

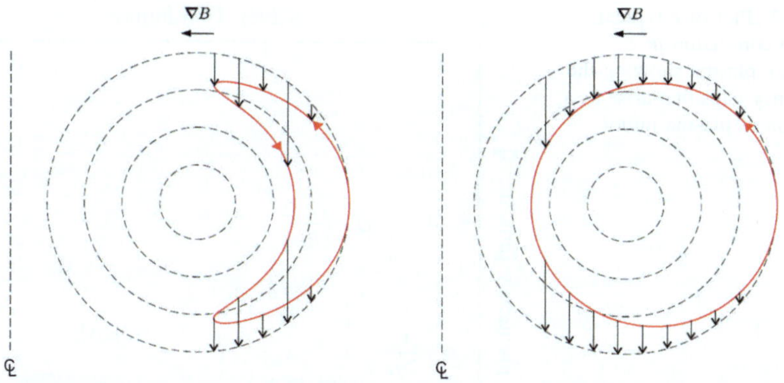

Fig. 9.1 Poloidal projection of typical energetic charged particle orbits in a tokamak. The particles experience a vertical drift dependent upon their velocity parallel to the magnetic field [16]

9.3 Alfvén Eigenmodes

9.3.1 Discrete Spectrum of Alfvén Eigenmodes in Toroidal Plasmas and Wave-Particle Resonance Condition

In toroidal plasmas, due to the geometric coupling of different poloidal harmonics, the usual spectrum of shear Alfvén waves forms some forbidden frequency zones (or frequency 'gaps') inside the Alfvén continuum. Under certain conditions, a discrete spectrum of weakly-damped Alfvén Eigenmodes (AEs) arises inside these 'gaps' [3]. Figure 9.2 shows an example of the Alfvén continuum in which the frequency gaps arising from the coupling of different poloidal harmonics can be seen versus plasma radius.

AEs that occur in the frequency gap that arises due to the coupling of neighbouring harmonics due to toroidal effects (namely the $\cos(\theta)$ variation of the equilibrium magnetic field around a flux surface) are called Toroidal Alfvén Eigenmodes, or TAEs, whilst those arising from ellipticity effects (coupling of m and $m + 2$ harmonics) are called Ellipticity induced Alfvén Eigenmodes, or EAEs. Modes in the next highest frequency gap (arising from the coupling of m and $m + 3$ poloidal harmonics) are called Non-circularity induced Alfvén Eigenmodes, or NAEs. Since the frequencies of AEs deviate from the frequencies of Alfvén continuum, the continuum damping effect is small for AEs and it is easier to excite these weakly-damped modes by small population of fusion-born alpha-particles.

Interaction between energetic ions and AEs is most effective if the waves and the particles can undergo a resonant interaction. The resonance condition for an energetic ion to interact with one of these Alfvén Eigenmodes is that the ion sees an almost constant phase of the AE along its orbit. This requirement translates into

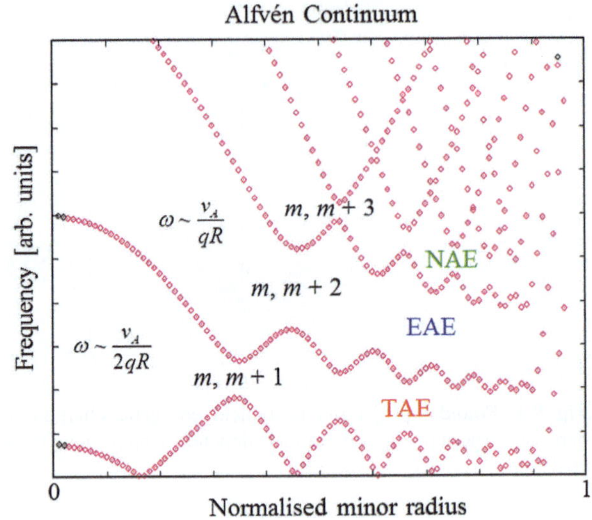

Fig. 9.2 Plot of a typical Alfvén continuum in a tokamak plasma showing the frequency gaps that arise as a function of plasma minor radius

a relation between the mode frequency, ω, and the particles toroidal precession, ω_φ, and poloidal transit frequency, ω_θ:

$$\omega - n\omega_\phi - (\mathrm{m} \pm 1)\omega_\theta = 0, \tag{9.6}$$

In terms of the passing particles' parallel velocity, the requirement for resonance with the Alfvén Eigenmode arising from the coupling of the m and $m + L$ poloidal harmonics is:

$$v_\parallel = \pm \frac{L}{2 \pm L} v_A. \tag{9.7}$$

This means that TAEs are resonant with fast ions travelling with $v_\parallel = v_A$ and $v_\parallel = v_A/3$, EAEs with ions travelling with $v_\parallel = v_A/2$, and NAEs with $v_\parallel = 3v_A/5$.

The drive for these instabilities comes from the radial gradient of the fast ion distribution (for toroidal mode numbers, $n > 0$) whilst the gradient in energy acts as a damping mechanism, since for the slowing-down distribution function of alpha-particles $\partial f_a/\partial E < 0$. A simple expression for the normalized growth rate of these modes is given by:

$$\frac{\gamma}{\omega} \sim \frac{\partial f}{\partial E} - \frac{n}{\omega e_\alpha}\frac{\partial f}{\partial \psi}, \tag{9.8}$$

where $f(E,\psi)$ is the distribution function of the fast particles in energy (E) and space as determined by the magnetic (poloidal) flux coordinate ψ. It should be noted that for a centrally peaked slowing-down distribution of energetic particles, both of the gradients in this expression are negative. This expression also indicates how it may be possible to control the stability of fast ion driven modes by tailoring the radial fast ion distribution, for example, by injecting neutral beams off-axis so their radial profile would be shifted with respect to the alpha-particle profile peaked on-axis. This mechanism is also discussed in Sect. 9.2.1 in connection with influencing the stability of sawteeth.

9.3.2 Damping Mechanisms of Alfvén Eigenmodes

In this section, the principal damping mechanism affecting energetic particle driven modes are presented and possibilities for possible control indicated.

9.3.2.1 Continuum Damping

Alfvén eigenmodes typically experience only weak continuum damping since they are located within the frequency gaps of the continuum and do not interact with it. Nevertheless, it is possible for the tail of the eigenfunction to extend out to a local Alfvén resonance surface where it can experience a small residual amount of continuum damping.

Since the radial variation of the centre of the AE frequency gaps arising from coupling between the m and $m + L$ poloidal harmonics is approximately described by

$$\omega = L\frac{v_A}{2qR} \sim \frac{1}{q\sqrt{n_i}}, \tag{9.9}$$

It may be possible to influence the level of damping by carefully engineering the radial profiles of the safety factor and ion density.

9.3.2.2 Ion Landau Damping

For typical tokamak plasma parameters a Maxwellian distribution of thermal ions contains a negligibly small number of ions capable of resonating with the AE at its primary $v_{\|} = v_A$ resonance. However, as described above, it may be possible for some of the sub-Alfvénic ions to interact with TAE through the $v_{\|} = v_A/3$ resonance. For a D-T plasma, this damping rate can be estimated as

$$\frac{\gamma}{\omega} \approx \frac{1}{243\sqrt{\pi}} q^2 \sum_{i=D,T} \beta_i^{-\frac{3}{2}} \exp\left(-\frac{1}{9\beta_i}\right), \tag{9.10}$$

where $\beta_i = 2T_i/m_i v_A^2$ is the ion beta.

Ion Landau damping is one of the most important stabilizing effects in ignited tokamak plasmas. Note, however, that this damping mechanism is exponentially sensitive with respect to β_i so that it is difficult to predict the level of damping with a high degree of reliability.

From a control perspective, both the safety factor q, and the ion thermal beta, β_i, could, at least in principle, be used to influence the degree of thermal Landau damping. It is also worth noting that for populations of deuterium and tritium in thermal equilibrium with each other ($T_{\mathrm{th,T}} = T_{\mathrm{th,D}}$) because of their larger mass, the tritium ions will be moving slower ($v_{\mathrm{th,T}} < v_{\mathrm{th,D}}$) than the deuterium ions and the deuterons will therefore provide a stronger source of Landau damping that the tritium. This points to using the fuel mixture as a possible control tool for stabilizing energetic particle driven modes. Fusion power production depends on the deuterium and tritium densities as $\sim n_D n_T \sim n_{D+T}^2 (n_D/n_{D+T})(1 - n_D/n_{D+T})$ which has a shallow maximum at $n_D/n_{D+T} = n_T/n_{D+T} = 0.5$. Small variations of the ratio of deuterons to tritons around this maximum have a small effect on the fusion power production but can enhance Landau damping in some conditions.

9.3.2.3 Radiative Damping of TAE

The small but finite coupling of the ideal MHD TAE with a kinetic Alfvén wave can be described by the dispersion relation:

$$\omega = \pm \underbrace{k_{\parallel m} v_A}_{\partial\omega/\partial k_\perp = 0} \underbrace{\left[1 + \left(\frac{3}{4} + \frac{T_e}{T_i}\right)(k_\perp \rho_i)^2\right]}_{\partial\omega/\partial k_\perp \neq 0} \tag{9.11}$$

where ω is the wave frequency, $k_{\parallel m}$ is the mth poloidal harmonic of the parallel wave vector, v_{A} is the Alfvén speed, T_{e} and T_{i} are the electron and ion temperatures respectively, $k_\perp$ is the perpendicular component of the wave vector and ρ_i is the ion Larmor radius.

The inclusion of Finite Larmor radius (FLR) corrections reveals a finite radial group velocity which carries energy away from the localisation of the TAE eigenfunction in the form of an outgoing radiative kinetic Alfvén wave (KAW). This effect is known as "radiative damping" [17] and, since its intensity is influenced by the ratio of electron to ion temperatures, it may be possible to influence its magnitude through careful use of external heating systems.

9.3.3 Diagnostic Potential

In scenarios in which the magnetic shear reverses, an extremum is formed in the Alfvén continuum and a new type of Alfvén Eigenmode may exist [18]. Such reversed shear configurations are often created in the early phases of a discharge when the plasma current is still ramping up and the current profile is evolving. In these cases, the Alfvén continuum and the Alfvén Eigenmodes associated with the shear reversal evolve in time and give rise to them being known as Alfvén Cascades because of their typically upward sweeping frequencies in time. They are sometimes also known as Reversed Shear Alfvén Eigenmodes (RSAEs) in recognition of the reason for their existence. Figure 9.3 shows an example of Alfvén Cascades in JET.

An example showing the evolution of the Alfvén continuum as current diffusion takes place is shown in Fig. 9.4. The change in mode structure from a single harmonic Alfvén Cascade mode (AC) to a 2-harmonic TAE as the mode enters the TAE gap is shown in Fig. 9.5.

Aside from the potential detrimental impact that these modes may have upon the fast ion population, their frequency evolution with time reveals information about the helicity of the magnetic field deep within the core of the plasma. Since these Alfvén Eigenmodes are only slightly displaced from the local Alfvén continuum, the dispersion relation describing their frequency is

$$\omega(t) \approx k_{\parallel}(t) v_A = \frac{1}{R_0} \left| \frac{m}{q_{\min}(t)} - n \right| v_A \tag{9.12}$$

By fitting this dispersion relation to their observed evolution, an accurate determination of $q_{\min}(t)$ can be made. If such a determination could be made in real-time, then these measurements could become important ingredients for plasma control, particularly for advanced plasma operating scenarios where the timing of the application of auxiliary heating systems in relation to $q_{\min}$ passing through integer values may be important for the creation of internal transport barriers.

9.4 Alfvén Eigenmode Stability Measurements

Even before various eigenmodes of the plasma are driven unstable through their resonant interaction with the fast ion populations present in the plasma, it is possible to measure their stability. This can be done by using external antennas to drive the plasma over a particular range of frequencies, for example, around the TAE frequency range, $\omega = v_A/2qR$, and by observing the plasma's response. Figure 9.6 shows an in-vessel view of one of the two arrays of TAE antennas in JET [19].

With the JET system, stable AEs in the plasma can be tracked in real-time allowing almost continuous measurements of their stability. If such modes were found to pose a significant risk to the operation of future fusion devices, then the ability to measure the proximity to stability limits would allow appropriate mitigating action to be undertaken in a timely manner.

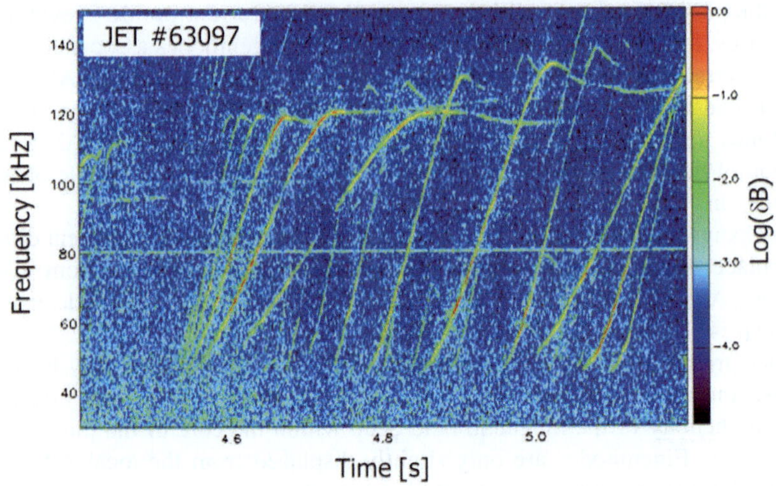

Fig. 9.3 Magnetics spectrogram showing the characteristic frequency sweeping evolution of Alfvén Cascades in JET pulse number 63097. The frequency sweeping timescale of 0.1–1 s is indicative of the timescale upon which the safety factor profile, q, evolves

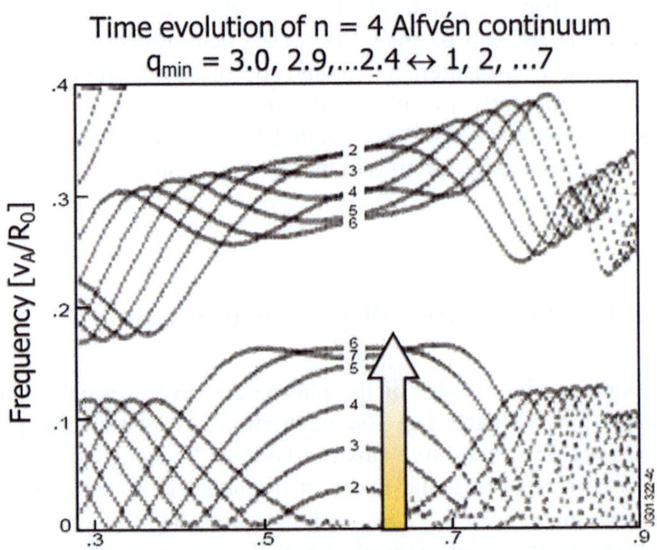

Fig. 9.4 Evolution of Alfvén continuum as current diffusion takes place and the minimum value of the safety factor profile, q_{min}, falls from 3.0 to 2.4. Each contour corresponds to a step in q_{min} of 0.1

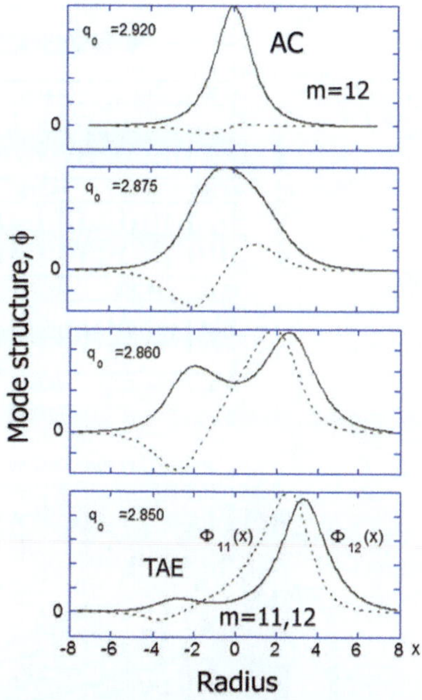

Fig. 9.5 Change in Alfvén Cascade (AC) mode structure as $q_{\min}$ drops as a result of current diffusion

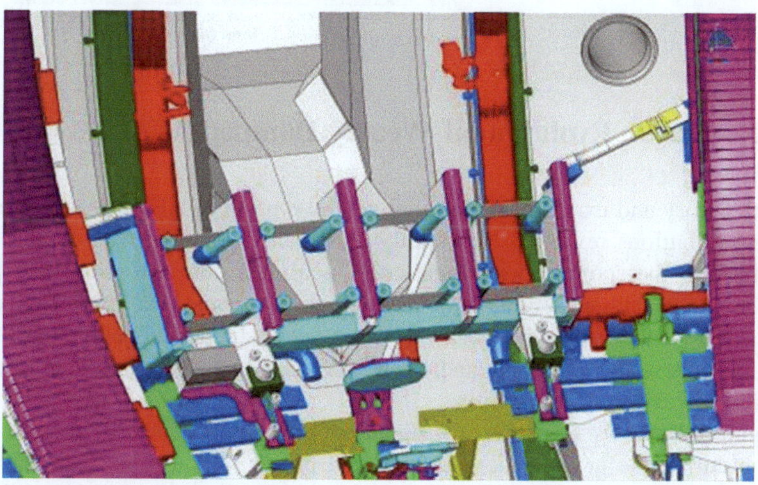

Fig. 9.6 In-vessel view of one of the two arrays of TAE antennas in JET

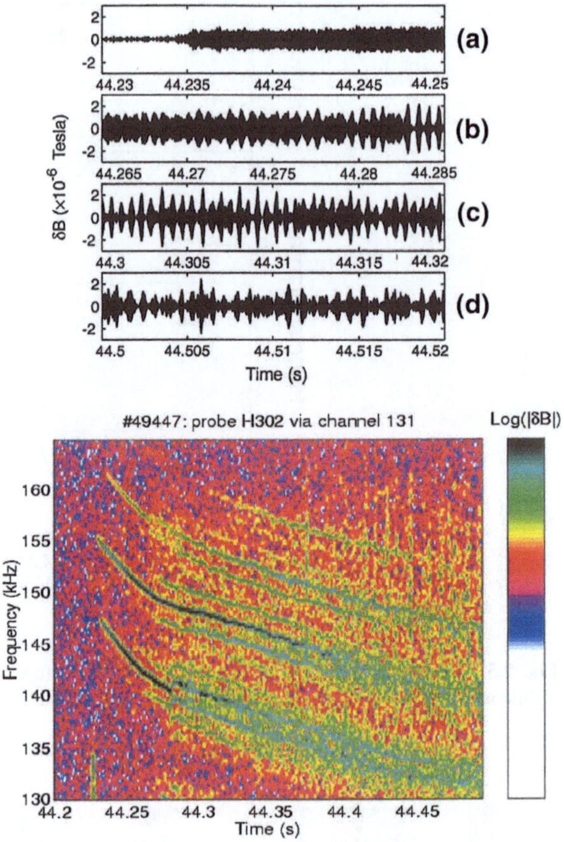

Fig. 9.7 For gradually increasing ICRH power with time, TAEs exhibit steady state, periodically modulated, and chaotic regimes. The *top* figure shows the magnetic probe signals. The *bottom* figure shows the magnetic spectrograms obtained via Fourier decomposition of the magnetic signal

9.5 Nonlinear Evolution of Weakly Damped AEs

In both theory and experiment, the nonlinear evolution of AEs and associated fast ion redistributions reveals a rich family of scenarios ranging from benign mode saturation to the spontaneous formation of nonlinear coherent structures (phase-space holes and clumps) with time-dependent frequencies [20–23]. This variety results from an interplay between the wave field which tends to flatten the distribution of resonant particles, and the relaxation processes which tend to restore the unstable distribution function.

Figure 9.7 shows how the raw signals and magnetic spectrograms of TAEs change during the gradual increase of ICRH power [22]. In this case, the distribution function of the fast ions resonating with the TAEs is formed via quasi-linear RF diffusion. The effective rate at which the distribution function is replenished, ν_{eff}, is therefore an order of magnitude higher than the rate associated with Coulomb collisions. In agreement with [22], such a diffusion-dominated distribution function gives a soft nonlinear regime of TAEs with steady or modulated amplitude.

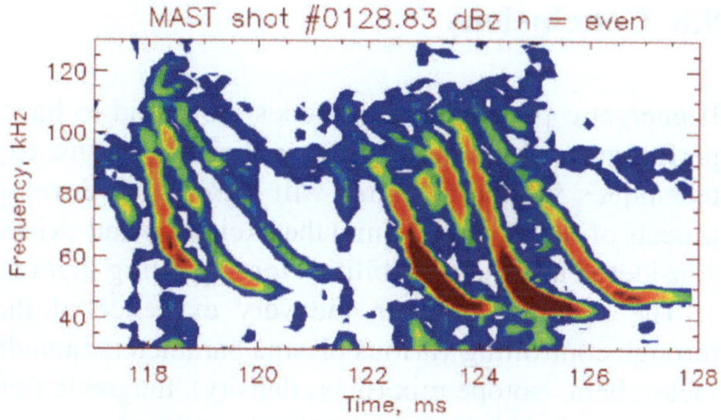

Fig. 9.8 Magnetic spectrogram showing NBI-driven TAEs with bursting amplitude and sweeping frequency on MAST [24]

On the other hand, TAEs driven by NBI-produced energetic ions, e.g. on MAST, very rarely exhibit a steady-state nonlinear evolution. Generally a bursting evolution of the mode amplitude with a sweep in the mode frequency is observed, as shown in Fig. 9.8.

Theory [25] explains that the difference observed may be due to the specifics of the fast particle relaxation mechanisms. Consequently, the earlier theoretical model [26] has been generalized by including dynamical friction (drag) as an additional relaxation mechanism. The upgraded model has revealed that only explosive behaviour is possible when the modes are near the threshold for instability when electron drag dominates over other relaxation mechanisms [27].

Interpretation of such phenomena requires a non-perturbative theoretical formalism, which has recently been developed in Ref. [28]. The underlying idea is that coherent structures with varying frequencies represent nonlinear travelling waves in fast-particle phase space. Given that the energetic particle density is usually much smaller than the bulk plasma density, it seems difficult for these particles to change the eigenmode frequency significantly. The way to resolve this difficulty is to take into account that, regardless of how small the energetic particle density is, a coherent group of these particles can still produce an observable signal with a frequency different from the bulk plasma eigenfrequency. The corresponding theoretical building block is then a nonlinear Bernstein-Greene-Kruskal mode, rather than a slowly evolving plasma eigenmode.

Frequency-sweeping nonlinear waves have been shown to generate larger fast particle transport levels than those from quasi-linear estimates [29]. This finding implies that the critical gradient estimates used for energetic particle-driven AEs [30] may not be applicable for frequency-sweeping nonlinear AE scenarios, and the type of nonlinear AE evolution should be assessed when a control tool for AE-induced transport is employed.

9.6 Conclusions

If energetic particle driven modes are found to have a negative impact upon the performance or operation of future burning plasma experiments such as ITER, then techniques for their control will have to be developed. In this chapter, various aspects of the physics behind the excitation and existence of such modes have been considered and the possibilities for exploiting them for control purposes indicated.

The stability, or even the very existence of the modes, can be influenced through controlling various plasma parameters including the plasma density, safety factor, beta, isotope mix (mass density), magnetic field, plasma flow (rotation) and the ratio of Alfvén velocity to the critical velocity below which ions start to dominantly experience pitch-angle scattering off other ions in preference to electron drag (v_A/v_{crit}).

The fast particle drive for the modes of concern can also be influenced by tailoring the energetic particle distribution function. For alpha particles this could be through the DT fuelling and fuel mix ratio and background temperature and density profiles of the plasma, for NBI via the beam injection geometry and for ICH, through the choice of heating scheme and position of the resonance layer. The magnetic field structure itself can also influence the population of resonant ions; the level of toroidal field ripple, 3D effects due to the application of 3D fields for example for ELM control, and even the aspect ratio can all play a role (although the latter is generally fixed within a single device).

The consequences of unstable modes may be controlled by influencing their nonlinear evolution near marginal stability; i.e. the transition between a steady-state saturated amplitude behaviour and more complex nonlinear behaviours may be controlled by influencing the role of dynamical friction at the resonance by altering v_A/v_{crit}.

Finally, even if it may not be practically possible to control all the energetic particle driven modes in the majority of cases, their observation may still have a broader use within control applications. The use of so-called MHD spectroscopy to help determine other plasma parameters which are desirable to control, e.g. the plasma safety factor, provides a strong motivation to continue to observe, interpret, and understand energetic particle driven modes in present day and next step devices.

Acknowledgment This work was part-funded by the RCUK Energy Programme [grant number EP/I501045] and by the European Union's Horizon 2020 research and innovation programme. The views and opinions expressed herein do not necessarily reflect those of the European Commission or of the ITER Organization.

References

1. J. Wesson, *Tokamaks* (Clarendon Press, Oxford, 2010)
2. A. Fasoli et al., Nucl. Fusion **35**, 1485 (1995)
3. C.Z. Cheng, L. Chen, M.S. Chance, Ann. Phys. **161**, 21 (1985)
4. B.N. Breizman, S.E. Sharapov, Plasma Phys. Control. Fusion **53**, 054001 (2011)
5. H.L. Berk, B.N. Breizman, H. Ye, Phys. Fluids B **5**, 1506 (1993)
6. S.D. Pinches et al., Comput. Phys. Commun. **111**, 133 (1998)
7. D.F.H. Start et al., Nucl. Fusion **39**, 321 (1999)
8. L.-G., Eriksson et al., Nucl. Fusion **39**, 337
9. J.-M. Noterdaeme et al., Fus. Sci. Tech. **53**, 1103 (2008)
10. S.E. Sharapov et al., Nucl. Fusion **39**, 373 (1999)
11. M. Keilhacker et al., Nucl. Fusion **39**, 209 (1999)
12. M.J. Mantsinen et al., Phys. Rev. Lett. **88**, 105002 (2002)
13. G. Duesling et al., Fusion Technol. **11**, 163 (1987)
14. S.E. Sharapov et al., Fusion Sci. Technol. **53**, 989 (2008)
15. A. Fasoli et al., Nucl. Fusion **47**, S1 (2007)
16. S.D. Pinches, PhD Thesis (University of Nottingham, 1996)
17. R. Mett, S.M. Mahajan, Phys. Fluids **B4**, 2885 (1992)
18. B.N. Breizman et al., Phys. Plasmas **10**, 3649 (2003)
19. A. Fasoli et al., Plasma Phys. Control. Fusion **52**, 075015 (2010)
20. B.N. Breizman et al., Phys. Plasmas **4**, 1559 (1997)
21. A. Fasoli et al., Phys. Rev. Lett. **81**, 5564 (1998)
22. R.F. Heeter et al., Phys. Rev. Lett. **85**, 3177 (2000)
23. S.D. Pinches et al., Plasma Phys. Control. Fusion **46**, S55 (2004)
24. M.P. Gryaznevich, S.E. Sharapov, Nucl. Fusion **46**, S942 (2006)
25. M.K. Lilley, B.N. Breizman, S.E. Sharapov, Phys. Rev. Lett. **102**, 195003 (2009)
26. H.L. Berk, B.N. Breizman, M.S. Pekker, Phys. Rev. Lett. **76**, 12567 (1996)
27. M.K. Lilley, B.N. Breizman, S.E. Sharapov, Phys. Plasmas **17**, 092305 (2010)
28. B.N. Breizman, Nucl. Fusion **50**, 084014 (2010)
29. M.K. Lilley, B.N. Breizman, Nucl. Fusion **52**, 094002 (2010)
30. W.W. Heidbrink et al., Nucl. Fusion **53**, 093006 (2013)
31. Progress in the ITER Physics Basis, Nucl. Fusion **47**, S1 (2007)

Chapter 10
Perspectives for Integrated Control

Piero Martin

Abstract A credible roadmap to fusion maximizes performance while still allowing for a safe, efficient and reliable operation of the plant. In this effort control of plasma quantities and off-normal events plays a very important role: the challenge will be not only that of controlling individual quantities, but also that of integration in a harsh nuclear environment. Operation of a fusion reactor will need complete mastering of the plasma. Real time control of MHD stability is a paradigmatic example. Successful control of MHD stability is based in fact on integrated control of both magnetic and kinetic quantities, influences component integrity, plasma-wall interaction and D-T burn and is a requisite for handling off-normal events like disruptions. This final chapter aims at presenting a global view of the open issues and of the potential solutions that characterize the challenge of integrating stability control in an effective fusion scenario.

10.1 Control Integration to Merge Performance and Reliability

Before becoming worldwide famous as aircraft pioneers in 1903, Wilbur and Orville Wright ran a shop specializing in bicycle repair and manufacture. At the turn of the century this was a growing business, as in 1887 the safety bike—i.e. that with two wheels of equal size, that was much easier to ride than the original high-wheel version—was introduced in the US from England. While almost everyone knows how to ride a bike, the physics of bike dynamics is much less well known, and still a subject of active research (see for example [1]). It is clear by

P. Martin (✉)
Department of Physics and Astronomy, University of Padova, Padua, Italy
e-mail: piero.martin@unipd.it

P. Martin
Consorzio RFX, Padua, Italy

V. Igochine (ed.), *Active Control of Magneto-hydrodynamic Instabilities in Hot Plasmas*, Springer Series on Atomic, Optical, and Plasma Physics 83,
DOI 10.1007/978-3-662-44222-7_10

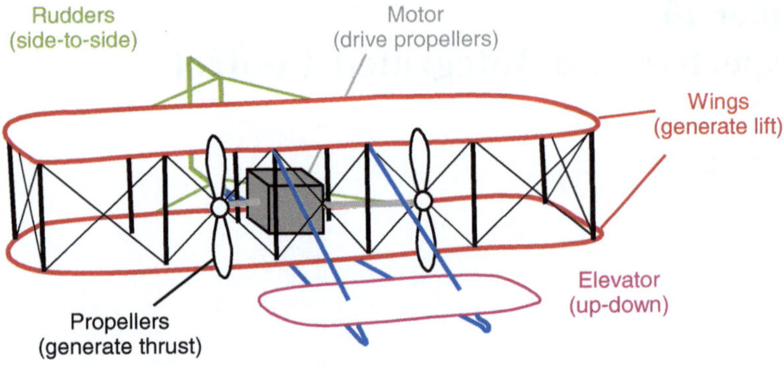

Fig. 10.1 schematics of the Wright 1903 airplane with its control systems

experience that bicycle stability is the outcome of a complex dynamic interaction between the rider and the bike, and that the bike is a highly controllable device.

It is therefore not surprising that the Wright brothers exploited their bicycle experience in designing their aircrafts, and in particular to realize that an aircraft intentionally designed to be unstable—relying for the flight on a high level of human control—was the way to go to fly. Counting on stability control more than on passive stability was the key factor that made the Wright's to win the competition for being the first to fly an airplane. A sketch of the 1903 wright airplane, with is main control systems, is shown in Fig. 10.1.

After the pioneering times, the need for longer flight times and for reducing control burden on the pilot led to more passively stable aircraft design. The boom of civil aeronautics—and the consequent requirements for comfort, safety and cost reduction—brought the aircraft design even more toward designs relying on passive stability. The introduction of computer control, in particular of the fly-by-wire system, has allowed to relax some of the passive stability requirements, leading for example to lighter, more efficient and cheaper airplanes. A passively unstable design is mandatory for high-performance aircraft, such as fighters or aerobatic airplanes, where high speed, high manoeuvrability and in general the capability of operating at the leading edge of performance is a must. Flying such an aircraft is guaranteed by an active control system, which provides stability in a closed loop cycle based on sensors, stability models, computers and actuators. Should such an active control system fail, flying would not be possible anymore.

Relying on an active control system—with the potential risks of its failure—is a price that in aeronautics must pay for high-end performance. In general, aircraft are an example of a system where the mission (e.g. safety vs. performance, comfort vs. manoeuvrability,..) determines the design and therefore the control requirements. The case of a magnetized fusion device has several analogies: the increase of performance—in terms of confinement and steady-state capabilities—and approach to the final goal, i.e. exploitation of fusion for electric energy production—is often accompanied by less stable scenarios. For these scenarios, as we

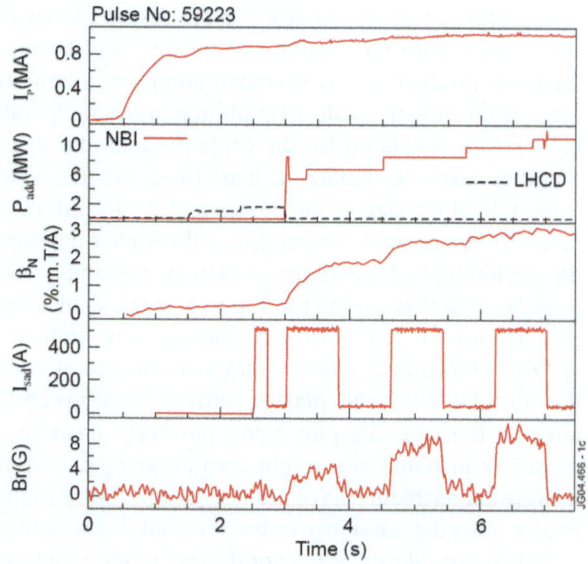

Fig. 10.2 Time evolution of plasma current, auxiliary input power (NBI and LHCD), β_N, current in the external coils (proportional to the applied magnetic perturbation) and plasma response (measured by the plasma edge radial magnetic field b_r) for JET #59223 (Figure from [2])

shall see for the DEMO case (the reactor prototype), the mission of the project strongly influences the control strategy. A paradigm of the link between stability and performance is shown in Fig. 10.2, which reports the outcome of a JET experiment [2] where an external perturbation is periodically applied to the plasma, and in the mean time the normalized pressure of the plasma itself, measured by β_N, is increased by adding auxiliary NBI power. As we have seen in the introduction and in Chap. 2 of this book, β_N is a good figure of merit for tokamak performance. We clearly observe how the plasma response to the applied perturbation, measured by the edge magnetic radial field, is directly correlated with β_N. The higher β_N, the stronger the plasma response to the externally applied perturbation, meaning that the plasma is closer to the stability boundary and external seeds for instabilities are more easily amplified.

The general point is that the higher the performance, the larger is the free energy available in the plasma, which can be released through instabilities driving relaxation towards a lower energy state. Magnetic and thermal energy, kinetic energy associated to fast particles, they may all be sources of free energy, as we shall briefly see in the next three examples.

As discussed by Zohm in the introduction, the fusion performance in a tokamak is measured by the triple product $nT\tau_E$, where n is the particle density, T the plasma temperature and τ_E the confinement time. In Chap. 2 we saw that moving from L to H-mode brings a confinement improvement by roughly a factor of two, with the build-up of a steep edge pressure gradient. But more free energy is available in the H-mode, which is characterized by MHD instabilities not present in the L-mode, like for example Neoclassical Tearing Modes (NTM, see chapter eight) and Edge Localized Modes (ELM, see chapter five). There is an optimal temperature of about 20 keV for the fusion global energy balance—identified by the balance between fusion reaction rate and bremsstrahlung losses. To increase the triple product further it is then important to maximize density and confinement time. Both roughly scale with plasma current. Operating with high plasma current is therefore beneficial for the triple product, but it is also a source of free energy. Stability analysis dictates a hard limit for the tokamak, the Kruskal-Shafranov limit. Given the size of the device and the toroidal field, this limit sets a maximum value for the plasma current. Since the higher the toroidal field the more severe are the technological challenges—which eventually set a limit on the maximum toroidal field that can be used in a tokamak—, the Kruskal Shafranov is a limit for plasma current. It is normally phrased as a limit in the edge safety factor q_a, $q_a \propto (a^2/R)(B_t/I_p) \geq 2$, where a and R are the minor and major of the device, B_t and I_p the toroidal field and plasma current respectively. As in the case of a fighter aircraft, that flies despite being passively unstable, also the $q_a < 2$ tokamak is passively unstable and might operate at $q_a < 2$ thanks to active control, as pioneering experiments have shown [3–5]. And as unstable aircraft gain in performance, also the passively stable tokamak does, being able to run at higher current.

When considering a reactor design, to the requirement of high confinement one may add the requirement of a fully non-inductively driven tokamak, i.e. a configuration that can be stationary and not relying on a transformer for current drive. The advanced tokamak (AT) regime presently the most likely option[1] for a steady-state tokamak (see [6] for a review of the steady state tokamak), since it provides high-pressure gradients also in the plasma core, and therefore high bootstrap current that can replace the inductively driven one. High-pressure gradients, though, provide free energy to drive external kink modes, or resistive wall modes (RWM) in the presence of a resistive wall.

As the devices approach fusion conditions, a larger population of fast ions is present in the plasma, and they can cause fast particle driven instabilities. The need for making room for blankets at the device periphery to convert the neutron flux—necessary in a reactor—also may act against stability as it increases the distance between the plasma and the wall.

Some of these instabilities, like for example ELMs, may lead to unbearable heat loads on the plasma facing components [7]. Others off-normal events like plasma disruptions that, besides causing premature plasma termination—already a bad event for an electricity production plant—might lead to very severe damage to the device.

A credible roadmap to fusion must therefore find the narrow route that maximizes performance while still allowing for a safe, efficient and reliable operation of the plant. In this effort control of plasma quantities and off-normal events plays a very important role: the challenge for it will be not only that of controlling individual quantities, but also that of integration in a harsh nuclear environment. Operation of a fusion reactor will need complete mastering of the plasma, that

[1] Besides of course the stellarator configuration, which is outside the scope of this book.

means real-time control of the plasma position and shape, of magnetic and kinetic profiles, of the plasma-wall interaction and of the exhaust, of the deuterium-tritium burn, and the ability of mitigating the consequences of off-normal events when there is no other option. This means integrating a large number of sensors and actuators with algorithms and first-principles and/or black-box models into an efficient control architecture, all in a challenging environment.

Real time control of MHD stability is a paradigmatic example of all this. Successful control of MHD stability is based in fact on integrated control of both magnetic and kinetic quantities, it influences component integrity, plasma-wall interaction and D-T burn and is a requisite for handling off-normal events like disruptions.

In the chapters of this book we have reviewed in great detail the physics behind the main MHD instabilities in a tokamak, how they set operational limits and the tools to control them. Based on that very detailed information, this final chapter aims at presenting a global view of the open issues and of the potential solutions that characterize the challenge of integrating stability control in an effective fusion scenario. We will first summarize in Sect. 10.2 the approaches to the control of individual instabilities presented in the previous chapters, highlighting in particular the "cross-talk" amongst them. Then, in Sect. 10.3, we will discuss the multi-faceted challenge of integrated control, to finally discuss in Sect. 10.4 how they are addressed in present and future research and in particular in ITER [8], with a final outlook on the present DEMO design.

A note for the reader: as this chapter touches almost all the subjects covered in the book, the number of references could be very large and many of them would duplicate others already quoted. Therefore we refer the reader to individual chapters and to the references therein quoted for a broad bibliographical coverage.

10.2 A Brief Summary of Present Control Tools

10.2.1 Sawtooth

The sawtooth instability was one of the first to be discovered in tokamaks [9]. They have beneficial effects, like preventing core impurity accumulation, but it took some time before the indirect implications for burning plasmas of this relaxation cycle were clearly identified. Sawtooth crashes cause a sudden perturbation of the plasma, which produces magnetic disturbances. If these disturbances are large enough, they may act as seeds of other dangerous instabilities, like NTMs.

The point is that in burning plasmas, the significant energetic ion population tends to stabilize sawteeth [10] and this is likely to result in longer sawtooth periods and therefore in more free energy being accumulated before the relaxation event takes place. This increases the likelihood for sawteeth to trigger other confinement-degrading instabilities and calls for technique to control the sawtooth oscillations, while retaining beneficial effects. At the moment the main goal of sawtooth control is to destabilise the internal kink mode and so stimulate a sawtooth crash when the sawtooth has not grown excessively (for a recent review on sawtooth control see [11]). An alternative approach is to deliberately maximise the sawtooth period. Indeed, this was originally considered the most desirable route to sawtooth amelioration in the 2007 ITER Physics Basis [2], but destabilisation is now generally the favoured option. This was driven in particular by concerns that grew about both the complexity of the NTM triggering mechanism—still relatively poorly understood—and the need for frequent expulsion of the on-axis accumulated higher-Z impurities that would otherwise radiate energy.

As discussed by Chapman in chapter four, the primary methods used to achieve this foresee actions on the quantities that control stability: energetic ions, plasma rotation and the local current density gradients can have a significant effect on the stability of the internal kink mode, thought to underlie the sawtooth phenomenon. Whilst the present explanation of the physics of sawtooth oscillations remains incomplete, the control dynamics is relatively well understood and numerical modelling has been able to explicate the sawtooth behaviour observed with different actuators in a number of tokamaks [11]. In particular Ion Cyclotron Resonance Heating (ICRH) or Neutral Beam Injection (NBI) are used to tailor the distribution of energetic ions. Electron Cyclotron Current Drive (ECCD), Lower Hybrid Current Drive (LHCD) or heating of the electrons inside the $q = 1$ surface with Electron Cyclotron Resonance Heating (ECRH) is used to control the radial profiles of the plasma current density and pressure, notably their local gradients near the $q = 1$ surface.

10.2.2 Neoclassical Tearing Modes

The NTM is an instability driven by plasma pressure and it is therefore more virulent in high performance regimes, where the thermal energy content of the plasma is higher [12, 13]. It needs to be avoided or, if that is not possible, controlled for efficiency and plant safety reasons. As shown in chapter eight by Maraschek, when an NTM is present in the standard H-mode it limits the maximum achievable plasma pressure, and therefore the reactor performance, causing an overall reduction of confinement of about 20 % for the (3,2) NTM, and even higher for the (2,1). In the hybrid H-mode scenario NTMs have a smaller impact on confinement. If the NTM grows large, in particular the (2,1), it can lock to the wall and cause a plasma disruption. NTM control will be particularly important in ITER because the onset beta value for NTM is observed to decrease with plasma flow, which in ITER will be significantly smaller than in present devices.

NTM control is presently based on two approaches: avoidance and removal or mitigation of unavoidable NTMs, both extensively discussed in chapter eight.

Avoidance of NTMs means eliminating the driving forces—i.e. modifying plasma stability properties—or the triggers for NTM. Since the main drive of the

NTM is the missing bootstrap current, a local reduction of the unperturbed bootstrap current close to the potential resonant surface should reduce the maximum saturated island size. The dominant part of the pressure gradient is the local density gradient: the reduction of the latter at the resonant surface then reduces the probability for NTM excitation. This can be done tailoring the density profiles, for example applying central ICRH, but is applicable only in a particular collisionality range. In addition, flattening of the core density profile might be detrimental in a reactor. Both these aspects make it problematic to apply this approach to reactor-scale devices. Avoidance—or at least significant delay in the NTM appearance—is obtained by applying ECCD locally at the (3,2) resonant surface before the mode appears.

A second approach for avoiding NTMs is based on avoiding the seed island by the reduction of MHD events, which may trigger NTMs. As we have just seen above, this concentrates mainly on the avoidance of large sawteeth and of fishbones at the $q = 1$ surface.

The removal or mitigation processes need to act when an unavoidable NTM is present in the plasma. ECCD application at resonant surfaces is an appropriate option [14], suitable to be used on real-time. Besides being used to destabilize sawtooth or in pre-emptive way, ECCD can in fact be applied directly at the island location (i.e. at $q = 3/2$ and $q = 2$) to replace the missing bootstrap current and therefore reduce—or heal—the island.

10.2.3 Edge Localized Modes

In the tokamak H-mode the steep plasma pressure gradient and the increased current density at the edge pedestal can exceed a stability threshold and drive a particular MHD instability referred to as an Edge Localized Modes (ELM) [15, 16]. The ELM appears in a cyclic fashion, and it causes the collapse of the edge pedestal, resulting in less steep pressure gradient within a few hundred microseconds. Heat and particles are expelled by the ELM from the confined plasma onto the plasma facing components. Then the edge pedestal recovers again, as in the pre-ELM phase. Each ELM is characterised by an increase in the radiation shown in the D_α line emission and by a burst of magnetic activity. ELM events can lead to large transient heat and particle loads on the plasma facing components as well as reducing the pedestal energy confinement by approximately 10–20 %. While in present devices this is acceptable, the extrapolation of their data on heat and particles deposited on the wall components for ITER leads to ELM energy loss ranging from approximately 5 to 22 MJ, where the uncertainty is due to the present gap in physics understanding. It is expected that approximately half of this energy will reach the wall and be deposited over a region of one square meter, known as the wetted area. Thus, the surface energy density could be 2.5–11 MJm^{-2} which is about 20 times higher than what is acceptable for the ITER first wall components, primarily made of tungsten or carbon fibre composites. ELM mitigation or suppression solutions are therefore mandatory for ITER.

Several ELM control methods have been developed and are presented by Liang in chapter five. They are based on a variety of actions made to the plasma, and follow three main routes: (a) weakening the loads on the plasma facing components by converting part of the ELM energy losses; (b) making them smaller and more frequent, to reduce individual ELM loads; (c) suppressing ELMs by controlling the pedestal pressure gradient or the edge current density to avoid exceeding stability thresholds.

Seeding the plasma edge with impurities like nitrogen or argon is a tool to increase radiation in the divertor during ELMs, but both experimental and modelling results show that reduction of ELM energy by this method is difficult for the large ELMs and may not work alone.

With fast vertical movements of the plasma column with pre-programmed frequency and amplitude—known as vertical kicks—the ELM frequency can be locked to the frequency of the externally imposed magnetic perturbation, enabling control of ELM frequency and size. The magnetic perturbation is induced by a set of vertical stabilization coils with controllable frequency and amplitude. This technique needs in-vessel coils to reach a high kick frequency. Recent results from JET [17] are promising, and this tool will be used for the ITER-like wall experiments on JET.

Launching trains of pellets allows for pacing ELMs at the pellet frequency. This can typically achieve a factor of two for the reduction of the energy per ELM. High frequency ELM pacing still needs to be demonstrated in large devices.

ELM suppression or reduction via applied resonant magnetic perturbations (RMPs) [18] is promising, although the physics mechanism is not well understood as yet. As discussed in Chap. 5, RMP fields offers an attractive method for next-generation tokamaks, e.g. ITER. The results obtained from DIII-D, JET, AUG, MAST, KSTAR, NSTX and EAST tokamaks have shown that magnetic field perturbations can either completely suppress ELMs [18], trigger small ELMs during ELM free periods, or affect the frequency and size of the type-I ELMs in a controllable way, preserving good global energy confinement [19–21]. Recent data show that RMPs impact on fast ion confinement and can cause significant fast ion losses [22]—something that might be not desirable in a reactor. Pellet pacing could work synergistically with RMPs, helping to compensate the density pump-out caused by RMP.

A joint effort from different devices and numerical codes will allow to extend the database and to understand ELM suppression.

10.2.4 Resistive Wall Modes

The advanced tokamak regime is a promising candidate for steady state tokamak operation, desirable for a fusion reactor. Constant power production is more convenient for energy conversion and avoids the power storage step required for pulsed operations. In addition, constant power loads on the plasma facing components allow simpler power handling, lower engineering efforts for reactor design, and smaller costs for the fusion power plant.

As discussed by Igochine in chapter six, the advanced tokamak regime is characterized by a high bootstrap current fraction—that eventually should completely substitute inductively driven currents—and a flat or reversed safety factor profile. This leads to operation close to the pressure limit (see Chap. 2) and as this limit is exceeded the external kink mode becomes unstable.

Without a conducting wall around the plasma, this mode would grow on an ideal time scale, which means microseconds, and in this condition would be a show-stopper. If a conducting wall is present in the device, the external kink is converted into the slowly growing Resistive Wall Mode (RWM) [23] (for a recent review on its control see [24]). The growth rate is then reduced to values of the order of the metallic wall magnetic field penetration time, typically of the order of milliseconds. This reduction makes it technically possible to act on the mode and to stabilize it with externally applied magnetic fields. Actively driven coils, which produce a magnetic perturbation cancelling that due to the mode, are the workhorse for RWM control and have been proved to work effectively both in the tokamak and in the reversed field pinch (RFP). RFPs, in particular, have demonstrated the possibility of full suppression of multiple simultaneous RWMs [25] and have studied how this depends on the active coil geometry and number [26].

Coils are used for integrated control and avoidance of RWMs and other MHD perturbations (e.g. error field, ELM..), which could act as RWM trigger. One available route is to avoid RWM growth using controlled coils with a wide-band feedback system that corrects slow (error field correction) and fast (RWM) dynamics simultaneously [27]. An alternative is based on a “three step strategy”, with increasing complexity in coil use. First one tries to avoid RWM by correcting error fields and ELM mitigation. If a mode grows anyway, the second step is to prevent coupling to the wall and suppress RWM in its infancy by active coil feedback. When also this fail, the third step—which is more challenging under many points of view—is to decouple the RWM from the wall, rotate and suppress it by feedback. Stronger feedback action is clearly needed in this case.

The feedback control of RWM has been recently particularly successful in allowing access to the unexplored operation territory of scenarios with edge safety factor close or below 2. High current, stable tokamak plasmas with edge safety factor below or around 2 are in fact attractive for magnetic fusion due to favourable high fusion gain and higher confinement. But they have long been considered inaccessible in modern devices owing to the unforgiving MHD instabilities. Even in devices with a resistive wall, the onset of an $n = 1$ resistive wall mode (RWM) leads to a disruptive limit at edge safety factor ≈ 2. Recently DIII-D and RFX-mod tokamaks have robustly overcome the edge safety factor $= 2$ limit by active control of plasma stability and demonstrated that operation below 2 is possible for hundreds of resistive wall times [4, 28, 5].

Interestingly, active RWM control might be helped by the interaction of the mode with plasma flow and fast particles, two players that contribute to RWM stability. The influence of the fast particles will be increasingly important in ITER and DEMO, which will have a large fraction of fusion born alpha's, while plasma rotation will likely be much smaller, if not absent, with respect to present devices. These interactions have kinetic origins that make the computations challenging. Correct prediction of the "plasma-RWM" interaction is an important ingredient, which has to be combined with the influence of external fields (resistive wall, error fields and feedback) to make reliable predictions for RWM control in a future reactor. In principle the stabilizing effect of fast particle in ITER might allow for a substantial increase in fusion performance with less demanding use of coils.

10.2.5 Disruptions

Disruptions are the most serious MHD off-normal events that may happen in a tokamak. So serious, that they can potentially cause fatal damage to the device. They are at the top in the list of problems that need to be solved to allow the safe operation of ITER. Disruptions are a rapid loss of the confined hot plasma and its current, often producing also a beam of runaway electrons. They cause heat loads to the plasma facing components and electromagnetic forces on the device structure. While disruptions are relatively common events in present tokamaks, the situation needs to change in ITER. As described by Hender in chapter seven, both energy dissipated to the wall and electromagnetic forces grow as L^3—where L is the linear dimension of the plasma [29]—which means that doubling the size of the tokamak (e.g. the step from JET to ITER) increases the loads at disruption by an order of magnitude. Maximum electromagnetic forces on the vacuum vessel are of the order of 4 MN in JET and are expected to be of the order of several tens of MN in ITER [30].[2] Disruptions and runaway electrons in ITER, if unmitigated, cause significant thermal loads on ITER plasma facing components and may lead to significant melting. If unmitigated, they can also cause mechanical loads on the so called safety important class (SIC) components, like vacuum vessel, port plugs and supports, cryostat support and other conductive in-vessel components.

To better clarify the extent of the problem, it is worth noting that four categories of mechanical loads are considered in the ITER design [31]: disruptions of category I are considered as a normal operational condition and 2600 events of this time are allowed for in the ITER lifetime; category II loads are allowed to occur only in a smaller number of events (400), while category III corresponds to severe disruptions, which should not happen more than 1-2 times during the machine lifetime. For example, a Vertical Displacement Event (VDE) due to loss of plasma magnetic control or a major disruption with very short current quench time ($\tau_{CQ} < 36$ ms) are category III disruptions. Category IV disruptions are considered

[2] A comparison that was suggested to me by Tim Hender: the weight force of an Airbus 380 airplane is about 5.5 MN, while the displacement (weight) of a medium size ship is about 50 MN.

extremely unlikely events—though not impossible—and none of them should happen ever (though the SIC components are designed to sustain one of them).

Finding effective means to control disruptions is therefore a must for ITER and for future devices. Control, in the case of disruptions, is a word that has several meanings, though.

First of all, control means avoidance. Avoiding a disruption is for example the outcome of an efficient control of those instabilities—NTM, RWM, etc....—that may eventually lead to a disruption. Control also mean predictability, i.e. the capability of detecting with sufficient warning time, and without false alarms, when a situation is evolving towards disruption. This is key for both avoidance and mitigation. Learning base methods like neural networks have been developed, but they have an intrinsic weakness since they need a training set of disruptions—obviously something not wanted in ITER—and do not work well outside the domain of their training. Ways to overcome these issues are being studied, but further efforts on models based on first principles, i.e. on physics, are needed. These are very difficult, since they need state-of-the-art non-linear, three-dimensional numerical simulations.

Disruption control means also mitigating their effects. In fact, even if much can be done to reduce the disruption probability, there will also be some of them that can never been avoided, like those caused by flakes of material falling off the wall, or by a power supply failure. The goal of the ITER disruption mitigation system is to reduce energy loads on PFCs and transform hard disruptions of Category II and III into milder Category I events. Chapter seven has shown that massive gas injection and killer pellets have both proved effective in giving the required level of mitigation of the electromagnetic forces arising from halo currents. These techniques are also effective in reducing heat loads, though issues of toroidal asymmetries, arising from the local nature of the massive gas injection or pellet injection, need to be fully understood. More problematic is at the moment the mitigation of runaway electrons. Theoretically the most effective method is boosting plasma density to such high level that the collisions inhibit secondary runaway growth, but such densities have not yet been achieved and pose significant problems. The research is now focused to understand whether such high densities need to be achieved to mitigate runaways, when all the mechanisms damping runaway growth are considered.

10.2.6 Fast Particle Driven Instabilities

A significant population of fast ions, i.e. ions with kinetic energies much higher that those of the thermal plasma, is implicit in a fusion-grade magnetized plasma [32]. Heating systems that produce fast ions will be used to initiate and control plasma burn and the D-T fusion reactions produce 3.5 MeV alpha particles.

The interaction of fast ions with MHD stability is bidirectional. As illustrated by Pinches and Sharapov in chapter nine, free energy associated with these fast ions has the potential to drive MHD instabilities that can cause both redistribution and loss of the energetic particle, therefore degrading performance. Expulsion of fast ion may also cause damage to the plasma facing components [33, 34]. Moreover, other MHD instabilities are affected by fast ions and/or the perturbations due to these MHD instabilities can cause fast ions anomalous transport. We have already seen the case for sawtooth and RWM stabilization. But—just to make other examples—sawteeth also lead to redistribution of fast ions [35], NTMs and magnetic field errors like those caused by ripples in the toroidal field at the plasma edge cause losses of fast ions [36], and also the RMP technique used for ELM control causes anomalous energetic particle losses. A positive feature is the possibility of tracking stable fast ion driven modes in real-time. This presently allows almost continuous measurements of the plasma stability [37]. If such modes were to pose a significant risk to the operation of future devices, then the ability to measure the proximity to stability limits would allow appropriate mitigating action to be undertaken in a timely manner.

Fast ion driven modes are also an opportunity to diagnose the plasma. For example, the frequency evolution with time of Reversed Shear Alfven Eigenmodes provides information about the helicity of the core magnetic field [38].

10.3 The Challenge of Integrated Control

As illustrated in details in the previous chapters, and briefly summarized here, there is bad and good news. The bad news is that several MHD instabilities are present in a tokamak and become more virulent as plasma performance is pushed towards fusion conditions. The good news is that we are learning about them, and in particular how to control them. We have seen in fact that several actuators are available to control, suppress or mitigate plasma instabilities. Looking into the future, the ultimate challenge is emerging: that of integrating several control actions into a reliable, efficient and as simple as possible strategy, with much more demanding constraints. Extra effort is in fact necessary with respect to the ability of controlling individually each instability in a relatively forgiving environment, as we often do now. Let us see why, starting from some of the issues that we will need to face.

The role of instabilities is not unique, and not necessarily bad. As an example, we have seen that sawteeth are potentially very dangerous as trigger of NTMs. Nonetheless their role in avoiding core impurity accumulation, and in burning plasmas in providing helium ash removals, is very important. Recognizing the strong links among these two opposite roles is what has steered the consensus towards techniques controlling sawteeth through their destabilization, more than the opposite.

NTMs are potential killers of the plasma, but in the hybrid regime—at a reduced amplitude—they may play a role in current redistribution and in clamping the core safety factor profile without spoiling significantly confinement [39].

Something similar, by the way, happens with their classical analogue—the tearing mode—in the reversed field pinch single helicity regime. Here a saturated resistive mode is the result of a plasma self-organization and leads to improved confinement [40]. ELMs are the result of a cyclical relaxation that help maintaining the edge gradients in the H-mode, but when scaled to ITER-class devices this plasma "breathing" needs to be faster and smaller than extrapolated to avoid material damages. A small (4,3) tearing mode might help in driving the plasma into the NTM Frequently Interrupted Regime, where the negative influence of (3,2) NTMs on confinement is weaker [41].

Instabilities and in general magnetic perturbation "talk" one to each other. This may happen directly, or indirectly—through their driving terms. Among the examples of direct interaction we have just mentioned sawteeth seeding NTMs. ELMs or fishbones can trigger metastable RWMs in high beta scenarios [42]. Magnetic field errors can seed NTMs or RWMs, but they are also by purpose applied to destabilize ELMs and make them harmless or, in certain cases, to make sawteeth smaller and more frequent.

The role of plasma quantities is not always unique in terms of stability. A population of fast particle may stabilize sawteeth and RWM, but also drives fast particle instabilities. The current density profile that provides a flat or reversed shear profile might avoid resonant surfaces for sawteeth and NTMs, but could give raise to other instabilities.

Stability control has a cost. This could be both direct—i.e. in terms of the capital cost of the control tools like additional heating or current drive systems, of the extra energy, which is needed to drive actuators and that needs to be taken into account in the overall energy budget—and indirect, since control actions may spoil plasma performance. These costs might be bearable, but need to be minimized.

A neat example of direct cost is NTM stabilization via ECCD. As discussed in chapter seven, with the application in ITER of 20 MW of ECCD power for NTM removal the pre-NTM energy confinement time is recovered. But the energy amplification factor Q, given by the ratio of the power produced by fusion to that given as input to the plasma, drops from 10 to 7 (8.5 for the optimistic case if only 10 MW are needed) [43]. The additional ECCD power has in fact to be included in the energy balance. Once the NTM has been stabilized $Q = 10$ is recovered by switching off the gyrotrons. When the next NTM gets triggered, the gyrotron are switched on again and the loop repeats. This—and other examples too discussed in the NTM and sawtooth chapters—mean that one has to be careful in applying ECCD. For example, a continuously applied pre-emptive ECCD will reduce the achievable Q-value, correspondingly. ECCD has to be used wisely and not wasted.

Another example—that applies to a reactor—is that of false alarms. Control system must be as reliable as possible. While shutting down a discharge because of a disruption false alarm in a present day tokamak is not an issue, the economic damage of switching-off a tokamak reactor for the same reason might be large.

Two examples of control indirect costs concern ELMs and NTMs. The most promising technique for ELM control is at the moment the application of resonant magnetic perturbations. The perturbation causes an artificial increase of edge transport in the pedestal region, thus reducing edge pressure. This reduction either completely suppresses ELMs, if edge pressure goes below ELM instability threshold, or at least make them smaller and more frequent. Obviously the artificial increase of edge transport causes also a small steady reduction of confinement, a price that can be paid in order to get rid of ELMs. Still not completely clear, as we have seen before, is the cost in terms of fast ion losses. Talking about NTMs, we have seen that a flat core density profile obtained with central ICRH may help for NTM stability; but flattening the core pressure profile is probably a price too high in a reactor, which makes this technique unlikely to be pursued in future experiments.

Control will be difficult in a nuclear environment. In ITER, and even more in DEMO, the nuclear environment and the consequent neutron flux will make much more difficult to operate sensors and actuators. Some of those presently used will simply not be applicable. And those, which will be there need to be extremely robust, since maintenance will be extremely difficult, if not impossible.

Opportunity for testing will be much less. While present day devices allow ample room for testing control techniques, with relatively small risk for the hardware, the next generations of experiments will be much less forgiving. This is challenging: training based methods, like for example disruption prediction tools, need in their present version a database, which can obviously not be provided in a device like ITER. And also the development of other active MHD control tools, will not be able to rely on experiments that can potentially fail, as happen today. Risks in experiments testing control tools need to be strongly minimize, which of course drastically reduce the input from experimental flexibility.

Noise can not be neglected. Given the safety standards required by the ITER, an issue like that of noise in the measurements will be very important. This applies for example to the axisymmetric control, like that of the vertical position of the plasma. The vertical stabilization system is crucial in ITER, given what a vertical displacement event can cause. A key parameter for this system to work is the noise in the dZ/dt measurements, where Z is the vertical position of the plasma. Noise is unavoidable, and its increase above a given level (at the moment 0.6 m/s) may produce unacceptable oscillations in the power taken from the grid for driving the vertical stabilization system or reduce the duration of the burn phase because of AC losses in the superconducting coils. The assessment of expected noise in ITER dZ/dt measurements, and the development of methods to reduce it, are still subject of investigation and are issues that need to be solved soon.

Not all can be planned. In addition to being able to coping with MHD instabilities that can be predicted or, the control system of a fusion device must to capable to handle exception, i.e. events not planned or desired—outside the standard function of the device—that can have significant impact on operation or device safety. This includes for examples faults of components and disruptions. Basic goals for exception handling systems are to minimize the probability of unexpected events and their consequences for the plant, and robustness. Handling could mean avoidance, recovery, switching to alternative regimes or, when nothing else is possible, safe landing through controlled shutdown.

10.4 Control Integration

Given the issues we have just analysed, and the importance that MHD stability has for a fusion reactor, it is clear that the development of an integrated control strategy is a grand challenge for exploiting fusion as an energy source. The strategy to address this challenge needs to be based on several pillars.

ITER will be the most important one. Its portfolio of sensors and actuators [44] gives the opportunity of explore the operation space, to integrate simultaneous MHD controls addressing the issues discussed before and to find the best operation points, which will be a precious input for DEMO.

Plasma axisymmetric magnetic control will be guaranteed by a combination of the central solenoid, of the poloidal field and of the internal vertical stability coils, with their power supplies. The inductive plasma current, shape and radial position control will have a settling time of about 5 s, while the faster vertical position control with the in-vessel coils will have a settling time of 0.1 s. This should allow or restoring the plasma vertical position after an uncontrolled vertical drift of about 16 cm, assuming a maximum noise in dZ/dt of 0.6 m/s.

The system of in-vessel coils is composed, in addition to the vertical stabilization coils, by 3 sets of 9 coils each, explicitly design for ELM control—via the application of resonant magnetic perturbation—and RWM control—via feedback control. A radiative divertor will act synergistically with RMP to mitigate ELM loads on the plasma facing components. Outside the vessel three sets of 6 coils each (top, bottom and equatorial) are present to deal with error magnetic field coming from equilibrium coil misalignments, feeds and from ferromagnetic materials, in particular in the Test Blanket Modules.

A combination of both ICRH and ECCD is planned for controlling sawteeth in ITER. It is predicted that a combination of 13.3 MW of co-ECCD from the equatorial launcher and 6.7 MW from the upper launcher would be able to reduce the sawtooth period by 30 %, while 10 MW of ICRH should be sufficient to negate the stabilising effects arising from the alpha population. ECCD will be a key player also for NTM stabilization. The available ECCD power of 20 MW at 170 GHz should be sufficient, and will be coupled to a system of steerable mirrors, all controlled in real time. Interestingly, ECCD might be coupled with the effect of applied magnetic fields following a scheme tested in DIII-D; NTM islands can in fact be dragged and locked in the best toroidal position for optimum application of the ECCD by use of magnetic active coils [45], a technique used for unlocking classical tearing modes also in RFPs [46]. To meet the $Q = 10$ goal of ITER baseline scenario, the ECRH power has to be turned off whenever it is not being actively used for mode control and the balance with the use of ECCD for sawtooth control needs to be found. This can be done either by following a prescribed sequence by switching-off central ECRH [47] or—instead of relying on smaller sawteeth—exploiting fast ions to deliberately stabilise the sawteeth, and to pre-emptively apply ECCD before each crash the near the $q = 2$ rational surface to stabilise the ensuing NTM [48].

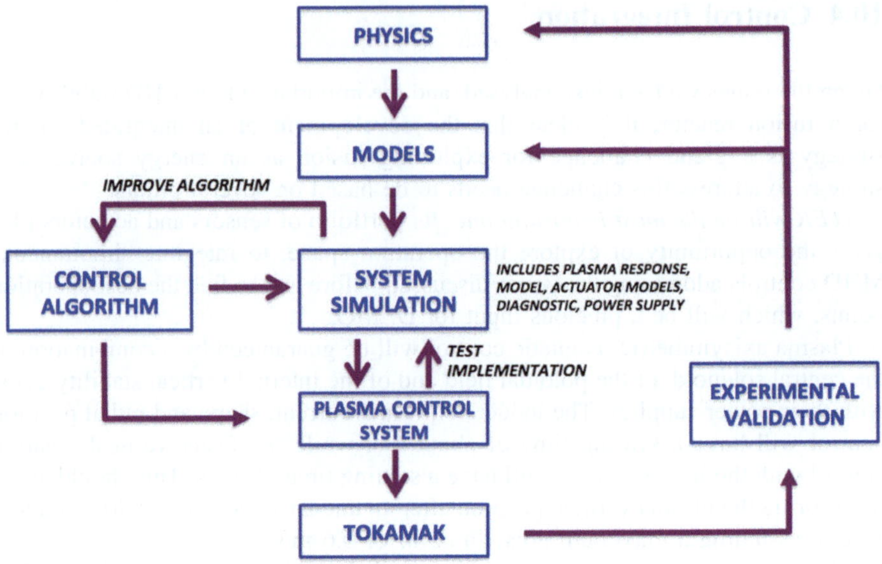

Fig. 10.3 Flow chart of an integrated control strategy (adapted from [49])

Modelling is crucial to develop efficient control. Modelling is clearly needed to better understand the physics underlying MHD stability. The case of disruptions is a paradigm from this point of view, since a detailed knowledge of the physics that drives disruptions would be very important to improve their predictability. And modelling is also key for high performance model-based control. The flow chart of a control development strategy, which includes the verification in simulations before experimental application is sketched in Fig. 10.3 (adapted from [49]) Models of the physics and of the machine boundary conditions, at various levels of complexity, are needed to develop the control algorithms and a robust system simulator, which includes models of plasma response, actuator, diagnostic, power supplies. As in aeronautics, the availability of flight-simulators are crucial to ensure a safe, error-free application of new control technologies.

A nice example of an integrated simulator for closed loop RWM control experiments has been developed and benchmarked in the RFX-mod RFP device [50]. The tool couples self-consistently a full 3D description of the machine boundary (with the Cariddi code), a 2D toroidal model of stability (MARS code) and a dynamic model of the control system cast in the state variable representation. Using actual PID gains and plasma equilibrium parameters such "flight simulator" successfully reproduces experimental closed loop RWM control, as shown in Fig. 10.4.

Figure 10.4a shows the time evolution of the amplitude of the radial field perturbation due to the ($m = 1$, $n = -6$) RWM measured in RFX and simulated by the integrated simulator. In Fig. 10.4b the growth rates of RWMs in RFX are plotted as a function of the proportional gain applied to the controller. If the gain is not high enough, only a slowing down of the RWM growth rate with respect to the

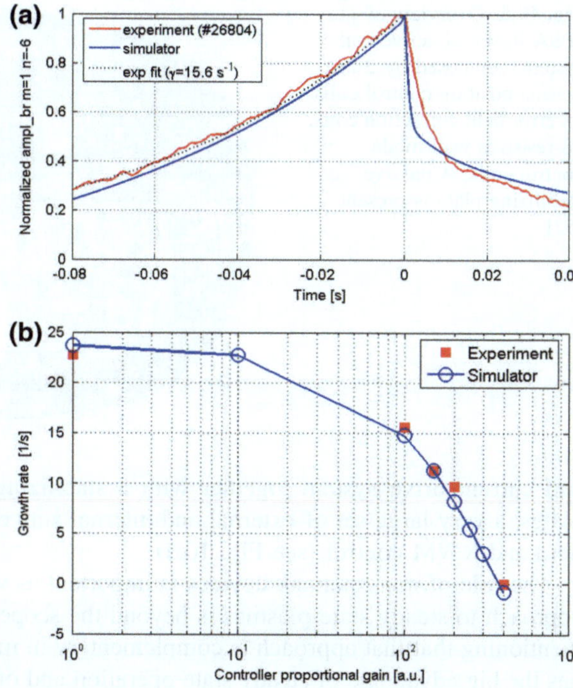

Fig. 10.4 RWM growth rates in RFX versus proportional gain in the coil-based control system. Stars are experimental measurements, open circles prediction obtained with the "flight simulator" (figure composed from [50, 51])

uncontrolled case (gain = 0) is obtained, while with the proper gain the growth rate is brought to zero. Stars are experimental measurements, open circles predictions from the flight simulator. We notice how precise is the agreement in both figures, meaning that the flight simulator is capable of completely predicting the stabilization physics for this device.

Accompanying experiments. The case of the RFX flight simulator is an example of how important is that construction and operation of ITER is accompanied by a large set of flexible experiments, that allow for developing and testing off-line control strategies to be later transferred to ITER. While these devices will not provide the same level of integration as ITER, they can scrutinize in depth and at low-cost, low-risk the building blocks of control strategies. How the present-day tokamaks have contributed—and will contribute—to the control grand challenge has been extensively described in the book. The new tokamak JT-60SA [52], which is scheduled to start its operation in 2019, will give an important contribution. JT-60SA is a fully superconducting tokamak capable of confining break-even equivalent high-temperature deuterium plasmas. JT-60SA will address the physics and technology of steady-state tokamak. It is designed to study power and particle handling for 100 s at high power with water-cooled divertors compatible with maximum heat fluxes of 15 MW/m^2 and it will be able to explore full non-inductive steady-state operation with 10 MW tangential NB current drive and 7 MW of ECCD. For the study of steady-state, non inductive scenarios, JT-60SA will be able to explore high beta plasma regimes by using a high power heating

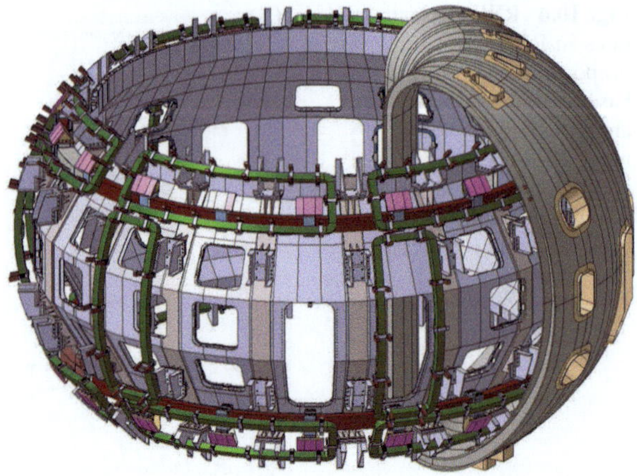

Fig. 10.5 Overview of JT-60SA in-vessel active coil system, composed by 2 fast plasma position control coils, 18 error field correction coils, 18 resistive wall mode control coils. A passive stabilizing plate is present [52]

and current drive system together with a stabilizing shell covered with ferritic plates, a very large set of external and internal active coils for fast position, error field and RWM control (see Fig. 10.5).

The role of non-tokamak devices is important as well. Covering the stellarator approach to steady state plasmas is beyond the scope of this book, but it is worth mentioning that that approach is complementary in many respects to tokamak and has the big advantage of steady state operation and of zero disruptivity. RFPs, has already mentioned, have provided pioneering results on MHD feedback control via active coils [53]. The size and the completeness of their coils system—RFX, for example, has 192 coils each independently driven—make their contribution important for the development of integrated MHD control.

DEMO. DEMO should be a step intermediate between ITER and the commercial power plant. To fulfil this goal, DEMO top level design requirements are (a) ability of net electricity production; (b) ability to be self-sufficient in terms of tritium production; (c) to provide a robust solution of all physics and technical issues still open at the moment; (d) to guarantee adequate availability/reliability operation over a reasonable time span and safe operation and minimization of radioactive waste.

At present there are two main DEMO design options: the first corresponds to a large, modest power density, pulsed device with a conventional plasma scenario, with a conservative design based on expected performance of ITER. The second to a device with higher power density and steady-state operation. Its design should at the moment be based on optimistic, advanced assumptions—therefore with higher risks than the first option, since extrapolations go much further.

Clearly control requirements for DEMO will not only be different with respect to those for ITER, but also they will depend on the detailed strategy chosen. Control needs for a steady state high beta device are very different in comparison to those for a pulsed machine. As for the case where we start this chapter, i.e. aircrafts, mission will define the design requirements for an integrated control system. DEMO will have a restricted operating space, based on the outcome of previous experiments and models, will have a very small set of sensors, both for the sake of easy maintainability and because of the hostility of the environment. Controls need to be not only integrated, but also reduced to the bare minimum. Therefore, it is expected that while control of ELMs, of fast particle driven MHD and reaction to off-normal events will be common to both, the pulsed approach will need more focus on NTM and sawtooth control, while in the steady-state, high beta DEMO RWMs will be key actors to address.

10.5 Conclusions

We started this chapter with the story of the Wright brothers, from a bike shop to flying the first airplane. The path to fusion has several analogies with that story and in general with the evolution of aviation. Being aware of the risks of oversimplification, it is nonetheless tempting to say that running present day tokamaks has some analogy to riding a bike. We have accumulated a large experience and we do it rather safely and achieve good performance—though in both cases we need to protect ourselves against unpredictable events and we still need to understand some details of the physics. But the next step to ITER, and even more to DEMO, will be qualitatively different. As the Wright first airplane, ITER will be the first fusion device to "fly", i.e. to demonstrate that fusion works. DEMO will bring fusion to its commercial exploitation, like modern airplanes do for aviation.

As it happened for airplanes, more than trying to completely avoid instability, it will be crucial to learn how to live with it and control it and whenever possible how to exploit it. This is why integrated control will need not only integration of various sensors and actuators, but also the ability of integrating experiments, simulations and modelling and of recognizing that, as for the airplanes, a broad experimental basis will be necessary.

References

1. http://bicycle.tudelft.nl
2. T.C. Hender et al., IAEA Fusion Energy Conference 2002—Paper EX/P2-22, http://www-pub.iaea.org/mtcd/meetings/pdfplus/fusion-20-preprints/EX_P2-22.pdf
3. P. Martin et al., Nucl. Fusion **51**, 094023 (2011)
4. J. Hanson et al., Accepted for publication in *Physics of Plasmas* (2014)
5. P. Piovesan et al., Phys. Rev. Lett. **113**, 045003 (2014)
6. C. Gormezano et al., Nucl. Fusion **47**, S285 (2007)
7. A. Loarte et al., J. Nucl. Mater. **313–316**, 962 (2003)
8. Hender T.C. et al., *Nucl Fusion* **47** S128
9. S. Von Goeler, W. Stodiek, N. Sauthoff, Phys. Rev. Letters **33**, 1201 (1974)
10. F. Porcelli, D. Boucher, M. Rosenbluth, Plasma Phys. Control. Fusion **38**, 2163 (1996)
11. I.T. Chapman, Plasma Phys. Control. Fusion **53**, 013001 (2011)

12. Z. Chang, J.D. Callen, E.D. Frederickson, R.V. Bud-ny, C.C. Hegna, K.M. McGuire, M.C. Zarnstorff, TFTR Group, Phys. Rev. Lett. **74**, 4663 (1995)
13. O. Sauter et al., Phys. Plasmas **4**, 1654 (1997)
14. H. Zohm et al., Plasma Phys. Control. Fusion **45**, A163 (2003)
15. F. Wagner et al., Phys. Rev. Lett. **49**, 1408 (1982)
16. J.W. Connor, Plasma Phys. Control. Fusion **40**, 531 (1998)
17. E. la Luna de et al.,in *23th IAEA Fusion Energy Conference*, Paper EXC/8-4 (2010)
18. T.E. Evans et al., Nature Phys. **2**, 419 (2006)
19. Y. Liang et al., Phys. Rev. Lett. **98**, 265004 (2007)
20. A. Kirk et al., Nucl. Fusion **50**, 034008 (2010)
21. J.M. Canik et al., Nucl. Fusion **50**, 034012 (2010)
22. M. García-Muñoz et al., in *IAEA FEC 2012*, Paper EX/P6-03 (*2012*)
23. A. Bondeson, D.J. Ward, Phys. Rev. Let. **72**, 2709 (1994)
24. M.S. Chu, M. Okabayashi, Plasma Phys. Control. Fusion **52**, 123001 (2010)
25. P. Brunsell et al., Phys. Rev. Lett. **93**, 225001
26. M. Baruzzo et al., Nucl. Fusion **52**, 103001 (2012)
27. Y. In et al., Plasma Phys. Control. Fusion **52**, 104004 (2010)
28. Zanca et al., PPCF **54**, 094004 (2012)
29. G. F. Matthews et al. *in Fusion Energy 2002 (Proc. 19th Int. Conf. Lyon, 2002) (Vienna: IAEA)* Paper EX/D1-1, http://www-pub.iaea.org/mtcd/publications/pdf/csp_019c/pdf/exd1_1.pdf
30. M. Sugihara et al., Disruption scenarios, their mitigation and operation window in ITER *Nucl.* Fusion **47**, 337 (2007)
31. S. Putvinski et al., in *2010 IAEA FEC*, Paper ITR/1-6
32. W.W. Heidbrink, G.J. Sadler, Nucl. Fusion **34**, 535 (1994)
33. H.H. Duong et al., Nucl. Fusion **33**, 749 (1993)
34. R.B. White et al., Phys. Plasmas **2**, 2871 (1995)
35. S.K. Nielsen et al., Plasma Phys. Control. Fusion **52**, 092001 (2010)
36. R.J. Goldston, R.B. White, A.H. Boozer, Phys. Rev. Lett. **47**, 647 (1981)
37. A. Fasoli et al., Plasma Phys. Control. Fusion **52**, 075015 (2010)
38. B.N. Breizman et al., Phys. Plasmas **10**, 3649 (2003)
39. C.C. Petty et al., in *Proceedings of 23rd International Conference on Fusion Energy 2010 (Daejon, Korea Rep., 2010)*
40. R. Lorenzini et al., Nature Phys. **5**, 570 (2009)
41. S. Guenter et al., Phys. Rev. Lett. **87**, 275001 (2001)
42. G. Matsunaga et al., Phys. Rev. Lett. **103**, 045001 (2009)
43. O. Sauter, M. Henderson, G. Ramponi, H. Zohm, and C. Zucca, Plasma Phys. Controlled Fusion **52**, 025002 (17 pp) (2010)
44. J. Snipes, in *MHD and Plasma Control in ITER, Talk Given at the 4th ITER International Summer Schhol on MHD and Plasma Control, Austin 2010,* http://w3fusion.ph.utexas.edu/ifs/iiss2010/iisstalks/Snipes_Joseph_talk.pdf
45. F. Volpe et al., Phys. Plasmas **16**, 102502 (2009)
46. R. Bartiromo et al., Phys. Rev. Lett. **83**, 1779
47. Goodman et al., PRL **106** (2011)
48. T.P. Goodman et al., Phys. Rev. Lett. **106**, 245002 (2011)
49. D.A. Humphreys, in *High Reliability Operation and Disruption Control in Tokamaks, Talk Given at the 4th ITER International Summer Schhol on MHD and Plasma Control, Austin 2010,* http://w3fusion.ph.utexas.edu/ifs/iiss2010/iisstalks/Humphreys_David_talk.pdf
50. G. Marchiori et al., Nucl. Fusion **52**, 023020 (2012)
51. Y. Liu et al., Plasma Phys. Control. Fusion **52**, 104002 (2010)
52. JT-60SA, *Research Plan version 3.1*, (Dec 2013). http://www.jt60a.org/b/index_nav_3.htm?n3/operation.htm
53. P. Martin, Fusion Sci.Technol. **59**, 602–616 (2011)

MIX
Papier aus verantwortungsvollen Quellen
Paper from responsible sources
FSC® C105338

If you have any concerns about our products,
you can contact us on
ProductSafety@springernature.com

In case Publisher is established outside the EU,
the EU authorized representative is:
Springer Nature Customer Service Center GmbH
Europaplatz 3, 69115 Heidelberg, Germany

Printed by Libri Plureos GmbH
in Hamburg, Germany